Local Planning
in Practice

Local Planning in Practice

Michael Bruton and David Nicholson

*University of Wales
Institute of Science and Technology, Cardiff*

Routledge
Taylor & Francis Group

LONDON AND NEW YORK

First published in 1987 by Hutchinson Education
Reprinted in 1988

Reprinted in 1990 by
By Routledge,
2 Park Square, Milton Park,
Abingdon, Oxon, OX14 4RN
270 Madison Ave,
New York NY 10016

Transferred to Digital Printing 2005

British Library Cataloguing in Publication Data

Bruton, Michael
   Local planning in practice — (The Built Environment series)
   1. City planning — England   2. Regional planning — England
   I. Title   II. Nicholson, David
   III. Series
   711'.4'0942   HT169.G7

   ISBN 0 7487 0373 X

Phototypeset in 10 on 12 pt Times Roman

Contents

3 The statutory local land use planning system in Britain I: The legislative and administrative framework

4 The statutory local land use planning system in Britain II: Local plans in practice

List of tables

List of figures

Preface

The official (central government) view of local plans is that such documents are restricted to a concern with the development and other use of land. The official view of the local planning process is that it is rational; focuses around the production of local plans and development control, and is largely administrative. The reality is that local plans and development control cannot be divorced from the much wider activity of local planning which is concerned to manage change in the environment within the context of attempting to secure social and economic change; the reality is that local planning is essentially a political process, characterized by complex inter-relationships between land use and social and economic processes; by the redistribution of resources; by conflicts of interest; and by bargaining and uncertainty.

This book aims to review local planning in practice, primarily in England and Wales, in a way which makes it comprehensible to students and practitioners concerned with planning and the development process. It sets out a range of theoretical or conceptual perspectives which contribute to an understanding of the way in which local planning is practised. Thus the complex, inter-related nature of planning problems is given emphasis; the significance of 'contingent factors' on planning situations is highlighted; the bargaining which is inherent in all redistributive processes is stressed – all of which point to a need for a flexible planning system which is capable of responding to a wide variety of complex and dynamic situations. Against this background, the system of local land use plans and local planning currently in operation is examined in a way which attempts to establish the strengths and weaknesses of the current system. The statutory local plans are evaluated and the use of non-statutory local planning instruments and the wide range of formal plans and programmes with a bearing on land use and development, such as GIAs, EZs and AONBs are reviewed in a way which highlights the influence of all these vehicles on decisions taken as part of the development process.

As always, a number of individuals have knowingly or unknowingly contributed to the production of this book. In particular a note of thanks is more than appropriate to UWIST for providing the resources which made this book possible; Anne Andress, Adri Stofaneller and Merle Jones who at different stages wrestled with various drafts of the text; Tracey Dinnock and Martin Morris who either drew, redrew or copied the diagrams, and to Lyn Davies who unknowingly pushed us into writing it. In different circumstances, this book or something similar would have been written by him rather than ourselves.

M.J.B. and D.J.N.

UWIST, Cardiff,
January 1986

13

Acknowledgements

We are grateful to the following for permission to reproduce copyright material:

1 F.S. Chapin and E. Kaiser and the University of Illinois Press for our Figure 2.1 from *Urban Land Use Planning*, 1979.
2 K. Christensen and the *Journal of the American Association of Planners* for our Figures 2.2, 2.3 and 2.4 from 'Coping with Uncertainty in Planning', *Journal of the American Planning Association*, Vol. **51**, No. 1, 1985.
3 *Town Planning Review* for agreeing to the use in Chapters 2 and 4 of material previously published by ourselves in 'Strategic land use planning and the British development plan system', *Town Planning Review*, **56**, No. 1, 1985, including our Figures 2.5 and 4.2; and M.T. Pountney and P.W. Kingsbury and *Town Planning Review*, for our Figure 7.2 from 'Aspects of development control, Part 1, in *Town Planning Review*, **54**, No. 2, 1983.
4 HMSO for our Figures 3.1, 3.2 and 7.3 from the *Development Plans Manual* (1970); DoE Circular 22/84; and H.W.E. Davies, D. Edwards, and A.R. Rowley, (1986) 'The Relationship between Development Plans, Development Control and Appeals'. Department of Land Management and Development, University of Reading.
5 Scottish Development Department for our Figure 3.3 from Planning Advice Note 30, *Local Planning*, 1984.
6 The School for Advanced Urban Studies, University of Bristol, for the use of material in Chapters 4 and 7, which was previously published by M.J. Bruton in *Policy and Politics*, Vol. **7**, No. 4, 1980.
7 *Planning Outlook* for agreeing to the use in Chapter 1 of material previously published by M.J. Bruton in *Planning Outlook*, Vol. **25**, No. 3, 1982.

1 Town planning, local planning and public policy

1.1 Introduction

The town and country planning system which has developed in Britain over the past 100 years or so clearly but simplistically establishes planners as producers of land use plans and controllers of development. Implicit in much of the legislation is the view that town and country planning is an apolitical, technical exercise concerned with physical development and the protection of the environment, operating within a loosely defined set of constantly changing policies designed to secure social and economic change. The inter-relationships between land use planning and policies for social and economic change are rarely made explicit. Indeed, until the 1970s, town planning practice operated almost as if it were self-contained and free-standing.

In reality, town planning is extremely complex, dealing with matters that are highly inter-related, surrounded by uncertainty and much affected by ideology. It is essentially '. . . a ruthless bargaining process' (Buchanan, 1968, p. 52) concerned with conflicts of interest and the distribution of limited resources. In the last 10–15 years, town planning practice has begun to adapt to this reality, and under the pressures created by structural changes in the economy is being shaped into a system which is concerned with the management of change in the environment. Such an approach inevitably accentuates the inter-relationships between social, economic, physical and political issues. In addition to the traditional roles of plan production and the control of development, it is also concerned to manage the allocation of resources in a way which achieves, as far as possible, policy objectives set out in plans. Although there have been legislative changes which, it can be argued, reflect this reality, e.g. the introduction of enterprise zones and urban development corporations, the statutory basis of the town planning system is largely unchanged. Town planners are still seen solely as producers of land use plans and controllers of development carried out by other agencies. At the same time, the introduction of new legislation and new policies, which explicitly reflect the ideology of the current (1986) Conservative government, is undermining the limited position of the development plan provided by statute.

17

Against this background, this chapter briefly reviews:

1   The evolution of the statutory town planning system which establishes planners as producers of plans and controllers of development.
2   The role of local planning within this system.
3   The changes being made to this system by the Conservative government in the early to mid-1980s.

1.2   The nature and evolution of the statutory land use planning system in Britain (Ashworth, 1954; Cherry, 1974)

Town and country planning as an activity of government initially developed out of a concern to improve social conditions in towns and cities through public health and housing controls. The public health problems associated with the population increase and rapid urbanization which were a feature of the Industrial Revolution, allied with a fear of social unrest, led to the introduction of nineteenth-century public health and housing legislation directed at the creation of acceptable sanitary conditions. As a result, local authorities were given powers to reconstruct insanitary areas and to make and enforce by-laws for controlling street widths and the height, structure and layout of new buildings. At the same time, the activities of social reformers such as Robert Owen, Cadbury, and Ebenezer Howard began to lead to the development of the view that through the medium of plans for town, country and suburb the social problems of overcrowding, poverty, ill-health, unemployment and insanitary and inadequate living conditions could be overcome. Indeed, the words of the then President of the Local Government Board, when he introduced the first town planning legislation – the *Housing, Town Planning, etc. Act 1909* – aptly summarize this view:[1]

The object of the Bill is to provide a domestic condition for the people in which their physical health, their morals, their character and their whole social condition can be improved by what we hope to secure in this Bill. The Bill aims in broad outlines at, and hopes to secure, the home healthy, the house beautiful, the town pleasant, the city dignified and the suburb salubrious.

The new powers introduced by the Act provided for the preparation of 'schemes' or plans by local authorities for controlling the development of new housing areas. The Act was succeeded by the *Housing and Town Planning Act 1919* which was concerned primarily with the provision of housing for the working classes and simplifying the administrative obligations for local authorities to prepare planning schemes introduced by the Act of 1909. Thus every borough or other district council with a population of more than 20,000 on 1st January 1923 according to the previous Census was required to prepare within three years a planning scheme for land in course of development or likely to be developed (Cherry, 1974, p. 72). Local authorities were urged to make their schemes simple and to concentrate on '. . . the essentials of town planning viz. the principal routes of communication and restrictions in regard to the character and density of buildings to be erected on the area' (Ministry of Health 1919–20, quoted in Cherry, 1974, pp. 82–83). Town planning it seems, even at this early stage, was already divorced from the concern to secure social and economic change which characterized the work of the social reformers and the introduction of the 1909 Act. Rather it was presented as being concerned with the preparation of land use plans and the control of development – the implicit assumption being that by producing plans and controlling development, implementation on the ground would naturally be in accord with policies contained in the plan and contribute to the achievement of social and economic change.

The *Town and Country Planning Act 1932* either repealed or consolidated almost the whole of the then existing general and local enactments relating to planning. It extended the powers of local authorities by enabling planning schemes to be prepared with respect to any land, whether urban or rural. At the same time, by making the preparation of such schemes voluntary, it removed the obligation which had existed for some authorities under the Act of 1919 to prepare schemes (Cherry, 1974, pp. 98–107). However, the 1932 Act reaffirmed the view of the earlier legislation that town planning was concerned with the production of plans and the control of development. In turn, the 1932 Act was itself superseded by a package of planning and planning-related legislation enacted in the period 1945–52 and which introduced a comprehensive and mandatory system of land use planning through the following:

1 *Distribution of Industry Act 1945*, which attempted to encourage new industries to establish in those parts of the country suffering the worst effects of unemployment and declining industry;
2 *New Towns Act 1946*, which set up the machinery for the implementation of new towns to accommodate overspill population from the congested inner areas of the major conurbations;
3 *Town and Country Planning Act 1947*, which required local planning authorities (*a*) to produce development plans for their areas defining patterns of future land use, and (*b*) to control new development in a way which took account of the provisions of the development plan;
4 *National Parks and Access to the Countryside Act 1949*, which allowed the designation of national parks and areas of outstanding natural beauty and improved public access to the countryside;
5 *Town Development Act 1952*, which enabled agreements to be made between different local authorities whereby development (mainly housing) could take place in one area to ease the problems existing in another area.

The thinking underlying this legislation was set out in three major reports:

the 1940 Barlow Report on the distribution of industrial population; the 1942 Scott Report on land utilization in rural areas; and the 1942 Uthwatt Report on compensation and betterment. Together these reports demonstrate a concern to secure social and economic change in post-war Britain, illustrate how land use planning might be used to contribute to the achievement of that change, and give an indication of the complex inter-relationships between land use planning and policies to secure social and economic change. Unfortunately, this thinking was not explicitly reflected in the legislation of the time, nor in the administrative procedures established to implement the legislation. The *Town and Country Planning Act 1947* was generally seen as being the major set piece of post-war planning legislation which simplified and strengthened the planning system by reducing the number of local planning authorities in England and Wales from 1441 to 145. However, it contained no explicit reference to the inter-relationship between land use planning and policies for social and economic change. Indeed, it is only when this Act is seen as part of a wider package of legislation which included amongst others those Acts listed above as well as the *Education Act 1944* and the *Health Act 1948*, that the concern with social welfare is apparent. Town and country planning was, at that time, seen by government as making a limited contribution to social and economic change by influencing the development and other use of land within a framework set by loosely articulated higher order policies for achieving that social and economic change. Similarly, the decisions (*a*) to administer the *Distribution of Industry Act 1945* through the Board of Trade rather than the Ministry of Town and Country Planning, and (*b*) to implement the provisions of the *New Towns Act 1946* through development corporations rather than the local planning authority, ignored the complex inter-relationships between social, economic, political and land use issues. At the same time, it reinforced the view that town planners were concerned solely with the production of plans and the control of development. This view was given greater emphasis by the way in which the requirements of the 1947 Act were amplified and implemented, through regulations, orders and circulars.

Briefly, the basic framework for land use planning required by the 1947 Act and associated regulations and circulars was the *development plan*. Each planning authority was required to draw up a plan showing the proposed pattern of land use for 20 years ahead. This development plan consisted of a loose hierarchical series of maps showing the proposed land uses; a written statement describing the proposals; programme maps showing the phasing of the development and reports of survey. The County Map, which was presented on a 1″:1 mile Ordnance Survey base, indicated the broad pattern of development in the area covered. Town Maps were prepared for urban areas on a 6″:1 mile Ordnance Survey base and showed the precise boundaries for areas zoned for the main urban land uses such as residential, industrial and commercial, as well as sites for specific uses such as schools and playing fields. Comprehensive Development Area maps (CDAs) were prepared for areas requiring development or redevelopment on a comprehensive basis. These were produced on a 25″:1 mile Ordnance Survey base and gave precise details of the proposed development. Each of these plans required the approval of central government, and provided the basis for considering individual applications for planning permission, which the Act required must be obtained from the planning authority before almost any building development or mineral extraction was carried out or any material change was made in the use of land or buildings. The 1947 Act also gave planning authorities powers to deal with specific problems of amenity including the preservation of trees and woodlands and of buildings of special historic or architectural interest, and the control of advertisements. Thus the development plan was essentially a plan for land utilization, concerned with policies of location and the use of land as a resource, which provided a clear framework for the control of development. The planner's role as a producer of plans and a controller of development was firmly established. The complex interactions between social, economic, political and land use matters were ignored. The simplistic view of the inter-relationship between the dynamics of change and the implementation of policies in the plan was confirmed.

The system when introduced was regarded as radical and comprehensive, and indeed it served well for a period of years (Buchanan, 1972). However by the early 1960s it was apparent that it was inadequate to cope with the rapidly changing problems facing society. In 1964, the Planning Advisory Group (PAG) was established to review the 1947 Act system and recommend changes. The Group found that the system was over-centralized, prone to delay, unable to influence the quality of design, inflexible and incapable of adjusting to rapidly changing circumstances (James, 1965). In its place it recommended a development plan system (PAG, 1965, pp. 8–9).

. . . which will –
(1) guide the urban development and renewal which is certain to take place;
(2) promote efficiency and equality in the replanning of towns;
(3) encourage better organization and co-ordination of professional skills so that town and country are planned as a whole;
(4) stimulate more purposeful planning of rural and recreational areas.

PAG worked on the assumption that local government would be reformed in the (then) near future and that a single tier of unitary authorities would be established. The land use planning system recommended for this new local government structure involved a development plan consisting of two tiers of plans – the *structure plan* and *local plans* (p. 26). The structure plan, which had to be approved by central government, was to set out broad policies indicating how the area covered by the plan should be developed.

The local plan, with a proposals map produced on an Ordnance Survey base, dealt with the detail of development in much the same way as the old Town Maps had done, and was to provide a framework for development control, which was to continue to operate unchanged. These local plans could be adopted by the local planning authority after a public inquiry into any objections was made. The land use planning system recommended by PAG was introduced by the *Town and Country Planning Act 1968* and is still operative although it has been amended subsequently, primarily by the *Town and Country Planning Act 1971* and the *Local Government Planning and Land Act 1980*. However, local government in the form envisaged by PAG did not materialize, and in place of the anticipated single tier unitary authorities a two-tier system of county and district council planning authorities was introduced in England and Wales by the *Local Government Act 1972*. Planning responsibilities were split between the two tiers with the counties being responsible for producing the structure plan and some development control and the districts being responsible for the production of most local plans and the bulk of development control. A similar but not identical land use planning system was introduced in Scotland by the *Local Government (Scotland) Act 1973*, administered by a system of Regional, Island and District Councils. In London and the six metropolitan areas in England the two-tier system established by the 1972 Act has recently been modified following the *Local Government Act 1985* by the abolition of the Greater London Council and the six metropolitan county councils on 1 April 1986. The London boroughs and the metropolitan district councils are now responsible for the preparation of unitary development plans, which combine features of both structure and local plans.

1.3   The role of local plans within the statutory land use planning system

1.3.1   Statutory basis for town planning

The Town and Country Planning Act 1971 (as amended) provides the statutory basis for the current land use planning system in England and Wales. It establishes (*a*) a development plan system involving the production of structure and local plans, and (*b*) a development control system which provides that, with certain exceptions, all development requires planning permission from the local planning authority. This Act is supplemented by a number of other acts, regulations and orders, together with circulars and advice notes issued by the Department of the Environment (DoE), which flesh out the details of the statutory land use planning system.

Briefly, the system when introduced was primarily seen as being concerned with land use; its objectives were described by the then Chief Planner in central government as (James, 1965, p. 22):

a. clarifying levels of responsibility, so that only major policies and objectives are brought before the Minister for approval, while matters of detail and local land use are settled locally;
b. providing more positive guidance for developers and development control . . . .;
c. increasing public understanding of the system and participation in the plan-making process;
d. simplifying planning administration.

Subsequent advice from central government in successive DoE circulars and memoranda dealing with structure and local plans in 1974, 1977 and 1979 clarified this role, and the relationship between land use planning and social and economic policies, when it expanded on the main functions of the structure plan, viz. (DoE, 1974a, 1977e, 1979a; para. 2.1):

to interpret national and regional policies in terms of physical and environmental planning for the area concerned. National and regional policies tend to be primarily economic and social . . . . : structure plans represent the stage in planning at which such policies are integrated with the economic, social and environmental policies of the country and expressed in terms of their effect on land use, environmental development and the associated transport system . . . .

The current development plans memorandum (Circular 22/84) identifies the functions of the structure plan as being (DoE, 1984g, para. 2.1):

a. to state the policies and general proposals for the development and other use of land . . . . ;
b. to take account of policies at national and regional level as they affect the physical and environmental planning of the area;
c. to provide the framework for local plans.

At the same time it makes it clear that in preparing the structure plan, consideration should be given to the relationship of planning policies and proposals to social needs and problems.

Acts of Parliament
The *Town and Country Planning Act 1971* has subsequently been amended and the system added to by a number of acts, the most significant of which are:

1 *Town and Country Planning (Amendment) Act 1972*, which makes certain amendments relating to development plans, the control of office development, building preservation orders and demolition of buildings in conservation areas.
2 *Local Government Act 1972*, which makes county councils the local planning authorities responsible for the production of the structure plan and strategic development control, and district councils the local planning authorities responsible for the production of local plans and development control.

3   *Town and Country Amenities Act 1974*, which makes further provision for the control of development in the interest of amenity.
4   *Inner Urban Areas Act 1978*, which gives additional powers to local authorities with serious inner-city problems so that they may attempt to participate more effectively in the economic development of their areas.
5   *Local Government Planning and Land Act 1980*, which introduced a number of changes to the development planning system, including the establishment of corporations to regenerate urban areas (urban development corporations) and enterprise zones.

Regulations and Orders

These Acts of Parliament are supported by Regulations and Orders, the most significant of which are:

1   *Town and Country Planning (Structure and Local Plans) Regulations 1982* (SI 1982, No. 555), which make provision for the form and content of structure and local plans and the procedures to be followed in their preparation, submission, approval or adoption.
2   *Town and Country Planning General Development Order 1977* (SI 1977, No. 289) and subsequent amendments, which provide for the grant of permission for certain classes of development and establish the procedures to be followed in making and dealing with planning applications and the making of appeals to the Secretary of State.
3   *Town and Country Planning General Regulations 1976* (SI 1976, No. 1419) as amended in 1981, which set out the procedure whereby a local planning authority may obtain a deemed planning permission for development which it intends to carry out.
4   *Town and Country Planning (Inquiries Procedure) Rules 1974* (SI 1974, No. 419), which apply to appeals to the Secretary of State in connection with applications for planning permission.
5   *Town and Country Planning (Use Classes) Order 1972* (SI 1972, No. 1385), which specifies certain classes of use of building, and provides that changes of use within the same class are deemed not to involve development and, therefore, do not require planning permission.

Circulars

In addition to Acts, regulations and orders, central government issues circulars providing advice on planning matters. The most significant circulars currently in operation relating to the development plan and development control systems are:

1   *Memorandum on Structure and Local Plans: The Town and Country Planning Act 1971 Part II (as amended by the Town and Country Planning (Amendment) Act 1972, the Local Government Act 1972, and the Local Government, Planning and Land Act 1980)* (DoE Circular

22/84, Welsh Office Circular 43/84), which provides basic guidance to local planning authorities on the development plan system of structure and local plans.

2 *Development Control-Policy and Practice* (DoE Circular 22/80, Welsh Office Circular 40/80), which sets out basic advice and guidelines as to how and with what objectives local planning authorities should exercise their development control functions.

3 *Town and Country Planning Act 1971: Planning Gain* (DoE Circular 22/83, Welsh Office Circular 46/83), which gives guidance on the ways in which 'planning gain' can be sought by agreement in connection with the granting of planning permission.

4 *Local Government Planning and Land Act 1980 . . . Town and Country Planning: Development Control Functions* (DoE Circular 2/81, Welsh Office Circular 2/81), which deals with changes in the allocation of responsibility for development control between county and district planning authorities.

5 *Crown Land and Crown Development* (DoE Circular 18/84, Welsh Office Circular 37/84), which sets out the arrangements by which government departments consult local planning authorities about their proposals for development.

6 *The Use of Conditions in Planning Permissions* (DoE Circular 1/85, Welsh Office Circular 1/85), which provides guidance on the use of the power to impose conditions on planning permissions.*

Town and country planning by statute is quite clearly seen as being concerned with the production of plans for land utilization and the control of development carried out by others; its contribution to the achievement of social and economic change is only implicitly acknowledged; the complex inter-actions between policy, resources, politics and interest groups which lead to development on the ground are ignored.

In Scotland the legislative basis for the statutory town and country planning system is provided by the *Town and Country Planning (Scotland) Act 1972*, as amended by, amongst others, the *Town and Country Planning (Scotland) (Amendment) Act 1972*; the *Local Government (Scotland) Act 1973*; the *Town and Country Planning (Scotland) Act 1977*; the *Local Government Planning and Land Act 1980*; and the *Local Government and Planning (Scotland) Act 1982*. These Acts are supported by a range of regulations, circulars and advice notes which, although different in detail, generally follow the same principles as the planning system in England and Wales.[2]

---

* A number of other important circulars relating to substantive areas of planning policies are reviewed in Chapter 7.

1.3.2   Development plans under the *Town and Country Planning Act 1971* (as amended)

The legislative and administrative basis of the town and country planning system is dealt with in detail in Chapter 3. It is touched on briefly here to set local plans into context, by way of introduction. The development plan system provided for by Part II of the *Town and Country Planning Act 1971* (1972 in Scotland) (as amended) consists of two main elements: structure plans and local plans.

The structure plan

The structure plan consists of a *written statement* which sets out the main strategic land use planning policies for the area and the most important general proposals for change. These proposals and policies for change are not site-specific; they relate directly to the development and other use of land, including measures for the management of traffic and the improvement of the physical environment, and must take account of a whole range of related matters which influence the future of the area as a whole, e.g. policies established under other legislation for housing, social welfare, economic development. The written statement may be illustrated by diagrams which are not map-based. The structure plan must be accompanied by an explanatory memorandum which sets out the reasoned justification for the policies and general proposals adopted.

Structure plans are prepared by county planning authorities in England and Wales, and the regional and island councils in Scotland for all or part of their area. They must be approved by the appropriate Secretary of State. In approving the structure plan, s/he can make modifications to it. The Secretary of State(s) must consider all objections or representations made in accordance with the regulations (the *Town and Country Planning (Structure and Local Plans) Regulations 1982*). Objectors do not have the right to be heard at an inquiry. Rather, the Secretary of State appoints a panel to hold an examination in public (EIP) into those issues on which he feels further investigation and discussion are required before he can approve, with or without modification, the structure plan.

The local plan

The local plan develops the policies and general proposals of the structure plan and relates them to precise areas of land. It consists of a written statement which sets out the planning authority's proposals for the development and other use of land in the area covered by the plan, and the reasoned justification for those proposals. The plan is supported by a proposals map on on Ordnance Survey base. Although planning authorities in Scotland are required to prepare local plans for all parts of their districts (*Local Government (Scotland) Act 1973*), there is no such requirement in England and Wales. Here, county and district planning authorities are advised to prepare local plans only where there is a clear need. There is a presumption in the 1971 Act (as amended) that local plans will normally be produced by district councils.

Each county planning authority is required to prepare a *development plan scheme* in consultation with their districts, setting out a programme for local plan preparation in its area; establishing which authority is responsible for producing each plan and identifying the relative priorities (Bruton, 1983a). A local plan must conform generally to the approved structure plan and, where it has been produced by a district council, that council must obtain a *certificate of conformity* from the county confirming that it accords with structure plan policies. If the county refuses to issue such a certificate, the matters in dispute are referred to the Secretary of State.

Statutory local plans may take one of three forms:

1   A comprehensive consideration of matters affecting the development of a particular area (prior to the 1982 Regulations referred to as a *district plan*);
2   A *subject plan*, which shows in detail the authorities' proposals on a particular issue or group of issues, e.g. mineral extraction;
3   An *action area plan*, which is produced for areas where substantial change is expected within 10 years, e.g. comprehensive development or redevelopment.

Local plans are normally adopted by the authority responsible for their preparation, although the Secretary of State has the power to require a local plan to be submitted to him for approval. The adoption of a local plan has the effect of automatically revoking the 'old style' development plan within the area to which the local plan relates, unless the Secretary of State makes a continuation in force order (DoE, 1984g, para. 1.9). The old development plan may also be revoked by the Secretary of State, but otherwise remains in force, so that in practice the 'development plan' for an area may comprise the operative provisions of the approved structure plan, any adopted local plans, approved and adopted alterations to such plans, and the old development plan. The first local plan to be completed under the *Town and Country Planning Act 1971* was adopted in 1975 by Coventry City Council (the Eden Street Action Area Plan); since then progress has been steady, reflecting the substantial completion of the structure plan framework by the early 1980s. By the end of March 1985, 358 local plans had been adopted in England, with a further 229 having been placed 'on deposit' in order to allow objections and/or representations to be made.[3] These adopted and deposited plans represent just over a third (38 per cent) of the number of plans that will have been produced once current preparation programmes, as set out in the development plan schemes agreed between county and district planning authorities, are completed. Further details on the scale and pattern of local plan deposit and adoption are given in Chapter 4.

In general terms, statutory local plans provide a detailed basis for development control and for the co-ordination of development and other land use; they include land allocations for future development and set out criteria against which applications for planning permission will be assessed. The perception of planners as producers of plans and controllers of development is very much to the fore. At the same time, the dynamics of change and implementation are by implication relegated to a position of insignificance. A simple system of statutory land use planning and control is established to deal with a situation of extreme complexity and variety. Reflecting this, many local planning authorities prepare *informal* or *non-statutory* local planning policy documents, either as well as or instead of statutory local plans, in order to be able to respond more flexibly to the planning issues they face. Central government advice on non-statutory local planning documents sees a continuing role for what is referred to as 'supplementary planning guidance' (development control notes and development briefs) which is recognized as a material consideration at appeals arising out of development control. Other types of informal material, notably non-statutory local plans and policy frameworks, are firmly discouraged, with Circular 22/84 cautioning that they 'can have little weight for development control purposes', and advising that land use policies and proposals should be established in a statutory local plan (DoE, 1984g, para. 1.13). This advice has been consistently given in local plan notes and circulars since 1976. However, the professional literature suggests that many authorities have a continuing commitment to the use of proscribed non-statutory local plans and policy frameworks (District Planning Officers' Society, 1978, 1982; Hayes, 1981; Shaw, 1982), and this is confirmed by more recent work which sets out the extent to which informal documents are used by local planning authorities (Bruton and Nicholson, 1984a; see Chapter 6). Local planning authorities have responded to the variety of local conditions facing them in their efforts to translate general policies for change into action on the ground by producing both statutory local plans and informal policy documents as judged appropriate.

1.3.3 Development plans and development control
The system of development control established under the *Town and Country Planning Act 1971*, which provides that all development, subject to certain exceptions, requires planning permission from the local planning authority, is of key importance for the application and implementation of statutory local plan policies. This is because the adoption of a local plan does not in itself lead to additional statutory powers accruing to the authority, nor does it guarantee the availability of resources to put plan proposals into effect, or directly bind other local authority departments or public agencies (Urwin and Wenban-Smith, 1983). Statutory local plans have legal effect primarily in the context of decisions taken by the local planning authority on

applications for permission to develop. This is by virtue of s.29 of the *Town and Country Planning Act 1971*, which requires local authorities to 'have regard to the provisions of the development plan, so far as material to the application, and to any other material considerations' in determining planning applications. A broad interpretation has been placed on s.29 by the courts, whereby the development plan – incorporating statutory local plans where these have been prepared and adopted – is considered to be merely one consideration among the range of considerations relevant to any given application. Moreover, in applying the provisions of the development plan in development control, local planning authorities must perforce have regard to central government's view of the proper relationship between the two. The influence of central government here stems from its ability to determine appeals against local authority refusals of planning permission, deemed refusals and conditions on permissions. Central government guides local authorities as to the relative weight it expects them to place on the development plan (including statutory local plans) through both actual appeal decisions and through advice in circulars, which effectively constitutes advance warning of the degree of support that central government is likely to give to development plan policies in its determination of appeals. This guidance is theoretically applied in appeal cases by individual members of the Inspectorate, which is responsible for handling all planning appeals. Basic guidance and advice is provided on the inter-relationship between development plans and development control (DoE Circular 22/84, *Memorandum on Structure and Local Plans*) and on how the development control system should be operated (DoE Circulars 22/80, *Development Control – Policy and Practice*; 1/85, *The Use of Conditions in Planning Permission*). More recently, the significance of the development plan, including statutory local plans, in determining applications for planning permission has been explicitly clarified in DoE Circulars 15/84, *Land for Housing*; 16/84, *Industrial Development*; and 14/85, *Development and Employment*. In effect, the development plan is only one of the material considerations that must be taken into account in dealing with planning applications; it should not be regarded as overriding other material considerations, and plan policies and proposals cannot in themselves provide sufficient grounds for refusal of permission. It seems that the role of planners as producers of plans is being diminished, whilst the role of planners as controllers of development is being confirmed. As far as non-statutory local planning documents are concerned, the view of central government is that while 'supplementary planning guidance' is an acceptable material consideration at appeal, the use for development control purposes of informal local plans and policy frameworks must be firmly rejected. Nevertheless, local planning authorities can still legitimately have regard to these documents as 'other material considerations' in determining planning applications, the risk being that decisions based on the provisions of such informal material will be overturned at appeal. However, there is some evidence of non-statutory plans and frameworks receiving support at appeal (Bruton and Nicholson, 1984c), indicating that Inspectors do not always act as compliant agents for central government policy.

1.4 Government and the town and country planning system[4]

1.4.1 Town and country planning and social and economic change

It is clear from the preceding sections that in the United Kingdom town and country planning is implicitly seen by government as part of a wider set of inter-related planning activities which operate in different areas (e.g. housing, health, education, social services, transport, employment); at different levels (e.g. national, regional, sub-regional, local) and through different agencies (e.g. ministries, *ad hoc* bodies, local authorities) to achieve the social and economic change sought by the founders of the town planning movement. The intention is that town and country planning will contribute to the achievement of this change by shaping and guiding development and the use of land. Thus, the role of the town and country planner in society is concerned solely with development and land use, operating implicitly within a framework set by higher order social and economic policies.

This somewhat narrow view of the role of town and country planning would be perfectly acceptable and workable in practice

- if there were a clearly established framework for planning through which national objectives for social and economic change could be coherently presented
- through which more specific sectoral objectives and policies relating to issues such as the economy, income, employment, industry, agriculture, natural resources, housing, social services, and transport could be presented in an integrated way at the national level;
- through which these national socio-economic and sectoral policies could be amplified in relation to the various regions of the country; and
- through which these 'regional' interpretations of socio-economic policy could be translated into a spatial structure, which in turn could be translated where appropriate into local detailed plans, and projects for development and implementation.

If such a framework existed, then the role prescribed in legislation for town and country planning would be feasible. Regrettably, no such framework exists in the United Kingdom (Bruton and Nicholson, 1985a). At the national level, policies for social and economic change are articulated in a disjointed way through White Papers, Green Papers, Acts of Parliament, departmental circulars and memoranda, and fiscal measures. With the exception of the short-lived National Plan in 1964 (Department of Economic Affairs, 1964), no attempt has been made to establish explicitly the following:

1 National objectives, policies and proposals for social and economic change.
2 More specific sectoral objectives and policies relating to issues such as the economy, industry, housing, natural resources, etc.
3 The allocation of resources to implement these policies.
4 The broad inter-relationship between these policies and regional and urban development.

At the regional level in England and Wales, attempts to relate national socio-economic and sectoral policies to specific regions have at best been half-hearted. For the period 1965–79, the Regional Economic Councils attempted to prepare broad advisory strategies for social, economic and physical development, although the quality of these strategies varied enormously (Glasson, 1978, pp. 253–299). In some parts of the country local authorities took the initiative in attempting to plan at this level and created standing conferences on regional planning, e.g. West Midlands Planning Authorities Conference (WMPAC), whilst in other parts of the country sub-regional planning studies were produced either by independent teams for a group of local authorities, e.g. Coventry–Solihull–Warwickshire or by teams commissioned by central government to examine the feasibility of large-scale planned expansion of selected areas of the country, e.g. South Hampshire.[5] Generally, however, these plans or reports failed to amplify the national socio-economic sectoral policies insofar as they affected their particular region, and (with exceptions such as WMPAC and the sub-regional studies) only rarely produced positive proposals for action. Thus the essential socio-economic framework which land use planning (or town and country planning) needs if it is to be used in a positive way to contribute to socio-economic change, is missing. Land use planning, with its narrow but clearly defined role, is forced to operate in a vacuum. How can it be used to give physical expression to national social and economic policies through structure plans and local plans when those policies are not clearly and coherently presented? The very role prescribed for town planning by legislation is one which it cannot presently achieve. This conceptual weakness which underpins our current system of land use planning has inevitably had a pronounced effect on the role of the planner in our society. It quite clearly prescribes a narrow land use based role for the planner. At the same time, by not providing a coherent higher order planning framework, it makes the successful operation of that narrow role almost impossible.

This imperfect situation is made even more difficult by the other implicit but fallacious assumption underpinning the legislation which provides for the current land use planning system, i.e. that planning as an activity is neutral and objective. If it were, then the idealized and comprehensive planning framework needed for a narrowly defined land use based planning system to operate effectively might solve the problems encountered for the planner in identifying his role. But planning is not neutral. Rather it is involved centrally in the distribution of scarce resources; conflicts of interest are an inevitable consequence of that distribution. Planning, far from being apolitical, is essentially a *political* activity concerned with the regulation of disagreements about matters of public choice (Rose, 1974). As such, it is inevitably partial, *ad hoc*, subject to pressures and much influenced and constrained by the political powers of the day.

In these circumstances, with an inadequate planning framework and the assumption that the planning function is apolitical, it is not surprising that the planner is confused and uncertain as to the role and purpose of the town and country planning system as provided by statute. On the one hand, the temptation must be to go beyond the boundaries of the statutory planning system in an attempt to address the complexities inherent in any consideration of economic and social issues. On the other hand, a position which is concerned solely with the use and development of land offers the security and professional legitimacy which stems from an established legislative base.

In Scotland, the situation is not so bleak – the existence of a clear regional and national identity; the recognition of the problems and opportunities facing the country; the semi-independent position of the Scottish Office – these factors appear to have encouraged (or, at least, allowed) the establishment of a planning machinery which is more explicitly designed to interrelate social, economic and physical planning. The National Planning Guidelines (Diamond, 1979); the Regional Report (McDonald, 1977); the liaison between local authorities and the Scottish Development Department (SDD); and the form of some of the structure plans suggest that a framework exists which allows town and country planning to operate approximately in the way central government implies it should.

1.4.2   Town and country planning: 1968–1979

In the years following the introduction of the *Town and Country Planning Act 1968*, planning practice operated as if a comprehensive and rational system of planning was in being. The structure plan was claimed by some to be '. . . a policy vehicle and not a means for expressing physical development proposals in detail. It had to set out the social, economic and environmental strategies for the area' (Drake *et al.*, 1975, p. 12). Advice from central government through various publications e.g. *Management Networks: A Study for Structure Plans* (DoE, 1971); the *Development Plans Manual* (Ministry of Housing and Local Government, 1970) encouraged this view, as did the insistence of the Planning Advisory Group that the new development plans should be based upon '. . . a much better understanding of the determinants of urban development, urban form and the functions of the countryside, and much greater knowledge of the social needs and aspirations of the community' (PAG, 1965, p. 53). At the same time, the Group pointed out that 'the problems of physical development and redevelopment . . . are often highly complex, involve investment decisions of great magnitude and extend across many related fields of policy. Many different agencies, public and private, are involved and many different interests are affected' (PAG, 1965, p.3). Planners attempted to produce plans based on a clear articulation of social, economic and environmental aims and objectives; alternative futures were forecast and evaluated using complex computer-based models; rationality and the systems approach dominated the profession.

Within a matter of years this broad approach was challenged and local authorities were advised in Circular 98/74 to be selective with regard to the policy content of structure plans, and to concentrate on the key issues of structural importance to the area concerned, i.e. to produce a mainly physical development policy vehicle which took account of social and economic matters (DoE, 1974c). This advice, whilst being consistent with central government's view of the role of town and country planning, ignored the difficulties of the system operating in a socio-economic policy vacuum.

During this period, the position of the development plan as almost the only policy formulation vehicle was challenged with the introduction of a number of new policy-making procedures in the public sector (see Chapter 5). For example, the *Housing Act 1969* introduced the General Improvement Area as a means of stimulating area-based improvement; Transport Policies and Programmes (TPPs) – in effect annual plans for investment into transport for a five yearly period – were introduced in 1975; Housing Strategy and Investment Programmes (HIPs) in 1978; the *Transport Act 1978* introduced the Public Transport Plan (PTP); the *Housing Act 1974* introduced the Housing Action Area (HAA). Industrial Improvement Areas (IIAs) followed. The *Water Act 1973* established the regional water authorities, who began to prepare their own plans and policies; the *Local Authority (Social Services) Act 1970* created separate departments for social services within local authorities; the *National Health Service Reorganization Act 1973* created the regional and area health authorities who saw the reorganization as an opportunity '. . . for health care planning to be comprehensive and co-ordinated with the planning of related local authority services' (Department of Health and Social Security, 1976, p. 3), whilst the inner-city problems have been singled out for special treatment under the urban aid programme and the *Inner Urban Areas Act 1978*. The net effect of these changes, which took place under both Labour and Conservative 'centrist' administrations, was to:

1 force town and country planning to retreat to the position earmarked for it by statute – a concern with the use and development of land; and
2 to introduce a short time scale, resource-based approach to planning characterized by a focus on particular areas or problems.

The indications are that these changes and additions to the system were introduced in an *ad hoc*, pragmatic way. By contrast, the changes to the system which have been introduced since the current Conservative government took office in 1979 would appear consistently and systematically to be seeking to change the very basis of the town and country planning system first introduced by the *Town and Country Planning Act 1947*.

1.4.3   Changes to the town and country planning system, 1979–1985

The underlying objective of the current (1986) Conservative government is to attempt to regenerate the economy and to make British industry more productive and competitive in trading terms. Shortly after their election in May 1979 the government made it clear that town and country planning is to contribute to this regeneration by becoming more positive and efficient; that the private sector is to become more involved in environmental and planning matters; and that the system is to be used to conserve the nation's heritage.[6]

The actions of the government since 1979 in modifying the town and country planning system confirm that their objectives for changing the emphasis of the system were more than just 'good intentions'. The cumulative effect of the changes introduced to date is a re-affirmation that town and country planning has a limited and specific concern with land use allocation, and a further limitation on the powers of local planning authorities through the formal involvement of the private sector and the centralization of important decisions on development proposals. Early in the life of this government, Stuart Gilbert (then DoE Under Secretary for Land Use Policy) suggested that town and country planning could make a contribution to the regeneration of the economy by not taking enforcement action against unauthorized small businesses, especially transport operations, unless a replacement site was available.[7] This informal advice was followed by DoE Circular 22/80, *Development Control – Policy and Practice*, which gave the same advice in more formal terms, as well as setting out the role that development control could play in encouraging business activity and the formation of small businesses. The 1977 General Development Order was amended in 1981 and increased the limits of permitted development for industrial applications, as well as exempting certain changes of use involving light industry and warehousing from the need to obtain planning permission.[8] Land in public ownership surplus to the statutory requirements for which it had been acquired was (and is) to be disposed of to the private sector, whilst in December 1981 the need to obtain an Industrial Development Certificate (IDC) to carry out certain types of industrial development in the more prosperous parts of the country (primarily the South-East and the West Midlands) was revoked.[9] These changes were designed to remove from industry some of the constraints imposed by the planning system, which were seen by the government as inhibiting the growth of the economy. The abolition of IDC controls effectively removed the redistributive teeth of post-1945 industrial location policy, and in doing so ensured that the social-welfare objective firmly established by Barlow, i.e. to secure a reasonable balance of industrial development throughout Great Britain, would henceforth be little more than a pipe-dream.

More recently, DoE Circulars 16/84 *Industrial Development* and 14/85 *Development and Employment* and the White Paper *Lifting the Burden*[10] seek to remove from the business sector what are seen by the government as unnecessarily restrictive land use planning constraints. Thus, the White Paper '. . . sets out the case for more freedom in the business sector and the need to deregulate' (para. 1.2); whilst the two circulars emphasize that development plan policies are only one of a range of material considerations to be considered in determining planning applications. More dramatically, Circular 14/85 argues that 'There is therefore always a presumption in favour of allowing applications for development, having regard to all material considerations, unless that development would cause demonstrable harm to interests of acknowledged importance' (DoE, 1985a, para. 3). It also points up what those 'interests of acknowledged importance' are, viz., 'Development proposals are not always acceptable. There are other important objectives to which the government is firmly committed: the need to preserve our heritage, to improve the quality of the environment, to protect the green belts and conserve good agricultural land' (*ibid.*, para. 3). A land use based planning system is pushed further towards its land use basis.

As well as modifying the statutory town and country planning system in these ways, the government has also introduced a number of policy initiatives which have been established largely independently of the formal land use planning system (see Chapter 5). Many of these measures demonstrate a concern to involve the private sector as a source of funding and expertise, and characteristically focus on major problem issues and/or areas in an isolated way, at the expense of a more comprehensive and inter-related approach. Perhaps the most obvious instance of attempts to secure private sector involvement was the establishment of the Financial Institutions Group (FIG) consisting of 26 managers seconded from financial institutions to the DoE for one year, following the 1981 Toxteth riots in Liverpool, to develop new ideas and approaches for securing urban regeneration. The proposals from FIG which have met with the most enthusiastic support from the government are those which involve co-operation between the public and private sectors. The best known of these is the Urban Development Grant, whereby public money is allocated to support urban development proposals put forward jointly by local authorities and private developers acting in partnership.[11] Other attempts to involve the private sector include Operation Groundwork, a programme of urban fringe land rejuvenation initially based in the North-West of England centred on a co-ordinated effort by public, private and voluntary agencies to achieve environmental improvement; the Derelict Land Act 1982, which attempts to involve the private sector as well as statutory undertakers and nationalized industries in the reclamation of derelict land by paying grants of up to 80 per cent of the net loss incurred in reclamation; and the appointment of two prominent businessmen to chair the Merseyside and London Docklands Urban Development Corporations, which was seen as a move to establish confidence and gain support from the private sector in restoring and regenerating two of the most neglected areas in the country.[12] All these changes can be justified in one way or another, e.g. using private sector money for the public good; using hitherto untapped private sector expertise in a novel way. However, the implications of the changes are that the town and country planning system is pushed gently into a more positive concern with land use and development matters. At the same time it becomes further fragmented and even more difficult to co-ordinate.

A number of new initiatives and measures illustrate the tendency to deal with particular problem issues and areas in an isolated fashion: Enterprise Zones, urban development corporations, task forces, the use of special development orders. Enterprise Zones for instance, designated by the Secretary of State under the *Local Government, Planning and Land Act 1980*, are seen by central government as a '. . . bold new experiment . . . where businesses can be freed from much detailed planning control and from rates' (Tom King, then Minister for Local Government and Environmental Services, quoted by Heap, 1982, p. 237). The prime aim of the Enterprise Zone initiative is to see how far industrial and commercial activity can be encouraged within their boundaries by the removal of tax burdens and some statutory planning and other controls. In size they vary from 50 to 400 hectares, and all contain land considered to be ripe for development. Although some planning controls have been maintained in the zones, their introduction marks a further reduction in the ability of the town and country planning system to influence the form of the built environment. The fact that central government can proudly advertise that '. . . Enterprise Zones are not part of regional policy, nor are they directly connected with other existing policies such as those for inner cities or derelict land' (DoE, 1981b, p. 3) supports the suggestion that the system is being modified to take positive but isolated action on what are seen to be specific problem issues and/or areas. A concern for the inter-related nature of social, economic and physical problems is noticeably lacking in this attitude – an impression which is reinforced by the decision of the Conservative government in 1979 to abolish the Regional Economic Planning Councils which were at least credible vehicles for attempting to plan at a regional scale. The perceived success of the relaxed regime of planning controls in Enterprise Zones has led to proposals in the White Paper *Lifting the Burden* (Cmnd. 9571; see also DoE, 1984a) to introduce new legislation to permit the setting up of 'simplified planning zones' – areas defined by local planning authorities where the requirement to obtain planning permission for specified categories of development would be waived. A further erosion of the statutory land use planning system appears inevitable, particularly given that (*a*) it is intended to give the Secretaries of State reserve powers to direct the preparation of proposals for a simplified planning zone, and (*b*) local planning authorities will be required to consider proposals for the establishment of simplified planning zones initiated by private developers.

The *Local Government, Planning and Land Act 1980* makes provision for the Secretary of State to designate an *urban development area* in order to secure its economic, social and physical regeneration, with an *ad hoc* body – *the ubran development corporation* – being appointed to achieve that regeneration. To date, two such areas have been designated and are the responsibility of the London Docklands and the Merseyside Development Corporations respectively. The objectives of these development corporations are to bring land and buildings into effective use; to encourage the development of new industry and commerce; to create an attractive environment; and to ensure that housing and social facilities are available to encourage people to live and work in the area. To achieve these objectives, the development corporations have extensive powers of planning and development control, and of land acquisition, management, development and reclamation, and disposal.

The concept of the urban development corporation was vigorously opposed by the local authorities, the opposition parties and the unions, largely on the grounds that the corporations would not be politically accountable to a local electorate and that the rationale for their introduction was based on a false premise, i.e. that administrative and organizational problems accounted for the decline of the areas designated as urban development areas, rather than structural economic and social factors. Moreover, the principle of the concept marks a severe threat to the nature and role of town and country planning, even though the two restricted areas designated to date do not significantly reduce or fragment the role and scope of the land use planning system as a whole. If further and more extensive urban development areas are designated, the effectiveness of the town and country planning system will inevitably be further reduced. *Ad hoc* bodies which are not politically accountable will be responsible for producing and implementing development and investment strategies for areas of the inner cities in a manner which ignores the wider implications of those developments. Once more positive but isolated action is being taken on what are seen to be specific problem areas and/or issues. At the same time, further power is concentrated in the hands of central government.

The same philosophy of taking positive action on particular problem issues or areas underpins the introduction of the *task force* as a new element in the government's attempts to wrestle with the problems of decline in our major towns and cities. Following his visits to Merseyside to explore at first hand the social, economic and physical problems in the area, the then Secretary of State for the Environment, Michael Heseltine, announced in

October 1981 that he would head a task force to be established in Merseyside. The role of this task force is described as being to '. . . bring together and concentrate the activities of central government departments and to work with local government and the private sector to find ways of strengthening the economy and improving the environment in Merseyside.'[13] The task force, the responsibilities of which extend beyond the boundaries of Merseyside to include the whole of the Merseyside Special Development Area and Runcorn and Skelmersdale New Towns, was criticized when it was introduced on the grounds that it created a new tier of government to co-ordinate the work of central government with that of the local authorities.[14] Whilst the task force concept challenges the traditional role of the local authorities in dealing with these issues, and concentrates more planning power in the hands of central government, the way in which it has been introduced in Merseyside suggests that by 1981 Heseltine was beginning to understand that the problems facing the declining industrial centres of the United Kingdom cannot be isolated and treated by relatively small *ad hoc* area-based approaches.

An even greater threat to the powers of local planning authorities is the potential use of *Special Development Orders* (SDOs) under s.24(3) of the *Town and Country Planning Act 1971* to grant planning permission. The consultation letter from the DoE to local authorities dealing with SDOs issued in 1981 states 'The purpose of making fuller use of SDOs would not be to make any general relaxation in development control, but to stimulate planned development in acceptable locations, and speed up the planning process'.[15] Critics of the proposal at the time commented that whilst the use of SDOs could lead to flexibility in implementing planning policies, it could also result in a loss of local control over certain categories of development, with an inevitable reduction in local democratic control over the planning process.[16] The way in which SDOs have been used, or their use threatened, tends to confirm that local democratic control of important development control issues will be removed. The first indication that this change in central–local government relations was likely to become a reality was the linking of an SDO with an architectural competition for the South Bank in London.[17] The use of an SDO to grant planning permission for the Mercury consortium's telecommunications network tended to blur the issue as it was generally accepted as being a sensible application of SDO powers.[18] However, subsequent events make it clear that this government intends to use SDOs as a means of securing the implementation of development proposals in the face of strong opposition from the local authorities directly concerned. For example, the Vauxhall Cross SDO, which granted planning permission for the winning design in an architectural competition, despite opposition from the local authority and the M.P. representing the area;[19] the threat of the Inspector holding the 1982 Inquiry into the Third London Airport that he might be forced to recommend to the Secretary of State that he use an

SDO to ensure that in the face of opposition from the mid-Herts and Essex authorities, housing land is released for development quickly enough if the Stansted expansion proposals are agreed;[20] the threat by the Secretary of State to use an SDO for the Hays Wharf development on the South Bank.[21] Once again, further power is being concentrated in the hands of central government.

Despite these apparently *ad hoc* and incremental changes, it can be argued that in practice the town and country planning system introduced by the *Town and Country Planning Acts (1968* and *1971)* remains largely intact. The county councils are still responsible for the production of structure plans, the district councils for the production of local plans and development control. Since 1979 the Secretary of State has only marginally modified the system, for instance by making provision for structure plan modifications to be approved without an examination in public; by reducing the period for public participation; by making provision to dispense with the local plans public local inquiry; and by transferring some development control responsibilities from county to district authorities. Thus it could be argued that this government has not seriously interfered with the town and country planning system introduced following the report of the Planning Advisory Group. However, policy changes introduced in DoE Circulars 15/84 *Land for Housing*; 16/84 *Industrial Development*, and 14/85 *Development and Employment*, which effectively remove much of the perceived authority previously associated with policies set out in the development plan, belie this argument. As regards employment-related development for example, Circular 14/85 makes explicit the presumption that planning permission should always be granted unless such developments '. . . would cause demonstrable harm to interests of acknowledged importance' (para. 3). The same circular also emphasizes that development plans '. . . should not be regarded as overriding other material considerations, especially where the plan does not deal adequately with new types of development or is no longer relevant to today's needs and conditions . . . .' (para. 5). Similar emphases run through Circulars 15/84 *Land for Housing* and 16/84 *Industrial Development*. In this context there has been a shift in the balance of power away from the development plan in favour of a more *ad hoc* approach in development control.

At the same time (*a*) the emphasis given to the importance of conserving the nation's heritage, improving the quality of the environment, protecting green belts and conserving good agricultural land in recent DoE circulars such as 14/84 and 14/85, and (*b*) the modifications the Secretary of State has made to structure plans submitted to him for approval, suggest that the narrow land use allocation basis of the system is to be given even greater prominence in the immediate future. In his decisions he is only allowing social and economic policies to be retained in the structure plan where they are used as reasoned justification in support of land use policies. Whilst this is entirely consistent with the legislative view of the role of the town and country planner, the absence in England and Wales of the requirement for the local planning or any other authority to produce a clearly articulated set of social and economic policies ensures that this is a variable feature of structure plans (Jowell and Noble, 1981). Other modifications prior to the approval of structure plans indicate that the Secretary of State is concerned to ease the constraints placed on potential developers by making policies less restrictive, e.g. in approving the Berkshire County Structure Plan he ordered that additional land be released to accommodate a further 8000 dwellings;[22] in approving the Somerset County Structure Plan he instructed that land allocated for housing should be increased to accommodate a further 4400 dwellings and the industrial land allocation should initially be increased from 153 hectares to 216 hectares, which was subsequently further increased to 246 hectares.[23]

The narrow land use basis of the town and country planning system presented by legislation is also emphasized by the most recent advice from the DoE on structure and local plans in Circular 22/84 *Memorandum on Structure and Local Plans* (DoE, 1984g, para. 4.2). Thus:

. . . non-land use matters, for example, financial support, consultation arrangements and proposed methods of implementation should not be included as policies or proposals in structure or local plans. These should, however, be included in the explanatory memorandum or reasoned justification if they are relevant to a full understanding of a plan's policies or proposals or provide a context for them.

The full significance of this section is made clearer when it is read in conjunction with an earlier draft version of Circular 22/84 (DoE, 1982, para. 4.2) which stated:

. . . non-land use matters, for example, financial support, subsidies, and car parking charges and matters arising from non-planning functions and powers of the local planning authority and of other authorities and public bodies should not be included in a structure or local plan as policies or proposals. These may be included in the explanatory memorandum or reasoned justification if they are relevant to the plan's policies or proposals, but only where such a reference will play a useful part in explaining or justifying the plan's policies and proposals.

Thus the range of methods which are essential to the management of change in the environment are denied the town and country planner. Similarly the review of the structure plan, which according to previous advice given in Circular 4/79 should be rolled forward every five years (para. 2.4), is now to be restricted to a concern with keeping abreast of major changes and should avoid unnecessary work (DoE, 1984g, para. 2.3). In the same vein, the advice in Circular 22/84 relating to local plans stresses that the need to produce a report of survey should be rare (para. 3.21); that serious consideration should be given to preparing one district-wide local plan rather than numerous part-district plans (para. 3.14), and that informal (non-statutory) plans should not be produced (para. 1.13). Whilst there is evidence to suggest that local authorities engaged in serious over-kill in producing early structure and local plans, the emphasis of the advice in the draft revision of Circular 4/79 was seen by many to mark '. . . a further stage in the limitation of any effective role for development plans . . . [and] . . . takes too restricted a view of the role of development planning.'[24] Certainly the restricted content of local and structure plans, allied with the prohibition of non-statutory local plans if adhered to by the planning authorities would ensure that plans produced within the town and country planning system would be simply land use allocation proposals.

Attempts have been made to speed up the processes of plan preparation and development control. In plan-making, public participation has been reduced to a minimum; the obligation in all cases to hold an EIP or PLI has been removed, and authorities are encouraged to reduce survey work to the necessary minimum. On the development-control side, amendments to the General Development Order in 1981 relaxed the limits of permitted development for industrial and residential purposes in certain cases;[25] local authorities are encouraged to improve the speed with which they process planning applications through the publication of a 'league table' of development control performance; DoE Circular 22/80 *Development Control – Policy and Practice* advises local planning authorities '. . . always to grant planning permission . . . unless there are sound and clear-cut reasons for refusal' (para. 3); and attempts were made to speed up the rate of handling planning appeals through the use of part-time Inspectors and the delegation to Inspectors of powers to determine appeals.[26]

The use of the town and country planning system to conserve and protect the nation's heritage has received considerable emphasis, despite the negative powers of control involved that have been explicitly relaxed in other areas of land use policy. In the words of the then Secretary of State for the Environment, Michael Heseltine,

I referred to conservation and protection of the environment, in which area I believe the planning system has achieved a very great deal. We have prevented much urban sprawl and ribbon development. We have saved buildings that need to be saved. The National Parks, the AONBs, the listed building machinery, the SSSIs, the Green Belt – to a greater or lesser extent, here lie the great successes of the planning mechanism. The ability to say 'no' – for that is what virtually all these policies involve – has contributed in an invaluable and vastly satisfying way. But you will have spotted the point. The successes I have identified involve the use of negative powers.[27]

Such powers continue to be used to protect the environment. For instance, the Special Development Order of 1981 restricted the extent of permitted development in National Parks, Areas of Outstanding Natural Beauty and Conservation Areas,[28] while more positively emphasis has been put on the need to reclaim derelict land, usually as part of a package of proposals designed to involve the private sector in regenerating areas of decline and decay in our major cities, e.g. Operation Groundwork North-West; the *Derelict Land Act 1982*. The Secretary of State has used his powers to protect the nation's heritage where major development proposals have threatened that heritage, e.g. he rejected proposals by the North-West Water Authority to raise the level of Ennerdale Lake; he rejected proposals by British Nuclear Fuels Ltd to take more water from Wast Water;[29] he 'saved' listed buildings from demolition in Stroud High Street;[30] he refused to provide grant aid for proposals to carry out drainage improvements in Halvergate Marshes, Norfolk, because of the threat the proposals carried for wildlife and natural vegetation in the area;[31] he invoked an Article 10 direction preventing Leicestershire County Council from giving planning permission to the National Coal Board's revised proposals for mining Belvoir;[32] he even went so far as to establish a new quango – the Commission for Ancient Monuments and Historic Buildings – with the intention of protecting the country's historic heritage.[33] Circular 12/81 *Historic Buildings and Conservation Areas* clearly articulates the government's policies to 'preserve the best of the nation's heritage' (DoE, 1981a, para. 3), whilst Circular 14/85 *Development and Employment* again re-affirms the importance of conservation, green belts, agricultural land and quality in the environment (*ibid.*, para. 2).

Two new pieces of conservation legislation have also been introduced: the *Local Government and Planning (Amendment) Act 1981* which has the effect of increasing local authority powers in relation to listed buildings and waste land; and the *Wildlife and Countryside Act 1981*, which provides for the protection of wild birds, animals and plants; public rights of way; the designation of SSSIs; and makes it possible for the Ministry of Agriculture to give grants for purposes other than agricultural production.

Finally, the decision to abolish the Greater London Council (GLC) and the metropolitan county councils on the 1st April 1986, together with the proposal that a system of Unitary Development Plans should be introduced for these areas, is yet another example of the erosion of the concept and practice of strategic planning. Rather than a strategic overview providing a framework within which lower level authorities can produce more detailed local plans, the intention is that the London boroughs and the metropolitan districts should produce their own hybrid unitary development plans incorporating both a general policy statement and the more detailed policies associated with a local plan. Under this system the Secretary of State for the Environment is likely to control strategic policy whilst the parochial interests of local planning authorities will almost certainly dominate local issues (GLC, 1985). Centralization and fragmentation of planning powers seem inevitable.

1.4.4   Overview

Since 1947 town and country planning legislation has quite clearly prescribed a narrow land use based role for the planner. No explicit reference is
made to the objectives for social and economic change which so motivated
the founders of the town planning movement. On the contrary, successive
circulars advising on how the system should be operated make it clear that
the structure plan is concerned to express social, economic and environmental policies established at the national and regional level in terms of their
effect on land use, environmental development and the associated transport
system. Local plans are concerned to develop the policy and general proposals of the structure plan, to provide a detailed basis for development control,
and to co-ordinate the development and other use of land.

The assumption would appear to be that there is a clearly established
planning framework through which national objectives and policies for
social and economic change are clearly articulated; through which the
regional implications of these social and economic objectives and policies
can be established to provide a framework for the structure plan, which in
turn expresses these social and economic policies in terms of their effect on
land use. The structure plan provides the guiding framework for the more
detailed local plans needed to guide and control development. If such a
hierarchy of planning levels existed then the role for town and country
planning established in legislation would probably be feasible and proper.

As it is, national policies for social and economic change are presented in
an *ad hoc* way, through legislation, regulations, circulars, White Papers,
Green Papers and fiscal measures. At the regional level, although attempts
were made to give some coherence to the establishment of socio-economic
strategies for the English Standard Regions, the results were largely ineffectual and in practice the structure plan operated in a vacuum. This position
was formalized in 1979 when the Regional Economic Councils were
abolished. Thus the town and country planner is expected to restrict his
concern to the development and other use of land without having clear
guidance as to the social and economic change that is being sought by the
government. The socio-economic policies and rationale on which the town
and country planning system should be based is uncertain; as a result the
planner faces a dilemma. Should s/he attempt as part of the structure plan to
articulate socio-economic policies within which general proposals for the
use and development of land can be established? Or should s/he make
assumptions about this socio-economic framework and focus on the narrow
land use allocation role? If s/he pursues the former course of action then any
socio-economic policies set out in the structure plan will be edited out by the
Secretary of State, or downgraded to the status of reasoned justification. If
s/he focuses on establishing policies for the use and development of land, the
very basis of those policies is undermined and the scope of the plan successfully to co-ordinate development is seriously impaired.

Despite these difficulties of operating a land use planning system in a socio-economic policy vacuum in the period between 1947 and 1979, central government quietly but consistently insisted that that was how it should be. The conceptual weaknesses underpinning the system ensured that it could not operate effectively. Since the return of the Conservative government in 1979 this narrow land use role of town and country planning has been re-affirmed – through modifications made to structure plans prior to approval; through new legislation, regulations and circulars. At the same time, the changes introduced into the system have begun to reduce the powers of the once comprehensive land use planning machinery by establishing *ad hoc* bodies to deal with particular problem issues or areas in isolation; by concentrating important planning decisions in the hands of central government, for example through the use of SDOs; by abolishing the Regional Economic Councils; by establishing *ad hoc* bodies such as the Merseyside task force; by involving the private sector more centrally in the process, and by encouraging a more *laissez-faire* attitude towards development to be adopted.

Many of the changes introduced are of themselves worthwhile. However to reinforce the land use basis of the town and country planning system without providing a clear framework for the articulation of socio-economic objectives and policies for change will do little to improve things. The uncertainties which have been inherent in the system since 1947 are inevitably reinforced. At the same time the cumulative effect of these changes is likely to reduce the powers of the local planning authorities to little more than a mechanistic process for controlling development, and protecting the nation's heritage – a process which has more in common with the way in which sanitary and land use controls were imposed in the last years of the nineteenth century and the early years of the twentieth century, than with the management of the complex and inter-related problems of our society.

1.5   Conclusions

The statutory town and country planning system is generally seen as the means by which physical development arising out of the activities of both the private and public sectors is planned and controlled in the public interest. Town and country planners produce land use plans of a form and with a content which is precisely defined; town and country planners control development in the light of policies and proposals in the land use plans *and* any other consideration material to the proposed development. Implicit in the legislation and the way the profession has operated is the assumption that town and country planning can somehow be separated from economic, social and political realities; that land use problems are simple and easy to resolve through the production of plans and the application of the development control process.

In reality, town and country planning forms one element in the wider field of public policy-making, which includes all actions of government from defence, through to education, welfare, housing and transportation. As a process '. . . public policy involves something to be affected in a particular way to secure a particular result, together with a statement as to how that result is to be achieved, and the necessary actions taken to implement the statement and achieve the desired results' (Goldsmith, 1980, p. 23). Public policies can be *distributive*, which generally benefit everybody; *constituent*, which in effect are about the rules of the game, e.g. election laws; *regulative*, which are designed to regulate behaviour, e.g. public health laws; and *redistributive*, which generally take from one group and give to another. As an outcome of implementing public policy, the consequences can be both intended and unintended. In the words of Goldsmith (1980, p. 23) public policies:

. . . may achieve what they set out to do, or they may have other consequences not all of which may be considered desirable. For example, many British housing policies have been designed to give private tenants both security of tenure and relatively low rents. One consequence of these policies has been that the private rental sector has virtually dried up, as landlords have found investment in private rental housing increasingly uneconomical, whilst another consequence has been to hasten the decline in the physical standards of much privately rented housing, as landlords have been unable or unwilling to undertake the necessary upkeep of their property. As a consequence, the need for urban renewal and rehabilitation has probably been higher than might otherwise have been the case, and it is doubtful whether these consequences of central government policy towards the privately rented sector were either foreseen or intended.

Town and country planning, as part of this wider field of public policy, is thus concerned with a range of complex, inter-related problems; problems which Rittel and Webber (1973) call 'wicked problems'. It is also concerned with policies which are distributive, regulative and redistributive. At the same time, town and country planning policies and proposals, as with public policy, are in part determined by the perceptions that local decision-makers have of their environment and their ideological stance. To quote Goldsmith again, the subjective understanding of the environment in which public policy makers operate '. . . is a closely interwoven mixture of perceptions, images, action precepts and so on, comprising ideology, attitudes and opinions, which, mixed together, determine the way policy-makers perceive and react to policy problems' (1980, p. 41). In effect, as well as dealing with complex 'wicked' problems, the town and country planning system will be operated in a way which reflects the local realities which, in management terms, are known as contingent factors. The work of Dearlove (1973) on Kensington and Chelsea; Young and Kramer (1978) on overspill housing in

Bromley; Jon Gower Davies (1972) on Newcastle and Dennis (1972) on Sunderland is evidence of this.

Against this background it is clear that town and country planning, as part of the wider field of public policy, is concerned with managing change in the environment through the production of land use plans and the control of development. As such, it is distributive, redistributive and regulative; it is thus political and concerned with conflicts of interest. Local planning, as part of the process of managing environmental change, has the same characteristics. This book is concerned to examine the operation of local planning in practice in the United Kingdom by reviewing the following:

1 The theoretical perspectives relating to complex 'wicked' problems; uncertainty, management and contingent factors; strategic planning; bargaining and conflicts of interest in the planning process (Chapter 2).
2 The operation of the statutory land use planning system (Chapters 3 and 4) and other formal area approaches and policy initiatives (Chapter 5).
3 The use of non-statutory local planning instruments (Chapter 6).
4 The inter-relationship between development control and local planning (Chapter 7) and between development processes and local planning (Chapter 8).

Throughout, land use planning and town and country planning are synonomous, and refer to the narrow definition of this activity as implied in the legislation, i.e. land use planning is concerned with the production of land use plans and the control of development. The land use planning system provided by the *Town and Country Planning Act 1971* thus embraces the preparation of structure and local plans and the operation of development control. Local planning, on the other hand, is seen as the management of change in the local environment – an activity which involves the establishment of plans and policies for change, which can be set out in statutory local plans, informal plans and policy frameworks, and in a range of other formal programmes, area approaches and policy initiatives with a bearing on land use and development; the provision of resources to implement the proposals for change; monitoring and controlling the change which takes place, and encouraging or influencing other bodies or individuals to contribute positively to the achievement of the desired change.

Notes

1 Parliamentary Debates on the Housing, Town Planning etc. Bill; *House of Commons Debates*, Vol. **188**, May 1908.
2 See the current *List of Department Circulars and Memoranda, Acts and Statutory Instruments* (Scottish Development Department and Scottish Office Finance Division) for details.

3 Department of the Environment schedules of local plan deposit and adoption in England up to the end of March 1985.

4 The content of this section appeared in a modified form in an article by M. J. Bruton in *Planning Outlook*, Vol. **25**, No. 3, 1982, pp. 81–88.

5 *A Developing Strategy for the West Midlands*, West Midlands Planning Authorities Conference, Birmingham, 1974; *Coventry–Solihull–Warwickshire: A Strategy for the Sub-Region*, Coventry City Council, Solihull Borough Council and Warwickshire County Council, 1971; Buchanan and Partners, *South Hampshire Study: Report on the Feasibility of Major Urban Growth*, HMSO, London, 1966.

6 'Tories pledge to simplify planning', *Planning*, No. 318, 18 May 1979, p. 1; 'Heseltine outlines his main priorities', *Planning*, No. 323, 22 June 1979, p. 1 for a summary of Conservative policy for town and country planning following their return to power. Also: Michael Heseltine on 'Positive planning', *DOE Press Notice*, 10 September 1981, speaking to the RTPI Summer School at Nottingham.

7 Quoted in *Planning*, No. 319, 25 May 1979, p. 5.

8 *Town and Country Planning General Development (Amendment) Order 1981*, SI 1981, No. 245.

9 *Town and Country Planning (Industrial Development Certificates) (Prescribed Classes of Building) Regulation 1981*, SI 1981, No. 1826.

10 *Lifting the Burden*, Cmnd. 9571, presented to Parliament by the Minister without Portfolio, July 1985.

11 'Fig seeds beginning to bear inner city fruit'. *Planning*, No. 496, 26 November 1982, p. 6.

12 'Businessmen appointed to chair UDCs', *Planning*, No. 355, 15 February 1980, p. 3.

13 'No Heseltine sparkle in Merseyside show', *Planning*, No. 440, 16 October 1981, p. 7.

14 'Heseltine warned on task force', *Planning*, No. 442, 30 October 1981, p. 3.

15 Department of the Environment, *Consultation Letter on the Introduction of Special Development Orders*, unpublished, DoE, London, 1981.

16 'Heseltine's Excalibur', *Planning*, No. 426, 10 July 1981, pp. 6–7.

17 'Uncertainty over policy on orders', *Planning*, No. 444, 13 November 1981, p. 1 and p. 3.

18 'Order on the right track', *Planning*, No. 468, 14 May 1982, p. 1.

19 'Commons back SDO', *Planning*, No. 475, 2 July 1982, p. 12.

20 'SDO threat at Stansted Inquiry', *Planning*, No. 492, 29 October 1982, pp. 10–11.

21 'Docklands SDO floated', *Planning*, No. 497, 3 December 1982, p. 1.

22 'The battle of Area 8', *Planning*, No. 477, 16 July 1982, p. 7 and 'Shaw letter adds to Berkshire confusion', *Planning*, No. 480, 6 August 1982, p. 16.

23 'Employment land boost', *Planning*, No. 455, 12 February 1982, p. 12.

24 'Memorandum to Marsham Street: Redraft plans advice', *Planning*, No. 494, 12 November 1982, p. 10.

25 *Town and Country Planning General Development (Amendment) Order 1981*, SI 1981, No. 245.

26 *Town and Country Planning (Determination of Appeals by Appointed Persons) (Prescribed Classes) Regulations 1981*, SI 1981, No. 804.

27 Secretary of State's Address to the Town and Country Planning Summer School, 1981.

28 *Town and Country Planning (National Parks, Areas of Outstanding Natural Beauty and Conservation Areas) Special Development Order 1981*, SI 1981, No. 246.

29 'Heseltine drowns lakes proposal', *Planning*, No. 451, 15 January 1982, p. 3.

30 'Heseltine saves Stroud High Street listed buildings', *Planning*, No. 452, 22 January 1982, p. 3.

31 'Major step forward in Halvergate saga', *Planning*, No. 495, 19 November 1982, p. 3.

32 'Heseltine puts brakes on Belvoir proposals', *Planning*, No. 495, 19 November 1982, p. 4.

33 'Heseltine backs monumental quango', *Planning*, No. 474, 25 June 1982, p. 8.

2 Theoretical perspectives on local planning

2.1 Introduction

In reviewing the development of the town and country planning system in the UK attention has been drawn to the fact that the statutory basis of the system gives emphasis to the role of town planners as producers of plans and controllers of development. The *Town and Country Planning Act 1947*, whilst establishing a loose hierarchy of plans and providing a strong basis for development control, made no attempt to develop a framework for planning which was capable of accommodating the complex problems and inter-relationships between land use planning and social, economic, political and financial factors. Similarly, the current planning system proposed initially by the Planning Advisory Group, whilst acknowledging that:

1 plan-making is a continuous dynamic and hierarchical process;
2 social and economic policies influence land use plans and policies; and
3 different problems in different situations need different solutions

has emerged in law as a system concerned almost entirely with the preparation of land use plans and the control of development, and lacking any systematic concern with the dynamics of change. Practice, by its actions in producing a wide range of statutory and non-statutory local planning instruments, is suggesting that the narrow and restricted role prescribed for town and country planning by the legislation is inadequate to cope with the current problems deriving from structural changes in society and the economy. It is the aim of this chapter to review the nature of the activity of planning and the various theoretical perspectives which contribute to an understanding of the process of planning and which could provide a framework within which the land use planning system might operate more effectively. In particular, emphasis is given to the following:

1 The relationship between land use planning and the wider socio-economic policy context.
2 The complex nature of planning problems.
3 The principles of strategic planning.
4 Planning as the management of change.
5 Bargaining and conflicts of interest in the planning process.

2.2   Planning and planners

2.2.1   General planning and physical planning
The *Concise Oxford Dictionary* indicates that a 'plan' is

a drawing; a diagram; . . . a large-scale detailed map of a town or a district; a table indicating times, places, etc. of intended proceedings etc; a scheme of arrangements; a project; a design; a way of proceeding.

The same dictionary defines the verb 'to plan' as

to design, scheme, arrange beforehand (procedure etc.)

Chadwick (1971, p. 63) more helpfully defines 'planning' as

. . . a process of human forethought and action based upon that forethought.

Although these definitions do not clarify *what planning is*, nevertheless they contain the essence of *what planning is about*. they also point to a fundamental distinction between the *general activity of planning* or policy-making and *physical planning*. By implication, these definitions see general planning as a procedure whereby one schemes or arranges beforehand and acts on the chosen scheme. Physical planning, on the other hand, is seen as referring to the physical design or plan of some artifact or building which might exist in the future. In this context planners involved with public policy claim a comprehensive concern with the inter-relationship between social, economic and physical change in city, town and region; with the relationships between residence, workplace, school and shopping; with the accessibility and other requirements of industry and commerce. They also justify their actions on the grounds that they serve the public interest (however that may be defined) and are needed to counteract the inadequacies of the market system. It is apparent that the majority of general plans have a physical component, e.g. a general plan for education produced by a local education authority will inevitably include a physical plan or plans locating and designing schools, and colleges. Similarly, a general plan for a firm setting out its manufacturing, market and financial strategy will also include a physical plan for accommodating the staff and 'hardware' involved in implementing that strategy. In both examples, the physical component of the general plan gives physical expression to more general policies, which are implemented through a process of management and financial investment.

Against this background, town and country planning in Britain is seen variously as being concerned to achieve '. . . the best use of land and the greatest possible improvement in human environment' (Town Planning Institute, undated, but pre-1964, p. 7); as being '. . . the art and science of ordering the use of land and the character and siting of buildings and communication routes so as to secure the maximum practicable degree of economy, convenience, and beauty' (Keeble, 1964, p. 9), whilst the Town

Planning Institute in 1967 defined it '. . . as a process, involving a recurring cycle of operations, for preparing and implementing plans for changing systems of land use and settlements of varying scale' (TPI, 1967, p. 6). Town and country planning was thus seen by the profession at that time as being concerned with the preparation and implementation of plans for systems of land use and settlement. By implication this activity is presented as free-standing. It is not seen as the physical element of a more general set of policies or plans to achieve social and economic change.

More recently, the Royal Town Planning Institute, in considering the future of the Institute, reinforced its concern with the physical environment of town and country but at the same time indicated that the process of physical planning should be related to the process of corporate planning (RTPI, 1971 and 1973). Thus, although there has been some broadening of view, it is clear that the profession sees town and country planning in more or less the same terms as the narrow land use base of the legislation. Planners are primarily producers of plans and controllers of development. On this view, a planning job is

one occupied by a person . . . involved in land use or physical planning at any level from local to national, and including a direct concern with the operation and implementation of planning legislation (Amos, 1982, p. 8).

Yet town planning in practice is much more than this. As well as producing plans for physical change, and controlling private and public development in accordance with these plans, town planners are also attempting to obtain finance to implement plan proposals, or seeking to influence other bodies to implement those proposals; town planners are attempting to relate the loosely defined sectoral national policies for social, economic and physical change to the social, economic, physical and political realities of the area in which they work. The planning task thus embraces 'all those forms of policy-making and implementation which have a significant bearing on spatial distributions of investments and the development and use of land at all scales' (Social Science Research Council, 1982, p. 9). The problem faced daily by the practising planner is that of integrating vertically oriented national sectoral policies with horizontally oriented land use and management programmes at the local level in an environment constrained by basic secular trends, conflicts of interest amongst groups, the values, attitudes and standards of the different interest groups, and the immediate circumstances. In the words of Chapin and Kaiser (1979, p. 23):

. . . because of the very tangible, complex and differential impacts of local decisions involving many persons and firms, what evolves is essentially an interactive progression of decisions rather than a single final decision on land use. Although practice varies from one locality to another, . . . in the land use planning process there is usually a recursive sequence in the way participatory, technical and political inputs to this process function to establish land use objectives, explore growth scenarios, test out development alternatives, and arrive at land use decisions.

As Figure 2.1 illustrates, the process and inter-relationships are complex.

In these circumstances it is clear that despite the legislation and despite the view of the profession, the town planner, if he or she is to function adequately, can no longer be merely a producer of plans and a controller of development. Rather the planner's role in practice is primarily concerned with managing change in the environment by applying in a general sense the basic managerial processes exercised by all managers in all types of organization:

- *Planning*, which involves the establishment of goals and objectives for the particular organization and determining ways of achieving these goals.
- *Organizing*, which includes the provision of human and physical resources needed to accomplish the plan objectives; the definition of the tasks of individuals and groups within the organization and the establishment of relationships between groups and individuals.
- *Controlling*, which is concerned to ensure that the implementation conforms to the plan requirement and that proper inter-relationships are established.
- *Leading*, which is the process of motivating and influencing others so that they contribute to the achievement of the goals and objectives set by the organization.

Town planners, or land use planners, if they are to be effective must become managers of environmental change.

2.3 Planning problems, 'wicked' problems and complexity

2.3.1 Characteristics of complexity

Real world problems are complex and highly inter-related, e.g. at a national United Kingdom level in 1984 attempts to deal with the energy (coal) problem were related amongst other things to the distribution of population, inflation, the trade unions, transport, civil rights, and law and order. Indeed, '. . . basically every real world policy problem is related to every other real world problem' (Mason and Mitroff, 1981, p. 4). Acknowledgement of this situation is of vital importance to planners for it means that attempts to resolve a particular policy problem must consider the potential relationships between it and other policy problems. In plan or policy formulation, it must be accepted that:

1 any plan or policy is or should be concerned to resolve a number of problems and issues;
2 these problems and issues will undoubtedly be inter-related to some degree; as a result policy to cope with one problem typically requires further policies to cope with all the other problems;

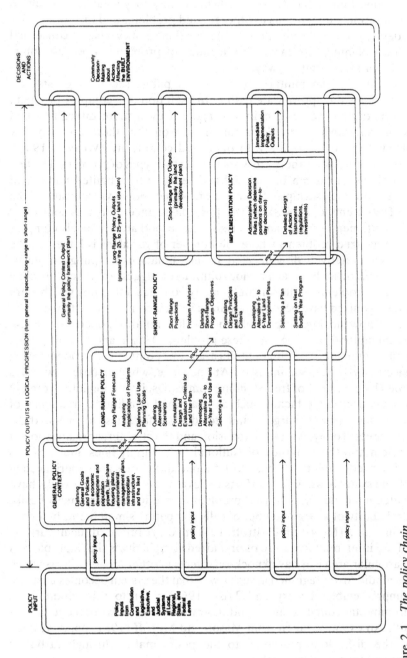

Figure 2.1 *The policy chain*
Source: Chapin, F.S. and Kaiser, E. (1979), *Urban Land Use Planning*, University of Illinois Press, Urbana, Figure 3.4

53

3   few (if any) problems can be isolated for separate treatment.

At the same time it should be noted that each policy solution is likely to create additional dimensions to other related problems and the policy solutions developed to deal with those problems. Thus policy-makers concerned with local planning are faced with a range of problems which are inter-related in a most complex way.

Experience in the military, business and public sectors suggests that organization is the appropriate path to the solution of a complex problem. However, organizing complexity can make it even more difficult to find workable solutions to problems for the reason that there are very few intellectual tools for coping with organized complexity (Weaver, 1948). What intellectual tools there are work best on two types of problem. The first are problems which can be isolated and reduced to a limited number of variables and relationships, and have a one-dimensional value system – the type of problems dealt with in the physical sciences through controlled experiment, measurement and analysis. The second are extremely complex but disorganized problems where the number of variables is large and the variables are relatively disconnected. In such situations sampling and statistical methods of analysis can provide solutions or suggest approaches to cope with these problems. Unfortunately, many complex situations which were once disorganized have become organized in their complexity with the result that it is now no longer possible to develop reliable policies or to use statistical methods to cope with these problems. Mason and Mitroff (1981, p. 6) cite a number of examples to illustrate this point, e.g. developments in the electricity-generating industry. At one stage, when electric power was provided by a number of free-standing companies, it was possible to predict reasonably accurately factors such as peak load conditions, frequency of demand, and power surges, and to design facilities to meet these requirements. Today, with the expansion and the structural organization of the system and the provision of National Grid supply lines, much of this statistical predictability has been lost. Similar difficulties are experienced in telephone systems where '. . . efforts to develop world-wide networks have presented many new and difficult problems in predicting frequency of calls, the probability of a large number of calls competing simultaneously for the same line . . . .' (Mason and Mitroff, 1981, p. 6). A further difficulty arising out of the inter-related nature of organized complexity is that a problem in one particular area can have knock-on effects in other areas – effects which can be modified or magnified in such a way that the system becomes unstable and unpredictable. Emery and Trist (1965) refer to this condition as 'environmental connectedness' and describe such an environment as 'turbulent'. An example taken from Mason and Mitroff (1981, pp. 8–9) illustrates the difficulties presented to the policy-maker through turbulent environmental connectedness, viz.:

For some years now the United States has been following a policy of farm price-support programs with the intention of helping disadvantaged farmers. The programs serve to reduce some of the uncertainty inherent in agriculture production and thereby permit farmers to streamline their decision making. One of the side effects of this change, however, was that large farmers were encouraged to mechanize, while small farmers suffered a reduction in their competitive position. Many small farmers liquidated their farms, resulting in massive layoffs of farm workers. Millions of farm workers, unemployed, fled to the inner cities where they became part of the urban crisis. Farm policy, economic policy, and urban policy, it turns out, were all part of a single, large, but highly inter-related system of organized complexity.

Policy-makers it seems are slow to realize that policies formulated to solve one problem have unintended consequences and create problems in other areas. Policy-makers are slow to appreciate that events which appear to be outwith their area of concern are connected with each other in such a way that often their own plans and policies are rendered obsolete or inadequate. In the field of public policy planners are not dealing with simple problems, or problems of disorganized complexity. They are faced daily with problems of organized complexity – problems which Rittel and Webber (1973) refer to as 'wicked problems'. Few recognized means have been developed to deal with these problems.

2.3.2 'Wicked problems'
'Wicked problems', or problems of organized complexity, are not wicked in the sense of being evil. Rather they are 'wicked' in that the more one attempts to solve them the more complicated they become. Rittel and Webber (1973) identify nine characteristic properties of such problems, viz.:

1 'Wicked problems' have no one definitive formulation.
2 Formulating or understanding a 'wicked problem' is synonymous with solving it.
3 There is no right or wrong solution to a 'wicked problem'. Solutions can only be good or bad relative to one another and the value system within which they are applied.
4 There is no way of knowing when a 'wicked problem' has been solved. As a result they require constant monitoring, and there is always scope for improving the solution propounded.
5 The possible range of methods which can be used to solve 'wicked problems' is unlimited.
6 There are many explanations for 'wicked problems'. Depending on the explanation chosen, so the solution differs.
7 It is never clear that 'wicked problems' are being dealt with at the proper level as they have no identifiable root cause, and can be considered as symptoms of other problems.

8   Once a solution to a 'wicked problem' has been attempted it cannot be reversed. What is done is done! Unlike the physical sciences there is no trial and error procedure which can be followed.

9   Every 'wicked problem' is unique.

Mason and Mitroff (1981, pp. 12–13) amplify the ideas of Rittel and Webber in the context of the planner or policy-maker who seeks to serve a social system by changing it for the better. In doing so they identify a number of characteristics of public policy 'wicked problems', which in their words 'spell difficulty for the policy-maker'. These characteristics are:

- *Interconnectivity.* This is where each problem is strongly connected to other problems and sometimes these connections form feedback loops. The opportunity costs and/or side effects of policies put forward to deal with a specific problem will often depend on events outwith the scope of the specific problem, i.e. contingent factors.
- *Complexity.* Given the numerous elements and inter-relationships which make up 'wicked problems' there are various points where analysis and intervention might focus and various approaches and policies for action might be applied.
- *Uncertainty.* 'Wicked problems' exist in a dynamic and uncertain environment. Therefore risk-taking, contingency planning and flexibility are essential ingredients in planning and public policy formulation.
- *Ambiguity.* There is no one correct interpretation of a 'wicked problem' or a solution to it. The value system of the individual determines how the problem/solution is seen.
- *Conflict.* Conflicts of interest among individuals or groups are inevitable in the competing claims for resources to resolve 'wicked' public policy issues. The interaction of these conflicts of interest determines how the solutions work out.
- *Societal constraints.* Social, organizational, technological and political constraints and capabilities have a marked influence on the policies selected to resolve 'wicked' public policy problems, and the success or otherwise of those policies in resolving the problems.

The implications of such public policy characteristics are profound for the land use planner working at the local level and seeking to function as a manager of change whilst operating within a much wider and more uncertain environment which is concerned to achieve more fundamental changes in society. In particular, difficulties are raised by the inter-relationships between specific proposals for land use and development and wider social and economic issues, at both the same 'local' level and at higher more general levels in the hierarchy, and between specific local development and resource allocation and politics.

2.3.3 Uncertainty in planning

Christensen (1985) in dealing with uncertainty in planning reinforces many of the points made by Mason and Mitroff (1981). She argues that, traditionally, planning practice has assumed that both the goals (or *ends*) of planning and the *means* of achieving those goals are known and accepted, i.e. the change being sought and the method(s) of bringing that change about are certain. In reality, the situation is much more complex (Christensen, 1985, p. 63), i.e.

Actual problems vary in uncertainty over means and ends. If people agree on what they want and how to achieve it, then certainty prevails and planning is rational application of knowledge. If they agree on what they want to do but do not know how to achieve it, then planning becomes a learning process; if they do not agree on what they want but do know how to achieve alternatives, then planning becomes a bargaining process; if they agree on neither means nor ends, then planning becomes part of the search for order in chaos.

In amplifying this statement, Christensen (1985) suggests a framework for clarifying the conditions relating to variable (or uncertain) planning problems. Briefly she

- establishes a two-dimensional matrix with the vertical dimension as the means or knowledge of how to do something, and the horizontal dimension as the goal or the desired outcome of planning (the former she labels 'technology'; the latter 'goal');
- dichotomizes the vertical and horizontal dimensions according to certainty/uncertainty. Thus, the technology dimension is dichotomized into known/unknown; the goal dimension into agreed/not agreed. Figure 2.2 illustrates this matrix.

Using the matrix, Christensen identifies four prototype variations of conditions that can characterize planning, viz.:

- known technology/agreed goal;
- unknown technology/agreed goal;
- known technology/no agreed goal; and
- unknown technology/no agreed goal.

Whilst this framework could be criticized on the grounds that it is apparently simplistic, nevertheless it does point very clearly to the complex range of variable situations which impinge on planning problems and their resolution.

In the first situation – *known technology and agreed goal* – both 'ends' and 'means' are certain. In such conditions public policy can be implemented through standard procedures which can readily be replicated, e.g. the provision of roads, water, drainage and other public services can be matched to local conditions at the same time as meeting the agreed desirable level of provision. However, even in this apparently stable situation, periodic uncertainty is inevitable if circumstances change, e.g. the certainty of the Welfare State in Britain has become less certain since the General Election of May 1979.

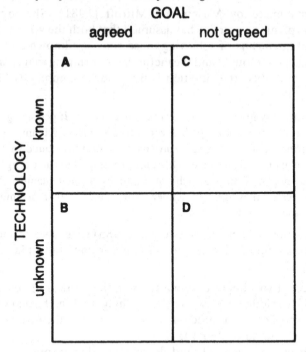

Figure 2.2   *Prototype conditions of planning problems*
Source: Christensen, K. S. (1985), 'Coping with uncertainty in planning', *Journal of the American Association of Planners*, Vol. **51**, No. 1, Figure 1.

The *unknown technology and agreed goal situation* generally arises when there is a public commitment to deal with pressing problems (an agreed goal) without the 'technology' (a proven solution) to deal with the problem, e.g. the agreed goal may be the regeneration of inner-city areas but the means of achieving that goal may be far from clear. Thus, the first priority is to obtain the missing knowledge about the solution. Practice frequently deals with such situations of uncertainty by a pragmatic, incremental, trial and error search for policies that work. Alternatively 'Agencies take on problems, then subdivide and delegate them to sub-units or external agencies in such a way that the problems are isolated. Thus, potential conflicts over goals are diverted, and attention can focus on technology' (Christensen, 1985, p. 65).

In circumstances of *known technology and no agreed goal* a process of bargaining surrounds the issue at hand, e.g. the British Airports Authority may know exactly how best to establish and operate a third London airport, but the residents of the areas likely to be affected by the development may well resist the proposal. In such situations, the bargaining may take a variety of forms although the expected outcome of the process is to accommodate multiple but conflicting goals. Thus, each bargain is tailored to its particular participants, is unique and, as a result, replicable results are precluded.

Finally, *unknown technology and no agreed goal* represents situations where there are multiple or unarticulated goals and no known effective means of achieving those goals. In the words of Christensen (1985, p. 65):

It is impossible to draw clear examples of uncertainty over both means and ends because these conditions are in chaos. Nevertheless, they are common. Goals are often nebulous and changing; facts are often ambiguous . . . . Moreover, the ways planning problems are defined focus our attention on some aspects of the problem and cause us to neglect others. These neglected aspects are left nameless and therefore untreatable.

It is evident that there is a fine distinction between, on the one hand, agreed goal and unknown technology and, on the other, no goal agreed and unknown technology, e.g. a concern to eliminate homelessness or to reduce poverty may well become a major policy issue. By defining these two problems in a particular way, and bringing political commitment to their resolution, agreement on a goal or goals might be reached and a search for the solutions undertaken. However, a continued failure to resolve the problem, allied with a continuing re-definition of the problem will eventually ensure that commitment to resolving that problem evaporates. Thus the problem reverts to the confusion characterized by uncertainty over both ends and means.

Figure 2.3 summarizes this analysis. However, it should be emphasized that real world problems may not necessarily fit into any single category, given that political and institutional forces shape the way problems are defined and treated, and a problem may shift its location as agreement over ends and means shifts. Nevertheless, the framework shows very clearly that:

- Three of the four sets of conditions that characterize planning problems contain uncertainty – about ends, or means or both.
- The set of conditions with agreed goals and proven technology is susceptible to periodic uncertainty as conditions change.

'In short, uncertainty characterizes all planning problems' (Christensen, 1985, p. 66).

2.3.4 The systems approach and the Law of Requisite Variety (Chadwick, 1971)

The previous section gives an indication of the variety and complexity of the problems with which public policy and planning have to deal. Indeed, the variety is infinite if the planner makes the decision to consider all aspects of the problem for '. . . by definition "all aspects" of the problem are infinite and variety in the problem must be matched by variety in the solver . . .' (Chadwick, 1971, p. 71). However, it is argued that by regarding the real

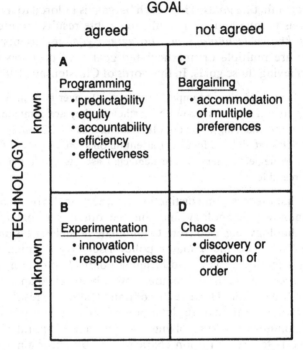

Figure 2.3   *Responses to prototype conditions of planning problems*
Source: Christensen, K. S. (1985), 'Coping with uncertainty in planning', *Journal of the American Association of Planners*, Vol. **51**, No. 1, Figure 2.

world as a system of systems, an order is introduced which constrains variety by studying only certain systems. At the same time by defining those systems through the medium of modelling, variety is further constrained. Thus by modelling a conceptual system of the real world it is argued that it is possible first to understand the processes of change; second to anticipate and evaluate those processes, i.e. the high variety of the real world is reduced to a conceptual system where the level of variety is capable of comprehension through the deliberate process of modelling. When it is desired to return to the real world in an attempt to anticipate and control the outcome of 'wicked problems', so the level of variety is increased to meet the circumstances of control. In the words of Chadwick (1971, pp. 63–82) '. . . planning is initially concerned with a conceptual framework which allows of the necessary processes of regulation of variety.' At the same time, the *Law of Requisite Variety* indicates that '. . . only variety can destroy variety' (Ashby, 1964), i.e. a system of given variety can only be controlled by a matching system which contains at least as much variety as there is in the system to be controlled.

The implications for planning are clear – the mechanism for managing the issues and processes with which planning deals should ideally possess the requisite variety to match the variety in the real world system at the level of variety necessary to secure the desired changes in the real world system. Again quoting Chadwick (1971, p. 81):

> ... the activity of planning ... must be seen as dependent upon the application of scientific method to the problems of the real world. In this endeavour, the insight provided by general systems theory, information theory and cybernetics – all inter-related, each calling upon and in turn shedding light upon the others – is remarkable: here is a means of ordering man's view of his place in Nature.

While the almost total reliance on scientific method advocated by Chadwick can be questioned, nevertheless the message is clear: *A complex system can only be guided and controlled by a system which is equally complex.*

The problems of complexity and inter-relatedness dealt with here are not unique to public policy issues or planning. They are problems experienced by managers of change in other fields. It is therefore appropriate to review the conceptual approaches to order and organization which have been adopted in the general field of management in an attempt to clarify ways in which public policy and planning might adapt to complexity.

2.4 Strategic planning and management

2.4.1 Organization theory and strategic planning

The concept of strategic thinking, together with the distinction between it and detailed tactical issues, is military in origin. In military terms, the strategy seeks to move and deploy forces in such a way that the enemy is required to fight at a pre-determined time and place; at this point, appropriate tactics are employed to defeat him in battle. In recent years, the business and public administration communities have borrowed the military concepts of strategy and tactics in an attempt to develop practical approaches to management and decision-making capable of dealing with an increasingly complex and unpredictable environment. This has led to views of the firm which differentiate between strategic and operational decision-making within a hierarchical framework (Ansoff, 1969; Simon, 1971).

Higgins (1980) succinctly summarizes the basic characteristics associated with this approach to managing complex change as follows:

> Any organization must possess some overall strategy before it establishes planning systems to serve it. The strategy will be a guide to action and determine the basic long term goals and objectives of an enterprise; it will establish the courses of action and allocate the resources necessary to achieve these goals. The formulation of this strategy is the responsibility of top management, and will largely determine the overall future of the organization.

Within the framework provided by the overall strategy, tactical or operational planning is undertaken to ensure that the resources of the organization will be allocated in the short and medium term to meet the organization's strategic objectives.

The great advantage of such a hierarchical arrangement is that a comprehensive but generalized overview of issues can be established at the top level and developed into more detailed policies and eventual implementation at the lower levels.

The major criticisms which can be levelled at this organizational approach are that it represents an over-simplified, machine-like view of the nature of planning and management and ignores the implications which might arise from the ways in which individuals and different organizations interact. In short, it attempts to codify a set of principles that can be applied universally to all organizations in all planning and managerial situations. A number of alternative approaches to this classical bureaucratic approach have been advocated and implemented, e.g. the behavioural organization theorists such as Elton Mayo and Kurt Lewin argue that organizational effectiveness is achieved by arranging things in such a way that people feel that they count and that their work is more meaningful. Whilst not rejecting the classical organization principles 'behaviouralists' believe that the formal hierarchical structure can be improved by making it less formal and less hierarchical, and by permitting more participation in decision-taking by the lower ranks.

2.4.2 Contingency approaches to management

More recent developments in management theory have attempted to take account of the realities of planning and managing highly complex and multi-dimensional organizations and systems. These developments are generally referred to as the contingency approach to management which '. . . views organizations as complex systems of interdependent parts operating within the context of an environmental suprasystem' (Newstrom, Reif and Monczka, 1975, p. xiv). This approach does not attempt to provide a code of universal principles which are applicable to most situations. Rather it argues that the 'right' approach to adopt in any situation will be contingent upon the specific conditions and circumstances within which decisions are made and implemented. Methods and approaches appropriate in one situation will not necessarily be appropriate in another situation, e.g. managing a university or a local authority is recognized as being a very different operation from managing an industrial or commercial firm.

Equally, similar organizations operating in different environments may need to plan and manage differently because of differences in the cultural, economic, political, social and legal climates within which they are situated. What the contingency approach offers is a conceptual framework with accompanying techniques and methods that can be used to identify the contingencies (which include both external and internal factors) likely to affect a particular situation; to develop and analyse alternative courses of action and to select the course of action that will best meet the requirements of the organization and associated interest groups. Newstrom *et al.* (1975) summarize these points most succinctly when they state that the contingency approach:

> . . . takes the position that theory acquires value only to the extent that it is successful in application and that theory must be adaptable to the needs and realities of the practitioner. It acknowledges that there are few universal principles that apply equally well in all situations. Instead, it emphasizes the conceptual framework, thought processes, and diagnostic and analytical skills that will enable managers to set objectives and develop the most appropriate means for achieving those objectives within the given situation (Newstrom *et al.*, 1975, p. xiv).

Christensen (1985, p. 66) advocates the adoption of a contingency approach to planning on the grounds that 'Planning processes can be understood as contingent because they are not pre-determined, but depend instead on problem conditions'. In amplification of this statement, she reviews the application of the planning process under the four prototype variable conditions summarized in Section 2.3.3, and shows that in *planning for known technology and an agreed goal* the planner knows the goal and attempts to achieve it through the application of a known and effective technology, e.g. the protection of existing shopping centres would be achieved by the application of policies precluding or restricting out-of-town shopping centres. These conditions form the basis of the traditional rationalist approach to planning which sees the planner as a highly expert professional carrying out specialized tasks, such as establishing the level of shopping that a town can support, or more generally selecting effective proven technologies (means) to achieve desired goals (ends). In this approach the planner is seen essentially as an optimizer who adapts to the contingent uncertainties and:

- selects the most appropriate means to achieve the ends;
- sets and organizes the various tasks needed to apply the means to achieve the ends;
- administers, manages and evaluates the application of the means and the achievement of the ends.

When the public concern is to solve a particular problem with no proven solution – *planning for unknown technology with agreed goal* – the planner's task is again to adjust to the contingent uncertainties and to seek a workable solution. In this situation, two approaches tend to be adopted – the first is a pragmatic, incremental, trial and error approach to find a solution which works. In the process, the planner is seen primarily as a pragmatist and an adjuster, who adapts the means to achieve the ends in the light of experience, e.g. if restrictive out-of-town shopping policies to protect existing shopping centres cannot withstand market and political forces, then the policies will be modified. The second approach attempts consciously to experiment with the unknown 'means' in an attempt to establish a workable solution. Planning is thus seen as a learning process which develops through the systematic application of alternative solutions. In this view of the world the planner is seen as a researcher, experimenter and innovator.

In conditions of *planning for known technology but no agreed goal*, planners invariably face confusion and disagreement over ends, which are only resolved through bargaining. In these political situations, the planner attempts to use processes to accommodate conflicting goals by:

– consensus building, where involved parties are encouraged to identify and solve their own problems (Susskind, 1981);
– mediating and traditional bargaining;
– advocacy planning (Davidoff, 1965).

An alternative approach is to divide the problem into subsets to avoid conflict and the need for resolution. In this situation, the planner avoids addressing the issue of uncertainty by dividing resources among different and often competing goals, e.g. the allocation of resources to improve the effectiveness of upland farming in Britain in the mid-1980s can and is conflicting with policies concerned to preserve the upland landscape. By treating these two issues separately, explicit conflict is partially avoided.

Finally, in conditions of uncertainty over both means and ends – *planning for unknown technology and unknown goal* – confusion and instability are inherent. In this situation, the planner adjusts to contingent uncertainties by becoming a problem identifier – formulating the problem in such a way that attempts to resolve it become attractive to government and the electorate, e.g. a combination of the Toxteth riots in 1981, and Heseltine's formulation of the inner city problem briefly reduced uncertainty about the 'ends' to be achieved in the inner city and allowed the planners to focus on technical aspects of how to solve the problem.

The conclusions drawn from this analysis (Christensen, 1985, p. 69) are:

. . . that planning processes can be understood as addressing different conditions of uncertainty. Thus planners must assess the actual conditions of uncertainty that characterize the particular problem they are confronting and then select a style of planning that suits those conditions. By tailoring planning to real world conditions, the planner is acting contingently. In doing so, the planner copes rationally with uncertainty.

In short, if planning is to contribute to change in the environment a contingent approach must be explicitly adopted. Figure 2.4 categorizes the planning roles and processes associated with the four prototype variable conditions

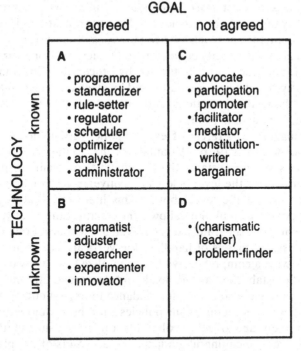

Figure 2.4 *Planning roles categorized by planning conditions*
Source: Christensen, K. S. (1985), 'Coping with uncertainty in planning', *Journal of the American Association of Planners*, Vol. **51**, No. 1, Figure 3.

reviewed above. However, there are potential dangers in moving entirely to a contingency approach, the most significant of which is that the objectives underpinning the overall strategy could be lost in the concern to adapt to complex and dynamic contingent factors at the tactical/local level. What is needed is an approach to planning which provides for a contingent approach set within a strategic framework.

Within the field of public administration, Friend and Jessop (1969) have applied ideas derived from both the contingency and the organizational approaches to management in developing their view that planning is a continuous process of strategic choice. This perspective shows that the difficulties of dealing with current decision problems can be reduced by considering them within a wider strategic context, which also embraces other related problems of present and future choice. Friend, Power and Yewlett (1974), together with Hickling (1974), have taken this work further and developed an approach to strategic choice and control which enables strategic policies to be formulated and implemented. The method accepts that such decisions are inter-connected, recognizes a hierarchy of policy formulation and choice, and establishes an approach to decision-making which recognizes the existence of uncertainty whilst allowing complex problems and policy options to be pursued. In the context of strategic planning in local government, the approach has been found to offer a useful model of the different kinds of uncertainty characterizing policy decisions, viz. technical uncertainty (about facts, usually tackled by officers); political uncertainty (about objectives and priorities, addressed by elected representatives); and uncertainty about the intentions and stances of outside organizations (Hunter and Trinnaman, 1983).

Commenting on the hierarchy of levels of policy formulation in relation to planning, Diamond (1979, p. 19) amplifies the relationship between the different levels when he states that 'each level of planning performs a strategic function for the level below and conversely is constrained by the strategic planning of the level above.' This hierarchical arrangement of choice and policy formulation allows the relationships between policy options to be pursued separately at each level, but within a framework which provides an explicit means of handling the vertical relationships between them. In this way, a comprehensive but generalized overview of issues and policies can be established at one level, and developed into more detailed policies at lower levels, so as to offer guidance to implementation agencies. Thus, for example, structure plan policies may be developed which are comprehensive enough to yield a coherent strategy, yet sufficiently detailed and specific to give meaningful guidance to such users of the plan as local planning authorities and development interests (Hickling, 1978).

A cautionary note, however, is sounded by Solesbury (1981), whose case study work on the application of the strategic approach to policy formulation and implementation in a number of fields indicates that the practical results of strategic planning have been disappointing. On the one hand, the planning process has sometimes failed to produce recognizably strategic policies, while on the other, such policies, once formulated, have frequently been ineffective in influencing subsequent events. Indeed, strategic policies have often been ignored by routine decision-making, despite formal commitment to them by public authorities. To meet this problem, Solesbury argues that attention needs to be diverted away from the planning of strategy towards its implementation, and that a pre-condition for such a re-orientation is the existence of a recognized authority at this level. Indeed, he comments that a strategic policy is unlikely to be implemented 'unless there exists a strategic authority responsible for it and committed to its implementation' (1981, p. 430). In effect, Solesbury is making a plea that strategic decisions and actions rather than strategic plans should be the ultimate output of the strategic planning process.

2.4.3   Strategic planning and land use planning[1]
In many cases, strategic policies aimed at securing social and economic change involve some form of physical development. If this development is to

be co-ordinated, it must be properly regulated in a way which contributes to the achievement of desired socio-economic change whilst, at the same time, taking account of physical environmental factors. Thus, physical land use planning is an integral part of the strategic planning process, at least in the public sector.

The above review of the theoretical basis of strategic planning makes it clear that although the process is applied in a range of commercial and public sector situations, the principles involved and the problems encountered are similar. However, the different sectoral concerns of public administration, such as education, health and transport, are qualitatively different from the concerns of business. The public sector is concerned to achieve a range of objectives relating to socio-economic change, often simultaneously through the activities of a number of agencies. By contrast the principal aim of the private sector is profit maximization. Any account of strategic planning in the public sector must accommodate this added complexity, and the associated range of contingent factors, by providing for a clear statement of socio-economic objectives and the articulation of sectoral policies in such substantive areas as industry, housing and transport, which are intended to make a contribution towards meeting these objectives. It should also clarify the role of physical land use planning in co-ordinating the implementation of developments associated with these policies, especially the linkages between strategic policies and local proposals for development. Additionally, given the work of Solesbury (1981) on the lack of success in implementing physical development proposals, any idealized framework for planning must take full account of the need to control the process of resource allocation in facilitating implementation.

These points suggest that three basic principles are necessary in establishing an idealized framework for the production of policies and plans. First, there should be a strategic planning framework with a hierarchy of levels of planning, where the level above constrains the planning of the level below and conversely is itself constrained by the level above. Only in this way can general policies for change be translated into more detailed proposals for action. Second, there should be clear links between strategy and implementation running from the higher order levels in the hierarchy to the operational levels. In this way, the vertical relationships between policy formulation and implementation can be clarified, and consistency of policy at different levels ensured. This is crucial, for policies must maintain strong links throughout the hierarchy if they are to be both valid and effective. Third, the framework must, at the same time, be sufficiently robust and flexible to accommodate the contingency approach to planning which is inevitably predominant in seeking to change 'wicked' social and economic situations through the activities of the public sector.

Moreover, any tendency to over-elaborate the hierarchy of levels of planning in order to account for the added complexity of public sector objectives must be tempered by our theoretical knowledge of implementation. The work of Gunn (1978), for instance, makes it clear that the greater the number of links in the process between policy formulation and implementation, the greater the chances of contingent factors impeding the successful implementation of those proposals. Indeed, Gunn's prerequisites for perfect implementation illustrate how difficult it is to guard against the likely effect of contingent factors, e.g. the existence of a single, relatively independent, implementing agency; complete understanding of the objectives to be achieved; a full specification of the tasks to be performed by each participant authority in moving towards these objectives; and perfect communication between these bodies. These conditions, whilst clearly not fully obtainable in practice, stress that the more agencies involved in the process the less likely is the successful implementation of plans and policies. Nor should the problems of establishing horizontal relationships between the various authorities and agencies working at the different levels in the hierarchy be overlooked. Though the existence of a hierarchy of levels of plan formulation makes it easier to identify the nature of the horizontal relationships required to achieve the overall objectives of national policy, the complexity of inter-organizational relationships must not be underestimated.[2]

2.4.4   An idealized framework for strategic planning and implementation in the public sector

In the light of this account of some of the theoretical aspects of strategic planning, an idealized conceptual framework for strategic planning and implementation in the public sector can be outlined (Figure 2.5). This framework clearly demonstrates the area of concern for local planning and provides for a hierarchy of levels of decision-making, with the level above constraining the level below; gives the opportunity of defining national social and economic objectives; and allows the co-ordination of the decisions and actions of a variety of agencies concerned with the implementation of sectoral policies through more detailed physical development plans and the programming of resource allocation. Though the details of the structure will vary in application depending on particular circumstances and needs, the fundamental character of the various levels in the hierarchy is clear.

Implicit in this structure is the acceptance of the fact that it provides no more than a way of thinking. If such a framework were accepted it would do no more than enhance the possibility of obtaining a more consistent and coherent interpretation of higher order policies through the adoption of more specific proposals for development and implementation. Fundamental to the successful implementation of policies and proposals at the different levels in the hierarchy is the need to allow the managers of change who are active at those levels to develop the most appropriate means of achieving the objectives set for them by higher order policies within the given situation

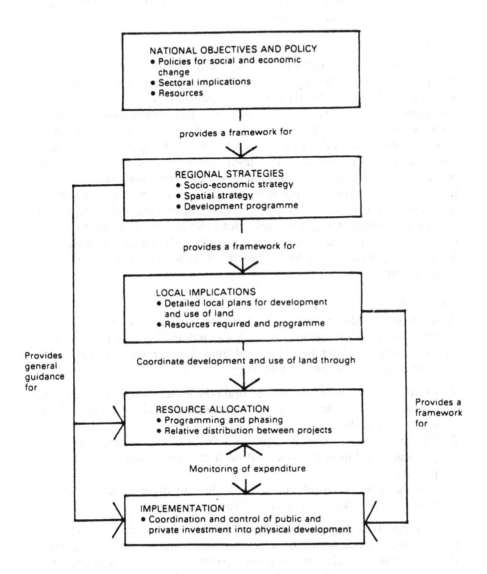

Figure 2.5 *An idealized framework for strategic planning and implementation in the public sector*
Source: Bruton, M. J. and Nicholson, D. J. (1985), 'Strategic land use planning and the British development plan system', *Town Planning Review*, Vol. **56**, No. 1, Figure 1.

facing them. Thus the framework assumes that in different situations different types of plan and resource allocation mechanisms will be used. However, all will be concerned to contribute towards achieving social and economic change, and all will have to take account of the contingent factors impinging on particular situations.

Level 1: National objectives and policies

At the top of the planning hierarchy, the nature of the social and economic change to be pursued throughout the policy formulation and implementation process should be clearly set out by government. In addition, more specific sectoral objectives and policies need to be stated in relation to issues such as the economy, income, housing and social welfare. The inter-relationships between these policies and urban and regional development would also be established at this level, together with an allocation of resources to implement policies on a sectoral and regional basis. An important function of these policy statements is to establish a commitment to an agreed set of national policies, in order to provide a clear 'input' to lower levels. These statements would ideally be produced for short periods of time (such as five years with a mid-term review) for which it is possible to foresee likely changes and the availability of resources necessary to implement the proposals.

Level 2: Regional strategies

At the next level down, strategic policies are needed to set out the way in which national policies relating to both sectoral and regional development can be expected to affect major regional areas. Strategies at this level should be concerned to establish the following:

1 More specific socio-economic sectoral policies for the particular region, as constrained by the more generalized policies of Level 1 and contingent factors in the region.
2 The broad spatial strategy required to co-ordinate the investment into physical development associated with the implementation of these policies.
3 A programme of development to implement these policies, which is related to the likely availability of resources.

The regional strategies would, therefore, comprise two distinct elements. One component establishes the socio-economic strategy for the area, while the other translates that strategy into spatial terms. The socio-economic element seeks to establish broad strategies for such sectors as employment, industrial development, housing, transport and education, within the constraints set by national level policies. Simultaneously, they would also be influenced by regional and local politics, together with a technical assessment of the feasibility of policy implementation. The second element, the spatial strategy, would then translate the socio-economic strategy into policies for physical development, identifying for example growth centres for additional population and employment, or the location of infrastructure improvements. Ideally, national coverage would be aimed for with a plan timescale of some ten to fifteen years.

Level 3: Local spatial implications

Such regional strategies, by themselves, cannot ensure that either their socio-economic or spatial policies are achieved. As Chapin and Kaiser (1979, p. 96) point out, it is only through a series of short-range programmes of development that higher level objectives can be achieved: 'there is no quantum jump into the future'. The third level in the hierarchy, which would form a sub-unit of the region within which it is located, therefore seeks to translate the general proposals and broad policies established in the level above into detailed local plans *where appropriate*. Full coverage of the region would not be expected. These plans would serve as a guide to developers and establish a basis for the co-ordination of public and private investment into development and re-development. They therefore potentially constitute an important lead-in to implementation and would be needed especially in areas with significant development pressures and/or where change is to be concentrated in the near future. In line with this emphasis on the progression of general policy towards implementation, the timing of proposed changes and the resources required would be worked out in detail; also, the timescale of such plans would be shorter than for those in Level 2, looking ahead a maximum of ten years. Again the form of these plans would be constrained by policies in Level 2 and influenced by more local contingent factors, and take full cognizance of the requirements of the *Law of Requisite Variety*.

Level 4: Resource allocation

Given that the failure to deal adequately with the resource aspects of strategic policies is a key factor in explaining the relative lack of success in implementing such proposals in the United Kingdom and elsewhere (Solesbury, 1981; Bruton, 1982), it is essential that account is taken of this factor. One solution would be the use of a series of sectoral rolling programmes for public investment into, for example, economic/industrial development, housing, transport and other elements of physical infrastructure and development requiring recurrent investment. Such programmes would be produced and rolled forward annually, and would outline spending plans for a relatively short period ahead, such as five years. These programmes would be constrained by the general policies and resource allocation proposals contained in the higher levels of the hierarchy, and would most likely be formulated for the regional areas defined in Level 2. However, given the range of potential policies and proposals, and the range of contingent factors in any specific situation, it must be accepted that a variety of ways of handling Level 4 would be developed.

Level 5: Implementation

Following resource allocation, the process moves into implementation, where 'planning and action become almost fused' (Chapin and Kaiser, 1979, p. 102). Here, the proposals contained in Levels 1 to 4 are converted into development through the co-ordination and control of public and private investment in the built environment. It is important to ensure through careful programming that the required combination of resources is available when required, and that development and expenditure are constantly monitored to ensure the early identification of any departures from schedule, together with the associated 'underspend' or 'overspend'.

2.4.5 Conclusions: town planning as the management of change in the environment

Planning problems are complex and 'wicked'. Town planners are concerned to manage change in the environment in a way which moulds and transforms 'wicked' problems for the benefit of society as a whole. As managers in the most general sense, town planners should be concerned to apply the basic managerial processes of planning, organizing, controlling and leading. Organization theories of planning suggest that the development of a hierarchy of planning and management levels is an effective way of coping with complexity. The contingency approach to management argues that no one approach is applicable to all problem issues in all situations. Rather, it claims that any framework for management should allow managers of change to select the approach and course of action which will best cope with the problem and the context within which the problem is located. Thus, if experiences in management are relevant in town planning, an idealized framework for managing change in the environment should ensure that higher order policies or objectives for social and economic change are transmitted vertically through the system in a way which co-ordinates the development of plans and their implementation at the different levels. At the same time, at the different levels in the system, managers of change in the environment must be free to develop approaches to, and solutions for, particular problems which respond to contingent factors which will include, amongst other things, policies set out in higher levels in the hierarchy.

The idealized hierarchical framework outlined here explicitly takes a policy-centred or 'top down' view of the process of implementing policy, and assumes that lower-level implementers are constrained by (or act as agents for) higher-level policy-makers. It is in the nature of idealized models to abstract from or simplify reality so as to allow analysis, and in this respect the schema developed here is no exception. It is important to acknowledge what is omitted from such models as much as what is included. The deficiencies of hierarchical perspectives on policy and implementation in local planning were first alluded to by Bruton (1980) and are conveniently summarized by Barrett and Fudge (1981, p. 4), who see the policy-action relationship 'as a process of interaction and negotiation, taking place over time, between those seeking to put policy into effect and those upon whom action depends' rather than viewing implementation as the hierarchical transmission of policy into a series of actions. The hierarchical concept of organization, whereby policy emanates from the top and is transmitted down the hierarchy and translated into more specific formulations, rules and procedures as it goes to guide or control action on the ground, gives little weight to power relations and conflicts of interests between those making policy and those responsible for implementation. In particular, Barrett and Fudge (1981, p. 12) are critical of two key assumptions of the hierarchical approach, namely that:

1   policy is formulated at the top (or centre) and is the starting point for action, and
2   that implementers are agents for policy-makers and are therefore in a compliant relationship to policy-makers.

The first of these assumptions is faulty, since policy may be a response to pressures and problems experienced on the ground; policy may be developed from a specific innovation, i.e. be preceded by action; and not all actions relate to specific or explicit policies. The second assumption ignores the fact that often those upon whom action devolves are not in any hierarchical association with policy-makers; 'implementation agencies will thus, in many instances, be autonomous or semi-autonomous, with their own interests and priorities to pursue and their own policy-making role' (Barrett and Fudge, 1981, p. 12). Where this is the case, action or implementation must be considered in terms of the nature of the power relations between and within the agencies involved. From this viewpoint, policy itself loses its absolute and fixed characteristic as an input to the process of implementation, and has to be seen instead as 'a series of intentions around which bargaining takes place and which may be modified as each set of actors attempts to negotiate to maximize its own interests and priorities' (*ibid.*, p. 25). Barrett and Hill (1984, p. 238), developing the ideas of Bruton (1980), suggest that 'the process of implementation is essentially a political process characterized by negotiation, bargaining and compromise' between the groups involved. This is to recognize that decision-making takes place in a socio-political context, where issues of power and influence and the ability to bargain can have a significant effect on the outcome. Indeed, such factors can often be more important in reaching decisions than social justice, technical evaluation or economics. For these reasons, it is important to review the nature of power, conflict and bargaining insofar as it relates to planning and managing change in the environment.

2.5 Planning, bargaining and conflicts of interest (Bruton, 1980; 1983b)

2.5.1 Consensus and conflict

The bureaucratic organizational structure for planning and managing complex issues implies that the process is logical and straightforward. It is assumed that technical analysis, reasoned argument and clear direction will deal with complexity. The behaviouralist view of management suggests that in reality the process is more complex involving human factors and their value systems. The contingency approach to management, allied with the view of public policy and planning problems as 'wicked problems', emphatically propounds the view that the planning and management process is highly complex; that no one approach is applicable in every situation; that contingent factors in the wider environment inevitably influence the process and that conflict rather than consensus is a likely component of the process.

Reference to official publications such as the *Town and Country Planning (Development Plans) Direction 1965*; the *Skeffington Report* (Ministry of Housing and Local Government, 1969); the *Report of the Planning Advisory Group (1965)*; the *Town and Country Planning Acts* (1968 and 1971) and the *Local Government Planning and Land Act 1980*, and associated DoE circulars dealing with the operation of the development plan system would seem to confirm that consensus is implicit in the official planner's perception of town planning. An alternative view which the planning profession is gradually coming to accept and which accords with the contingency approach is that planning is concerned with the distribution of resources; that conflicts of interest arise in reaching planning decisions, and that such decisions are inevitably political (Pahl, 1970; Simmie, 1974; Blowers, 1980; Goldsmith, 1980).

Perhaps the most fundamental weakness of the official (government) view of town planning is its failure to take account of the way in which society is organized, decisions are made, and planning takes place. The implications of the official view are that planning is an apolitical activity which operates in a culturally and politically homogeneous society, where the only source of conflict between planners and the public results from the latter's ignorance of planning matters. Yet British society is structured in such a way that the methods used to cope with the distribution of resources inevitably produce conflict between groups, not only about the way in which current resources are distributed, but also about the way in which resources will be distributed in the future. Given that this distribution of resources is manifested spatially as well as socially, it is inevitable that conflicts of interest will arise in reaching planning decisions (Simmie, 1974; Pahl, 1970). Indeed, Goldsmith (1980, p. 126) claims that '. . . planning, since it is primarily concerned with the use of physical space is an activity central to the conflict about the allocation of space'. Conflict is part of the wider issue of power relationships, and power has been defined as a form of social relationship whereby the behaviour of one group in society is dependent on the. behaviour of other groups. This wide ranging and abstract definition can be delimited firstly by *describing power* in terms of magnitude (how much?) distribution (who controls?), scope (what is controlled?), domain (whom is controlled?) and secondly, by *explaining power* by reference to the resources, skills, motivation and bargaining ability to establish relationships possessed by the groups concerned (Schelling, 1960). Obviously, the extent of a group's power and the resources and skills it has available are important factors in the establishment of any power relationship.

Following on the realization that planning decisions are distributional, political and involve conflicts of interest, the same profession has reluctantly but gradually begun to accept that in many instances conflicts of interest over planning issues are resolved through some form of bargaining process. As long ago as 1968 Buchanan suggested that the real nature of planning is '. . . a very ruthless bargaining process' (Buchanan, 1968, p. 52). More recently the work of Jowell (1977), Grant (1978), Willis (1982), Hawke (1981a, b), Heap and Ward (1980), and Ward (1982) on 'Section 52' agreements establishes quite clearly that bargaining for 'planning gain' forms an integral part of the development control process. This view is supported by Healey *et al*. (1982, p. 82) in their work on the relationship between development plans and the processes of implementation as the following quotation indicates:

Development Plans may be of particular importance in areas where there is conflict of interest over land, and where planning authorities seek to implement policies which restrict the opportunities for some of these interests through a 'regulatory-negotiative' approach.

The implication is that both the local authority and the developer start from an implicit bargaining position and negotiate through to a decision which is acceptable to both sides. However, neither the work on Section 52 agreements nor the work of Healey *et al.* give any indication of how this bargaining takes place, although theoretical accounts of bargaining processes have been provided by Schelling (1960) and Bacharach and Lawler (1981).

2.5.2 Distributional bargaining and conflict

Schelling (1960) fully develops the theme of conflict and bargaining. He argues that in any power relationship conflict should be taken for granted with the controlling and responsive groups in society attempting to 'win' – but to win relative to their own value system rather than to win by 'defeating' the opposition. Winning is achieved by distributional bargaining (whereby a better bargain for one means less for the other), mutual accommodation and the avoidance of mutually damaging behaviour. Bargaining is guided mainly by the expectations of what the other will accept, and the final bargain is struck when one of the parties makes a sufficient and final concession. This form of conflict is characterized by the following features:

1 *Commitment.* This is the ability to commit oneself to a course of action or outcome, and communicate this commitment to other parties in the negotiations. Invariably in planning, commitments are adopted by groups concerned to oppose development proposals put forward by central or local government. The theory of distributional bargaining suggests that the opportunity exists for all parties involved in the conflict to adopt commitments of their own, in an attempt to counter other adopted commitments. In planning situations where public sector decisions on major development proposals tend to be taken by elected representatives (usually in Committee), it is more difficult for the responsible authorities to adopt a commitment in opposition to a commitment apparently endorsed by large numbers of the population whose interests the representatives have been elected to protect.

2 *Communication of commitment.* Having established a commitment, it is essential that it is communicated to the other parties involved in the negotiations. Thus in planning situations, public meetings, the press and other media, and official channels of communication are used by all parties to ensure that the nature and extent of the adopted commitment is made clear to the other parties concerned.

3 *The use of a bargaining agent.* This is a common feature of distributional bargaining whereby a body (or bodies) is established by the parties concerned through which negotiations are conducted. The use of such bargaining agents allows the groups controlling them to give instructions which cannot apparently be changed thus increasing the pressure on the other parties in the negotiations. In planning situations bargaining agents tend to be groups formed through the initiative of a number of individuals to oppose specific developments.

4 *The use of intersecting negotiations.* This is an integral part of distributional bargaining, used to strengthen an adopted commitment, e.g. by pointing to a range of contingent situations where an established position would be prejudiced if a concession were to be made on the issue currently under negotiation. Given the nature of planning, with decisions being taken by elected representatives allegedly in the public interest, it is unlikely that intersecting negotiations will be used openly by the authorities to strengthen any commitment adopted.

5 *The use of restrictive agenda.* If negotiations are underway simultaneously on two or more issues then it may well be advizable for one party to ensure that certain other matters are not negotiated at the same time or on the same agenda.

6 *Compensation and/or arbitration.* These are essential elements of a distributional bargaining situation. In planning the role of the Inspectorate and the public inquiry thus takes on an added significance.

Table 2.1 summarizes the main characteristics of distributional bargaining. Whilst this approach to bargaining goes some way towards explaining how conflicts of interest are expressed and resolved, it can be criticized on the grounds that it pays too little attention to the characteristics of the parties involved in the negotiations, e.g. their values and resources;

Table 2.1 *A summary of the approach to bargaining taken by Schelling*

Assumptions

1 In any power relationship conflict should be taken for granted.
2 Bargaining strength derives from commitment (giving the impression of a strong and inflexible position), and bargaining skills.
3 Groups attempt to win relative to their own value system rather than by defeating the opposition. Winning is achieved by distributional bargaining.
4 The ability of one participant to gain his ends is dependent to an important degree on the choice or decisions that other participants make and inter-dependent decision taking, in which strategic behaviour is concerned with the other party's choice.
5 Bargaining is guided mainly by the expectations of what the other will accept.
6 A final bargain is struck when one of the parties makes a sufficient and final concession.
7 The bargaining process is seen as one of logical behaviour within a broad strategy in which actors converge towards agreement.

Characteristics of the bargaining process

1 *The adoption of a commitment.* All parties may do this to put the others under pressure and to counter their commitments. Thus bargaining becomes explicit.
2 *The communication of the commitment*, either directly or through third parties. It is essential to keep communication channels open to ensure continuous negotiation.
3 The use of *bargaining agents* with bargaining skill. This allows an apparently immovable position to be adopted.
4 The use of *intersecting negotiations*, where other issues of high value to the parties may be introduced into the bargaining process in order to strengthen an adopted commitment.
5 The use of a *restricted agenda*, i.e. controlling the issues brought into the bargaining arena to avoid prejudicing other related issues.
6 *Compensation and/or arbitration* are used where the concessions offered are not sufficient to produce agreement by themselves.

Deficiencies of the approach

This approach to bargaining pays too little attention to the characteristics of the parties involved; their inter-relationships and the inevitably uneven distribution of power.

---

the inevitably uneven distribution of power between those parties; their inter-relationships and the fact that parties may co-operate in attempting to achieve their objectives.

2.5.3 Power dependency and bargaining

Bacharach and Lawler (1981) have developed a power dependency approach to bargaining which partially overcomes these shortcomings, although it is similar in some respects to Schelling's theory of distributional bargaining. Basically Bacharach and Lawler concentrate on bargaining

Table 2.2 *A summary of the approach to bargaining taken by Bacharach and Lawler*

Assumptions

1 Bargaining is the active component of conflict and the main form of tactical action.
2 The dependence relationship between parties is central to bargaining, i.e. an actor's outcomes are contingent on what other actors do simultaneously and/or in response to his behaviour.
3 Dependence varies with the alternatives available to each actor to achieve the same or equally satisfying outcomes, and depends on the value (or commitment) given to a particular outcome.
4 Tactical action is based on the power distribution within a relationship. Actors base their tactical action on the perceived power of the opposition and the likelihood of them using it. It is designed to project an impression of their own power.
5 Groups attempt to win relative to their own value system and by altering the dependence relationship in their favour.
6 The bargaining process is seen as an information manipulation game in which the actors are uncertain about each other's goals, aspirations and intentions.

Characteristics of the bargaining process

1 Tactical action concentrates on:
   (a) *reducing one's own dependence* on the opposition by increasing one's own alternatives, and giving an impression of a small commitment to an outcome.
   (b) *increasing the dependence of the opposition* by reducing his/her alternatives and making him/her put a higher value on the issue at stake.
2 Tactical action is designed to induce concessions from the opposition and increase one's ability to manipulate the behaviour of the opposition.

Deficiencies of the approach

This approach to bargaining assumes that tactical action is purposeful and can be identified. It also provides little guidance as to how the different components of tactical action can be identified.

power which they claim is derived from the dependence relationships between the parties involved in the negotiations. These dependence relationships set the stage for bargaining; generate the issues in conflict and influence the bargaining outcome. Bacharach and Lawler also see the use of 'tactics' as the key dimension of the bargaining process, and suggest that by analysing tactical action, bargaining power (which they call 'context') can be related to bargaining settlement (which they call 'outcome'). They stress the need for a theory of tactics which captures the dynamics of the use of power and which identifies the links between different dimensions of power.

The characteristic of the power-dependency approach to bargaining are summarized in Table 2.2. However, although this approach focuses attention on the characteristics of the parties involved in the process, accommodates the co-operative approach to bargaining and takes account of the unequal distribution of power between those parties, there are difficulties in applying it to specific situations, e.g. it provides little guidance as to how the different components of tactical action can be identified. A comparison of the theories of Schelling, and of Bacharach and Lawler, reveals a number of common features as well as unique characteristics. Thus both approaches:

- see mutual dependence as a pre-requisite of explicit bargaining;
- focus on the visible components of bargaining (Schelling's strategy of conflict; Bacharach and Lawler's tactical action);
- assume that parties wish to avoid mutually-damaging behaviour.

At the same time differences between the two approaches suggest ways in which they can be amalgamated to create a framework which takes account of the characteristics of the parties involved and their inter-relationships; the uneven distribution of power and co-operative approaches to bargaining against which bargaining situations can be analysed. Thus the view of Bacharach and Lawler that the bargaining settlement can be predicted on the basis of bargaining power is particularly helpful in situations where the exchange of information is imperfect; where the bargaining undertaken is more implied than explicit; or where it proves impossible to reach an agreement which is acceptable to all sides. By contrast, Schelling's approach has significant advantages where the bargaining is more explicit and is concerned with distributional issues. Without doubt the way in which he presents the expected characteristics of bargaining make it easier to apply in practice. These theoretical approaches to bargaining have been utilized by Bruton (1980, 1983b) in undertaking empirical work which gives some indication of the importance of power and bargaining in local plan preparation (see Chapter 4) and development control (Chapter 7).

2.6 Contexts for local plans and local planning

The contingency approach to management emphasizes the contextual features impinging on the activity in question. For local plans and local planning, these features will be physical, economic, political and social in nature. The precise and relative contribution that these factors make to the operating context within which local plans are prepared and used can be expected to vary between localities. Planning contexts at the local scale will thus be characterized by extreme variety. To the extent that this is so, generalizations about the effects of any particular contingent factor will be difficult and not necessarily valid. It is possible to identify contingent features that will usually be relevant (e.g. the strategic/structure plan context), but not to predict what the influence of these will be for any particular area. This is so notwithstanding the fact that generalizations *can* be made about the nature of the processes of bargaining which arise from the role of local planning as an inherently distributional activity taking place in a socio-political context. However, what Strauss (1978) calls the 'contexts for negotiation', and which provide the 'rules of the game' and set constraints making bargaining parties unequal or limiting the scope for negotiation, are less amenable to analysis but no less relevant (Barrett and Hill, 1984). The difficulty of assessing in any general way the contextual or contingent factors impinging on local planning can be clearly seen in the study of speed, economy and effectiveness in statutory local plan preparation and adoption undertaken for the DoE by Fudge *et al.* (1983). Here, the context for local land use planning receives considerable attention, especially insofar as particular contextual features affect the individual stages of plan preparation. Local planning authorities are regarded as confronting a series of tensions in determining both the most suitable means for local intervention and the need for statutory local plans in their overall strategy. Most notably, tensions arise from attempts to accommodate the three principles identified by Regan (1978) as underlying the land use planning system. These principles are:

1   *Judicial*, emphasizing the rights of individuals in a system which intervenes in private property rights, and leading to the establishment of agreed processes for the determination of disputes.
2   *Professional*, stressing technicality and rationality and embodied in the desire for long-term comprehensive approaches with policy justification stemming from technical analysis.
3   *Political*, emphasizing the role of public opinion and of the political party in control, and implying an adaptive stance able to shift in line with changing priorities.

The land use planning system in general (and the local plan in particular) attempts to give a major role to these three components, even though each represents a different way of carrying out local planning and producing local plans. Attempting to reconcile these differences within a single administrative process of plan production ensures that the tensions involved are reflected in the preparation process itself. In 'choosing' or evolving an overall local planning strategy local authorities must face and resolve these (and other) tensions, and Fudge *et al.* suggest a checklist of questions that offer guidance in determining such an approach. This list considers the following issues: the operational context; the type of planning guidance required; the type of statutory plan needed; the role of the plan; consultation and participation; and the preparation process. However, this checklist has only general rather than specific relevance. First, 'there is a complex inter-relationship between a whole range of factors that produces a unique solution in each authority; even where the characteristics of areas appear superficially similar, quite different approaches have been adopted with some justification' (Fudge *et al*, 1983, p. 19). Second, 'it is inappropriate to offer specific guidance about the type of plans that should be prepared in different circumstances because each local authority's "operating context" is likely to be unique' (*ibid.*, p. 28). In an attempt to overcome this problem of generalization, and to account for (and explain) the form and operation of local planning, Healey (1983, p. 245) sets out to relate local planning as a government activity and local plans as a specific form of that activity 'to the wider sphere of the nature and articulation of political and economic forces as they affect both central and local government; the operation of firms in general, and their development activities in a particular locality'. Such a 'political economy' approach to the role of the state and to particular sectors of production as they affect locational and land needs however has its own methodological problems. In particular (*ibid.*, p. 245):

Coherent theoretical structures are at such a broad level of generality that they cannot be tested against actual events, while in investigating events, it is difficult to substantiate precisely the connections between these and wider forces, even though a good interpretive guess can be arrived at by assembling various forms of evidence.

One way round this difficulty is to use analyses of social and political structures to identify 'general organizing ideas which suggest the key relations and tendencies which affect detailed events' (*ibid.*, p. 246), rather than attempting to trace every link in the chain between theoretical structure and the object of analysis. The research task is defined as follows (*ibid.*, p. 260):

. . . to relate the way the state actually operates and what the role of different parts of the state is at any point in time and in any locality, to interests broadly linked to the economic tendencies of a late capitalist economy and to the political demands this creates.

Following O'Connor (1973) and Saunders (1981) this approach recognizes a fundamental distinction between production and consumption interests in the economy, which are nevertheless inter-related. In the field of land use and development, both production and consumption interests are addressed by local planning. Healey (1983, p. 261) outlines the following 'organizing ideas' to structure analyses of how and why production and consumption interests become involved in land use planning:

1   State intervention in land ownership, use and development has to take account of demands for land for the production of goods and for social needs.
2   In capitalist societies these demands are likely to be in potential conflict.
3   In the United Kingdom further demands for land, and hence potential for conflict, are created by the house-building and property development sector and by cultural traditions which stress the heritage value of certain landscapes and buildings.
4   State regulation of development benefits firms producing goods, firms producing property, conservation interests, and the community at large from social services and facilities. All also benefit to some degree from more direct intervention in providing and servicing land and buildings.
5   Since in these interventions the actions of the state have to appear legitimate to all interests, procedural devices are likely to emerge to demonstrate that the actions of the state are in the general public interest. However, the more powerful interests will seek to ensure control over such procedures so that these operate to their advantage.
6   It is unlikely that state land policies can operate effectively when defined in isolation (either as a distinct sector of state activity or at a particular level within the state) since land use and development is a consequence of economic and social activities, including state policies, and because where land is required this involves the consideration of specific sites and localities.

Production and consumption activities concerned with the development and other use of land are also necessarily interested in particular sites and areas. Hence both sets of interests must be concerned with *local* planning policies as formulated by the local planning authority. In seeking to explain why particular local planning strategies have been taken up, Healey's approach emphasizes the need to relate such strategies to the wider social, political and economic contexts in which they have been developed. It also suggests that the relationship between policy and implementation is likely to be interactive in the sense that different and competing interests will be involved. The outcome of this process may be seen as being derived through negotiation or bargaining between interests (e.g. Bruton, 1980). In essence Healey's work reinforces the views expressed in Sections 4 and 5 of this chapter concerning the need to adopt a contingency approach to planning; an approach which acknowledges the role of bargaining in the process of plan preparation and implementation. By implication once again planners are seen as managers of change in the environment.

2.7 Conclusions

The statutory town and country planning system in Britain is restricted to a concern with the use and development of land. Town planners are seen by government as producers of plans and controllers of development and the means of producing those plans and controlling development are narrowly prescribed. In reality the type of problems faced by town planners are complex and highly inter-related. They are 'wicked' problems characterized by inter-connectivity, uncertainty, ambiguity, conflict and societal constraints, where policies to solve one problem often have unintended consequences and create problems in other areas; where policies and events in other areas are often connected to apparently unrelated problems in a way which influences the outcome of policies to deal with those problems. Thus it is unrealistic to view problems relating to land use in isolation – they are an essential element of the social, economic, and political context within which they are located. In this situation planners in practice have gradually transformed their role from being producers of plans to that of managers of change – producing policies for change; organizing the resources needed to implement those policies for change; controlling that implementation and motivating other bodies to contribute to the achievement of that change. In doing so, local planners have sought to strengthen their approach to local planning issues by shifting from plan-makers to catalysts and negotiators, an approach suggested as long ago as 1974 by MacMurray. The decline in the British economy, with limited economic growth occuring only in certain restricted areas of the country, has contributed to this transformation of the town planner's role.

Management theory suggests that only by structuring complex organizations in a hierarchical way will the problems of complexity and inter-connectivity be satisfactorily handled. The British town and country planning system has moved some way towards this by adopting a loose hierarchy of land use plans but it has not fully realized the need to establish a framework which provides for both the following:

1 The vertical linkages between general (political) policies for social and economic change and implementation on the ground.
2 The integration of the 'vertical' policies with the horizontal (spatial) relationships needed to secure social and economic change and physical development on the ground.

In short the relationship between general policies and specific plans has not been fully thought through.

At the same time, the contingency approach to management informs us that there is no one 'right' solution to a problem. Rather, the correct solution to adopt in any given situation will depend on the specific conditions and circumstances within which decisions are made and implemented. Thus in a local planning context it would seem reasonable to expect that the 'right' approach to adopt will be, for example, contingent upon the state of the local economy; the local political situation; the nature of the strategic planning framework; resource constraints; county–district relations; the degree of uncertainty and so on. Given the wide range of possible contingent factors and circumstances it is unlikely that the narrow and inflexible land use plans prescribed by government will be capable of meeting all situations. The *Law of Requisite Variety* suggests that this will not be the case whilst the range of non-statutory local plans and policies in use by practice and the variety of special area work such as GIAs, HAAs, EZs, confirm that this is not the case.

Similarly, it is clear that implicit and explicit bargaining is integral to the process of plan preparation and development control, although the nature and extent of the bargaining will vary with the contingent and contextual factors. To quote Chapin and Kaiser (1979, p. 23) once again:

. . . because of the very tangible, complex and differential impacts of local decisions involving many persons and firms, what evolves is essentially an interactive progression of decisions rather than a single final decision . . . there is usually a recursive sequence in the way participatory, technical, and political inputs to this process function to establish land use objectives, explore growth scenarios, test out development alternatives, and arrive at land use decisions.

Notes

1   The content of this section appeared in a modified form in an article by M. J. Bruton and D. J. Nicholson published in *Town Planning Review*, Vol. **56**, No. 1, January 1985, pp. 21–41.
2   A series of papers illustrating the complexity of inter-organizational relationships is included in *Town Planning Review*, Vol. **51**, No. 3, July 1980, pp. 257–338.

3 The statutory local land use planning system in Britain
I: The legislative and administrative framework

3.1 Introduction

The discussion in Chapter 2 emphasizes the importance of situating both statutory local plans and the wider local planning effort of local authorities firmly within the relevant policy and operational contexts. The hierarchical strategic planning framework, for instance, identifies local planning as an activity which is both theoretically subordinate to higher level policy initiatives and itself acts as a controlling influence over the lower implementation level. Moreover, it is also argued that such an understanding of the vertical limitations placed upon local plans and planning must be supplemented by an appreciation of the operating environment in which local plans are prepared and local planning strategies are evolved. Contingency factors arising from this environment will have a varying influence upon the development of approaches to the local planning task taken by individual authorities, and so can be used to aid an understanding of why such approaches differ within a common legislative framework.

This chapter, together with Chapter 4, focuses on the development and use of the statutory local plan, but these points as to the importance of the context in which such plans are produced and used must be kept in mind. This is because the pattern of production of plans, the objectives they seek to fulfil, and the issues they address cannot be satisfactorily discussed solely by reference to legislative procedural requirements. The legislation actually embodies considerable administrative discretion, even though the DoE through policy directives and advice has subsequently sought to limit the boundaries of such local autonomy. This chapter begins with a review of the statutory development plan context within which local plans have been produced, concentrating on the various statutory (development plan schemes, certification) and informal (local plan briefs) devices which have been introduced in an attempt to ensure the conformity of local plan proposals with structure plan policies. The legislative provisions for local plans and the manner in which these provisions have developed and changed over time, both through amending statutes or evolving DoE policy, are outlined. The statutory procedures for plan production are then reviewed, together with the way in which these requirements have been received and worked out in practice; attention is concentrated on the public local inquiry as a means of conflict resolution and plan legitimation. Chapter 4 then develops this review of the statutory local land use planning system by considering local plan production to date (March 1985) and the content of adopted and deposited plans. These two chapters together allow an evaluation of the effectiveness of the statutory local plan against the idealized hierarchical model of policy formulation set out in Chapter 2. The relevance of this exercise is that such a model is largely assumed by the development plan legislation with which we are here concerned.

3.2   Local plans: the statutory planning context

Following their review of the land use planning system set up under the *Town and Country Planning Act 1947*, the Planning Advisory Group proposed a two-tier system of land use plans, comprising strategic 'county' and 'urban' plans (later enacted as structure and urban structure plans, respectively) which were to be subject to ministerial approval, and more detailed 'local' plans designed to implement the intentions of the policy plans; these latter would not be submitted to central government for approval, but 'must conform' with the policies laid down in the urban plan or county plan (Planning Advisory Group, 1965, p. 10). These 1965 proposals were largely enacted in the *Town and Country Planning Acts* of 1968 and 1971. Initially, it was assumed that the new system would be operated in a manner similar to that set up under the 1947 Act. Here, the responsibility for development plan preparation outside London had been vested in 79 county boroughs, covering the cities and larger towns, and 45 counties. These were unitary authorities, although there was provision for the delegation of some development control functions by county councils to the urban and rural district councils in their area. The Planning Advisory Group took this unitary pattern of responsibility for plan preparation as a framework for their proposals, so that the issue of conformity between structure and local plans was not seen as problematic. The expectation was that once the 'local planning authority' (county or county borough) had evolved a strategy in an approved structure plan, local plans would follow as required (Ministry of Housing and Local Government, 1970). Nevertheless, the 1971 Act required authorities to ensure that local plans conformed generally to the relevant structure plan (s.11(9)).

However, the operation of the new development plan system was complicated by the reorganization of local government in April 1974 following the *Local Government Act 1972*. This legislation established a two-tier structure of metropolitan and non-metropolitan county and district councils which in a modified form remains in operation today.[1] The new development plans system was split between the two tiers, with the districts being given major responsibility for the preparation of local plans.[2] This was a logical and attractive allocation, but the move created a number of problems for the practical implementation of the new planning legislation for counties and districts alike. Not the least of these was the pressing need to define what constituted a 'strategic' issue, and to establish how far structure plans could reasonably go in scope and depth of treatment (Finney and Kenyon, 1976). In turn, this difficulty immediately raised the possibility of conflict between tiers, for clearly the more the county-prepared structure plan established matters of detail, the more the district councils were constrained in terms of the degree of discretion open to them in producing policies for their local plans. Structure plans, in fact, assumed a new (unofficial) function, 'as documents prescribing the power relationships in planning between the two tiers' (Grant, 1982, p. 48). The boundary of acceptable detail in structure plans has been set in practice by the Secretary of State's modifications to structure plan policies considered to be too detailed. In addition, the new allocation of planning functions meant that the problem of conformity between structure and local plans took on a new dimension in the mid-1970s, since each was to be largely prepared not only by *different* authorities, but also by authorities of different *tiers*. Since such authorities varied in terms of geographical area, functions and (possibly) political composition, differences between structure and local plan priorities and policies were anticipated. The need for formal machinery to ensure conformity was evident (Royal Town Planning Institute, 1972), and the *Local Government Act 1972* made provision for two separate devices designed to ensure the smooth consistent operation of the new system, the development plan scheme (DPS) and the certificate of conformity. In addition, a third non-statutory device was added by the DoE, the local plan brief, while Circulars such as 74/73 *Co-operation between Authorities* emphasized the need to establish constructive relationships between both county and district planning authorities and development plans and development control work (DoE, 1973a).

Development plan schemes have to be prepared by county planning authorities in consultation with district authorities under s.10C of the 1971 Act[3], and must be kept under review and amended if thought fit. Agreed schemes must be sent to the Secretary of State: although not subject to his approval, schemes may be amended by him if a district authority makes representations that they are dissatisfied with the county's proposals. DoE Circulars such as 74/73 have emphasized that these powers to settle disputes will be used only as a last resort. The legislation is explicit on form and content; schemes should:

1 designate the authority responsible for plan preparation and adoption;
2 specify the area to be covered by each plan;
3 specify the title, nature and scope of each plan;

4   set out a programme for the preparation of local plans;
5   indicate the relationship between plans; and
6   identify any plans to be prepared concurrently with the structure plan.

Subsequent circulars and advice have reiterated these requirements as to content, also giving emphasis to the need to prepare schemes quickly and to propose local plans only for areas where there is a clear need and where the staff resources exist to complete the work in a reasonable time scale.[4] The objectives of development plan schemes have not been so clearly spelled out, but a range of circulars and other documents suggest that the main aims are to secure effective co-operation between county and district, and to establish a realistic local plan programme with clearly established priorities. Circular 74/73, for instance, recommended that local authorities use development plan schemes to promote effective co-operation and minimize delay, dispute and duplication (DoE, 1973a), whilst DoE Circular 58/74 (which dealt with the development plan provisions of the *Local Government Act 1972*) suggested that schemes should be prepared quickly with the immediate aim of avoiding uncertainty and bringing out the priorities for the more urgent local plans. In allocating responsibility for plans, the availability of planning staff was important (DoE, 1974a). Local Plans Note 1/76, although now cancelled, reaffirmed the view of the development plan scheme as a co-ordinating device, and drew attention to the potential of schemes as a management tool for concentrating resources, deploying staff efficiently, and ensuring that the most appropriate type of plan was produced to meet the needs of the area (DoE, 1976). Later development plan circulars and memoranda have stressed that priorities for local plans should be realistic in the context of staff resources and local planning problems, and that schemes should also consider whether or not proposed plans are needed (DoE, 1979a, 1981c, 1984g). There is also evidence to suggest that, in addition to these official aims and objectives, schemes have been used by county and district planning authorities as one of several instruments whereby the relative planning role and responsibility of the various authorities was established. Bruton, for example, has shown how the preparation of a development plan scheme and the production of local plans was used as part of a process of distributional bargaining to establish the relative levels of responsibility between district and county (Bruton, 1980). Leach and Moore have stressed the role of various legal and quasi-legal documents and agreements as 'bargaining counters' in the development of a county/district relationship, where such documents 'can be used as a power resource in the development and deployment of strategies for coping with the interdependency with the other authority' (1979, p. 168). In many respects, the successful use of the development plan scheme to resolve questions of the relative responsibilities and planning roles of county and district authorities is a necessary pre-condition for the achievement of the official aims and objectives.

It is hardly surprising, therefore, that an early study of a limited number of agreed and draft schemes by DoE research officers Mabey and Craig (1976) found many of the official objectives to be as yet unfulfilled: The schemes examined showed a desirable intention to co-operate, but exhibited some confusion over responsibility for plan preparation, as well as over the scope, type and area of the proposed plans; there was little rigorous assessment of the time or manpower implications of the suggested programmes. More alarmingly, Mabey and Craig suggested that the outcome of their analysis implied the intended preparation by English local authorities of in excess of 3500 local plans, a total programme regarded as unrealistic given the resources of time and staff that would be required. Mabey and Craig attributed the scale of local authority intentions to such factors as a tendency to favour a large number of small area district plans, and the intended use of subject plans to deal with topics better dealt with in district plans or sup-plementary guidance. Local Plans Note 1/76 (now cancelled) expressed concern at the 'proposals to prepare large numbers of district plans, often covering very small areas and for subject and action area plans inappropriate circumstances', going on to refer to the problems associated with producing large numbers of local plans and thus by implication favouring the product-ion of fewer plans (DoE, 1976, para. 9). A more recent study of schemes (Bruton, 1983a) has shown that the early fears of Mabey and Craig were not realized, the first round schemes proposing a total of 2606 local plans (Table 3.1). Nevertheless, this still represents a massive commitment. By 1981, development plan schemes showed a total of 1511 local plans being pro-posed, a reduction of 42 per cent on those included in the initial schemes. This fall is partly attributable to the influence of the consistent Departmental advice that 'the making or review of a scheme is not the occasion for including every possible plan' (DoE, 1979a, para. 3.17). The large reduction in the number of action area plans (-83 per cent) had been anticipated by Mabey and Craig, and can be attributed to the need for local authorities to

Table 3.1 *Local plans proposed in first round DPS, DPS operative in 1981, and DPS February 1982, England**

	First Round DPS		DPS Operative 1981				DPS February 1982			
					Change from previous round				*Change from previous round*	
Type of plan	*No.*	*(%)*	*No.*	*(%)*	*No.*	*(%)*	*No.*	*(%)*	*No.*	*(%)*
District	1468	(56)	1227	(81)	-241	(-16)	1149	(83)	-78	(-6)
Action area	588	(23)	102	(7)	-486	(-83)	78	(6)	-24	(-24)
Subject	550	(21)	182	(12)	-368	(-67)	155	(11)	-27	(-15)
Total	2606		1511		-1095	(-42)	1382		-129	(-9)

Source: Bruton, 1983a, Table 1.
* Excluding Greater London.

be more realistic in their proposals for this type of plan as the structure plan for their respective areas was submitted to the Secretary of State for approval. The comparatively small reduction in the number of district plans proposed (−16 per cent) can be related to the need perceived by district planning authorities for a more detailed guidance framework for development control than provided by the structure plan. It is also likely that the proposed production of local plans, especially district plans, was seen by district authorities as a means of establishing their position in the planning hierarchy and challenging what had been until 1974 the dominance of the counties (McAuslan and Bevan, 1977). No doubt in the early years following reorganization, development plan schemes were as much concerned with establishing planning roles and responsibilities as with setting out a co-ordinated and realistic work programme. There are also suggestions that many local plans were proposed for internal authority political reasons (Duc, 1979; Williams, 1978), and that planning officers used the number of local plans proposed to justify part of their establishment. Following the request in Circular 23/81 (now cancelled and replaced by Circular 22/84) that county planning authorities review their development plan schemes with the objective of ensuring that there is a clear need for proposed local plans, and that the work required to enable the plan to be placed on deposit could be carried out within about a year (DoE, 1981c), there was a further reduction in the number of local plans proposed to 1382 (9 per cent less than in the 1981 schemes). Much of this reduction, however, can be attributed to the completion and adoption of plans rather than to the exhortations of the Circular.[5] These most recent schemes also show that by 1982 district and action area local plans had clearly become the province of the district planning authorities, with the counties having prime responsibility for subject plans (Table 3.2). This suggests that the respective local plan-making responsibilities of the two tiers have been generally resolved.

Disputes referred to the Secretary of State have been rare, presumably

Table 3.2   *Local plans proposed in development plan schemes, February 1982, by type of authority, England**

Type of plan	Country		District		Total	
	No.	(%)**	No.	(%)**	No.	(%)**
District	16	(1)	1133	(99)	1149	(83)
Action area	3	(4)	75	(96)	78	(6)
Subject	126	(81)	29	(19)	155	(11)
Total	145	(10)	1237	(90)	1382	(100)

Source: Bruton, 1983a Table 2.
* Excluding Greater London.
** Expressed as percentage of total of type of plan.

because county and district authorities share a common interest against the central determination of local planning strategies. Nonetheless in some cases the DoE has found itself involved as an arbiter, seeking to resolve through informal consultations and advice or formal directives such disagreements as those between Tyne and Wear MCC and the constituent districts, Avon CC and Northavon DC,[6] and Greater Manchester MCC and Manchester City Council (Healey, 1984). Several of these disagreements have centred on the question of whether various open land/urban fringe/ Green Belt issues, such as the fixing of Green Belt boundaries, are best settled by a single county adopted subject plan or a series of district adopted local plans. Counties viewing Green Belt as a strategic planning tool have tended to favour the former, while districts anxious not to have their scope for expansion too severely curtailed have favoured the latter. Ironically, however, the DoE has itself been accused of taking a less than consistent stance on the question of whether Green Belt is a suitable topic for subject plan treatment (Munton, 1983). Unreported informal disputes over development plan schemes, which may result in a scheme not being updated if the contents cannot be agreed, are probably more common.

Recent schemes[7] vary in form and content, but all establish priorities for local plan production, establish the nature and scope of each plan, and clarify responsibilities. Few, however, consider in detail the need for a local plan, while there is little evidence of the use of schemes as devices to deploy or manage staff, or of account being taken of staff/resource availability. Current schemes, therefore, appear to meet most of the *legislative* requirements as to form and content, although there is less evidence of concern to satisfy DoE *policy* requirements. The majority of schemes for instance contain proposals for plans which will clearly not be placed on deposit in the timescale envisaged formerly by Circular 23/81 and now by Circular 22/84, i.e. within one year from the start of production.

For county planning authorities, the development plan scheme provides little control over the actual policies set out by districts in agreed local plans. A more explicit policy control for the county is provided by the certification procedure introduced by the *Local Government Act 1972*.[8] This strengthens the requirement for 'general conformity' between the structure plan and local plans (both during the formulation of local plan proposals and at adoption) that was already part of the new planning legislation following the proposals of PAG.[9] Briefly, district councils preparing local plans now have to submit the plan to the county for certification as being in general conformity with the structure plan before placing the plan on deposit for public inspection.[10] As with development plan schemes, the Secretary of State retains powers of determination in the event of a dispute, but again these are regarded as powers of the last resort. Introducing the new procedures, DoE Circular 58/74 argues that since both the conformity requirement and the idea of co-operation and consultation between county and district authorities in plan-making is not new (cf. Circular 74/73) 'the need to obtain a certificate is therefore not an onerous addition to the process, but simply a safeguard'. Any doubts over conformity are to be resolved through the normal course of consultations; irreconcilable local disputes should be referred initially to the Department on an informal basis before embarking on a formal application to the Secretary of State (DoE 1974a, para. 11).

In practice, although the scope for problems to arise at the certification stage is potentially great, only a few cases of dispute have been reported; there is no doubt that the majority of local plans that have been prepared conform generally to the structure plan. This is partly due to the processes of consultation recommended by circulars, whereby legitimate conflicts of interest between strategic and local authorities have been resolved through constructive dialogue between the parties (Briscoe, 1985). Additionally, where there is a clear conflict between approved structure plan policies and draft or emerging local plan proposals, the district concerned may well prefer to abandon the statutory procedures and use the plan informally, thereby avoiding conflict over certification. This point suggests that county planning authorities will encourage districts to prepare statutory rather than informal local plans, since the conformity of the former to structure plan policy can be ensured while that of the latter cannot. In any event, the certificate only establishes a general conformity, and this is a vague and flexible phrase. DoE Circular 4/79, for instance, advised that ' "conforms generally" should be regarded as embracing "does not conflict with" ' (DoE, 1979a, para. 3.62). This flexibility can be exploited to reconcile apparently conflicting proposals. For example, in his report on the Fareham Western Wards Action Area Plan in 1978 the Inspector responded to objections to approved *structure* plan growth proposals by recommending a reduction of one-third in the number of dwellings to be proposed in the *local* plan. This was not considered to endanger conformity since the structure plan allocations were for a longer timescale than that addressed by the action area plan (Grant, 1982). Such recommendations do not remove the statutory requirement on the local planning authority to ensure conformity at the time of adoption, although the Inspector's recommendations should be capable of implementation 'without disturbing the general conformity' between structure and local plan (DoE, 1984g, para. 3.71). The Fareham case, however, does illustrate that the conformity requirement can bear a very loose interpretation indeed.

Despite these factors, a number of disputes over conformity have arisen which have gone to central government for arbitration. These disagreements have arisen from county judgements that district-prepared local plans have failed adequately to embrace structure plan policies, commonly in respect of Green Belt or urban restraint issues but also on industrial and shopping topics. These disputes have highlighted the practical difficulties of setting limits to 'general conformity' which are acceptable to both county and district authorities. The emergent response of the DoE to these problems has been to encourage the county to certify the disputed plan (failing which the Secretary of State may certify the plan himself), leaving disagreements to be settled at the public local inquiry stage.[11] The Bromsgrove District Plan, for instance, is reported to have only achieved certification by Hereford and Worcester CC following informal Departmental pressure.[12] Tyne and Wear MCC, however, refused to follow Departmental advice to certify Newcastle upon Tyne City Council's City Centre Local Plan, which was subsequently certified by the Secretary of State using his 'reserve' powers. Here the DoE noted that the statutory requirements for conformity were met 'if the proposals of a local plan are broadly in keeping with the policies and proposals of the structure plan'. In this case, structure plan modifications agreed with the DoE limited additional retail provision within the city centre to 5000 m^2; local plan proposals allowed for an additional 17,000 m^2 of retail floorspace.[13] Elsewhere, the DoE has asked Oxfordshire CC and Humberside CC to certify disputed plans and settle the issues at the public local inquiry.[14] Not all such disputes have been settled in favour of the districts, with Leeds City Council recently having been directed to revise the Morley Local Plan following county refusal to certify.[15]

Nevertheless, the principles adopted in these cases by the DoE effectively reduce the powers of the counties to achieve conformity between district-prepared local plans and the county structure plan, especially on detailed issues such as site allocations. Not only is a liberal interpretation being placed on general conformity, but the reserve powers of the Secretary of State are being threatened to encourage certification. The ability to fight the local plan at the inquiry is no compensation. Counties have no special powers of objection. Moreover, objecting to a previously certified plan clearly places counties in a contradictory position. If the objection relates to a proposal with strategic implications contrary to structure plan policy, why was the plan certified as in conformity? A further danger here, which counties will wish to avoid, is that accepted strategic policies will once more be thrown open to debate. Finally, there is no compunction upon the district council concerned to accept the Inspector's recommendations, even assuming that he/she accepts the county's objections and recommends modifications to the local plan. A county which had thus complied with Departmental requests and certified a disputed plan could theoretically see its objections upheld by the Inspector but legitimately ignored by the district, which could then proceed to adoption. More recently, the DoE has clarified the position by advising that a certificate should be refused where there is 'significant conflict' between structure and local plan provisions, but that where there are other difficulties (such as matters of detailed interpretation or proposed implementation) which do not disturb general conformity the county planning authority should issue a certificate and resolve the outstanding issues through the objection procedures (DoE, 1984g, para. 3.43). At the same time, Briscoe (1985) reports that the DoE has attempted to define more closely the boundaries of 'general conformity', operating a rule of thumb that local plan provisions which vary by more than 10 per cent from approved structure plan policies are not in accord. Briscoe suggests that where approved strategic policies are up to date this is unlikely to cause problems, although difficulties may arise should the county wish to certify local plans as being in accord with revised and submitted (but not approved) strategic policy. However, expedited procedures apply when proposals for the alteration, repeal or replacement of the structure plan have been submitted to the Secretary of State. Under these procedures, the Secretary of State may where appropriate give a direction enabling the county planning authority to issue a certificate of general conformity as if the proposals for structure plan replacement, repeal or alteration had already been approved (DoE, 1984g, paras. 3.48 to 3.53).

The final device intended to secure integration between the two tiers of plans is the local plan brief. The focus here is explicitly on the policy issues and the proposals to be included in local plans, but the local plan brief is a DoE policy invention only and has no legislative foundation. The idea first seems to appear as the 'planning brief' in Circular 74/73, which suggests that briefs be prepared by the county council in association with the districts in order to supplement the structure plan 'by setting out the thinking behind the plan and relating its aims, objectives and policies to the more detailed treatment appropriate to the local plan' (DoE, 1973a, para. 12). In advance of structure plan approval, interim planning briefs were suggested to help with the production of informal local plans, which at the time were being actively recommended by the DoE to meet perceived shortages of land for housing.[16] By 1977, interim local plan briefs were to be used, not to assist in the preparation of informal plans, but to help with those local plans being prepared concurrently with the structure plan (DoE, 1977e). Current advice envisages the preparation of briefs by the county planning authority to provide the structure plan context (DoE, 1984g). Little is known in detail about the content of local plan briefs, but as a type of informal or non-statutory local planning instrument the scale and manner in which they are used is taken up in Chapter 6.

This review of the statutory context for the use of local plans has emphasized the operation of two particular external contingent factors, the strategic context (currently embodied in the county-prepared structure plan) and the national policy directives emanating from the DoE. Integrating and co-ordinating devices provided for by statute have been interpreted by DoE advice and henceforth worked out in practice. In some cases, these devices have provided a focus for inter-authority conflict rather than a channel for integration, but these instances are relatively rare. There is no doubt that development plan schemes and certification have, in the majority of cases, simply acted to formalize working arrangements for consultation and co-operation that would have developed in any event. In doing so, they have reflected the strategic planning principle that policies established at one level should be influenced and constrained by those developed at the level above. At the same time development plan schemes have been used as 'counters' in bargaining and negotiations between county and district authorities as levels of responsibility for plan-making have been defined.

3.3   The type, form and content of local plans

In its review of the development plan system that had been established by the 1947 Act, the Planning Advisory Group (PAG) identified several key weaknesses with this legislation as implemented. In particular, the system was considered to be 'too detailed for some purposes and not detailed enough for other purposes', while the distribution of responsibility between central and local government was judged to be imbalanced in that many issues of local interest and with no policy significance were brought before the Minister for approval (PAG, 1965, p. 8). The 1947 Act had *required* the preparation of town maps by county boroughs, and of county development plans by the counties; here, town maps could also be prepared for urban areas. Town maps could also be used to define the boundaries of Comprehensive Development Areas (CDAs). A hierarchy of plans is evident here, but in practice the format and content of these plans was similar, with each being prepared on a map base (of varying scales) and each subject to Ministerial approval. The system was characterized by an apparent comprehensiveness and precise land use allocations, especially within urban areas. The detailed nature of the 1947 Act development plans not only meant that substantial delays occurred in obtaining central government approval, partly as a consequence of the large number of objections that were generated, but also meant that once approved the plans dated quickly. In addition, it proved difficult to adapt the plans to emergent planning problems and the policies being evolved to deal with them, so that they became increasingly irrelevant as vehicles for local, strategic and regional planning policies. For instance, PAG noted how both the town map and the CDA had proved to be inadequate means of controlling the growth of private development activity in town and city centres in the early 1960s; these problems with the statutory system had in fact led central government to promote the concept of a non-statutory town centre map to guide renewal (Ministry of Housing and Local Government, 1962a). One of PAG's main concerns was the failure of the 1947 system to deal with economic growth and its development implications, a shortcoming which was seen as particularly pressing since the Group anticipated a period of rapid population growth and rising standards of living, leading to 'a surge of physical development on a scale that this country has not previously seen and this will occur, overwhelmingly, in and around the towns' (PAG, 1965, p. 8). As we have seen, to guide this urban development and renewal the Group proposed a split in formal plan preparation between strategic urban and county plans, to be 'primarily statements of policy illustrated where necessary with sketch maps and diagrams and accompanied by a diagrammatic or "structure" map designed to clarify the basic physical structure of the area and its transport system', and local plans, intended as 'a guide to development control and a basis for the more positive aspects of environmental planning' (*ibid.*, 1965, pp. 9 and 10). This contrasted with the 1947 system in that PAG sought to distinguish clearly between strategic policy issues and detailed land use proposals/site allocations by allocating these different planning concerns to appropriate policy vehicles which would differ both in *format* and in *content*.

Local plans, by rendering general policy and broad proposals into detailed local terms, would 'serve as a useful guide to developers and as a basis for co-ordinating public and private action in development and redevelopment'. As such they were not viewed 'primarily as a control mechanism, but as a positive brief for developers, public and private, setting the standards and objectives for future development' (*ibid.*, p. 31). Here, the economic context of growth (with its consequent urban expansion and renewal) which PAG assumed that their new system would need to cope with is clear. In this situation, local plans were to serve as devices for the guidance and co-ordination of development, with the emphasis on promotion rather than control. The main purposes of local plans would thus be (PAG, 1965, p. 32):

1   to implement the intentions and fill in the details of the policies and proposals in the submitted plans;
2   to provide planning authorities with a recognized means of planning at the local level . . . .;
3   to encourage authorities to undertake comprehensive environmental planning . . . .;
4   to provide the basis for development control, for detailing land use requirements and sites for public buildings, to help to answer questions on land searches and to provide developers with positive guidance on development standards;
5   to help the public understand and take part in the detailed planning of their town.

A number of types of local plan were identified:

1   Action area plans

These were regarded as the most important category since they provided the main device for channelling the anticipated urban growth and redevelopment. This development activity was to be handled in a *comprehensive* manner; action areas were those areas, to be defined in the strategic plans, 'which are to be planned and developed, redeveloped or improved in a comprehensive manner over the next ten years or so'. As such, 'they represent the main programme of action for the town' (*ibid.*, p. 32). As a vehicle for comprehensive development proposals, action area plans were intended as replacements for the inadequate town maps and CDAs, particularly as a basis for the use of other statutory powers such as compulsory purchase orders.

2 District plans

Recognizing the need for detailed local planning in urban areas outside of proposed action areas, the Group suggested 'district plans' to provide a more detailed basis for development control than the strategic urban plan itself, and to apply the techniques of traffic and environmental management in a co-ordinated way. District plans would thus serve to provide a link between the strategic plan and specific action areas; 'eventually a mosaic of local plans can be built up for the town as a whole, setting the action areas in a wider context and relating one action area to another' (*ibid*., p. 33).

3 Town plans and village plans

Outside of the urban areas, where the strategic county plan was proposed as the broad policy framework for development, the Group suggested that detailed local plans for small towns and villages could be prepared as required 'as and when staff resources allow'. These plans were to be 'of a very simple character and the small physical scale of the subject matter must not lead to over-meticulous planning or planning in excessive detail' (*ibid*, p. 34). PAG clearly envisaged the gradual preparation, according to resources and priorities, of a large number of local plans, each dealing with a single settlement.

In addition to these defined types, the Group also referred briefly to 'other types of local plan' which 'may be prepared for dealing with special problems in rural areas', such as mineral workings or recreational development (*ibid*., p. 34). These plans would thus deal with single policy topics where the in-depth treatment appropriate to a local plan was justified, although it was expected that in general the county plan itself would provide sufficient policy guidance for the rural areas.

PAG proposed that only the first tier strategic plans should be submitted to the Minister for approval, and that responsibility for local plans should be assumed by local planning authorities. While authorities would be *required* to submit urban and county plans to central government for approval, the position on the preparation of local plans was much more flexible. PAG recognized that it would not be possible or desirable to prepare local plans for every town or village in a county, or for all parts of towns covered by urban plans. Local plan preparation was to be undertaken in the light of local planning problems and priorities, and authorities 'should not feel obliged to undertake the preparation of more local plans than they can reasonably cope with' (*ibid*., p. 32). This emphasis on local discretion extended to the form and content of local plans, which was to be 'left largely to local option'; though a general consistency would be desirable, 'there is no need for rigid uniformity, either as between planning authorities or as between local plans for different towns . . . the plans can be designed to meet the needs of the case and the kind of planning job required' (*ibid.*, p. 35).

These proposals were largely accepted by central government in a 1967 White Paper which contains only a passing reference to the need at the detailed level for action area plans and other local plans (Ministry of Housing and Local Government, 1967, para. 18). The lack of attention given to local plans at this stage perhaps indicates an acceptance of PAG's arguments for local discretion, encouraged no doubt by the desire of central government to avoid the problems of delay caused by the requirement of the 1947 Act that each and every development plan be submitted for Ministerial approval. This interpretation is supported by statements made by the Minister for Housing and Local Government during the passage of the Bill for what became the *Town and Country Planning Act 1968*:[17]

One of the main purposes of separating structure plans from local plans [is] to rid the central government of the responsibility for a mass of detail which is better dealt with and decided at local level.

In line with this philosophy,

I wish to make a plea . . . for flexibility . . . local plans, as we envisage them, will vary considerably in their type and scope, and we should be losing some of the benefits of the change that we are making if we tried to prescribe too rigidly the requirements that would apply to all local plans.

Central government was here signalling its desire to make only the minimal statutory provisions for local plans, particularly in terms of form and content. Nevertheless, the same Ministerial statements also show the development of central government thinking about possible legitimate types of local plan, although the subsequent legislation (the 1968 Act) only provided for 'action area' and 'local' plans at the detailed level. Thus,

Local plans will be either plans with a varying degree of detail about a particular part of the area of the planning authority or plans which have a particular function or cover a particular aspect of planning for a rather wider area.

In addition to action area plans, the 'area' plans also included

district plans covering a larger part of the area of the planning authority. The basic purposes of these will be to state more fully the working features of the locality . . . and to show the principal land uses as a guide to development control.

There was also a possibility of 'a different kind of local plan which would be in the nature of a special policy or project plan'. This seems to be a development of PAG's proposal for local plans dealing with single topic policies. Relevant topics might be mineral development, recreation, or access to the countryside. These early refinements to the concept of local plans found no echo in the 1968 Act, which simply stipulated that local plans were to consist of a map and written statement which formulated in appropriate detail 'the authority's proposals for the development and other use of land in that part of their area'.[18] Action areas were defined as a type of local plan, and were to be designated in a structure plan. However, the emergent central government thinking during the passage of the Bill was later formalized in the publication *Development Plans: A Manual on Form and Content* (Ministry of Housing and Local Government, 1970) and regulations produced in 1971, which summarized the various functions and types of local plans described in the manual, as well as outlining the procedures to be followed in producing local plans.[19] The manual provided advice and illustrative examples of how the new two-tier development plan arrangements could be worked out in practice, with the strategic county or urban structure plans providing the framework for the preparation of more detailed local plans (Figure 3.1). Following PAG, the functions of local plans were defined as follows: to apply the strategy of the structure plan; to provide a detailed basis for development control; to provide a basis for the co-ordination of development; and to bring local and detailed planning issues before the public (Ministry of Housing and Local Government, 1970, para. 7.4). These four functions still define the official role for statutory local plans (DoE, 1984g, para. 3.1). Three different kinds of local plan were distinguished in the manual: *district plans* for comprehensive planning of larger areas expecting piecemeal change; *action area plans* for comprehensive planning of areas shown in the structure plan for improvement, redevelopment or new development; and *subject plans* for dealing with particular planning aspects in advance of a comprehensive plan or where such a plan was not needed. This classification of types retained action area plans as defined by statute, but strengthened PAG's 'district plans' by including the town and village plans within this category; topic subject plans (on such matters as land reclamation, mineral development or linear developments such as a motorway corridor, river valley or coastline), which had been hinted at by the PAG and during the passage of the Bill, were now explicitly introduced. There was no indication that these would necessarily be confined to rural areas.

The initial operation of the new development plans system was complicated by local government reorganization, which took place in England and Wales in April 1974 following the *Local Government Act 1972*. This legislation made the counties responsible for structure plans and gave the districts primary responsibility for local plans, and made provisions for development plan schemes and certification to ensure conformity. To take these changes into account, new regulations were issued in 1974,[20] reiterating the three possible types of local plan (district, action area and

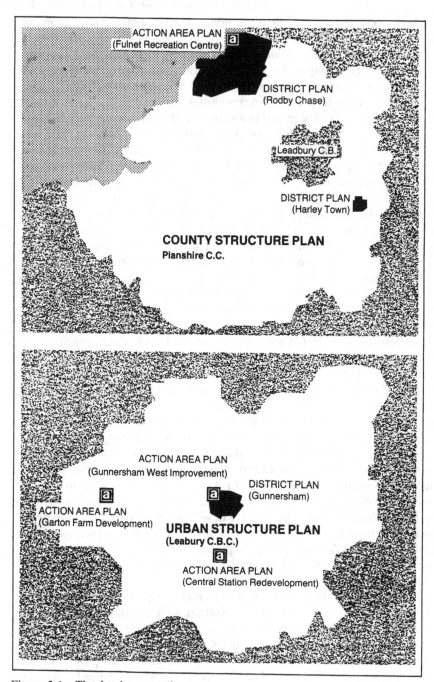

Figure 3.1 *The development plan system*
Source: *Development Plans: A Manual on Form and Content* (1970), Ministry of Housing and Local Government, HMSO, London.

100

subject) and the requirement that the map component for the plan be on an Ordnance Survey base. The regulations also spelled out in topic form the various matters to which local plan proposals should relate, e.g. housing, industry and commerce, shopping, education. Most local plans have since used this topic division of proposals as an organizing principle. Finally, the regulations specified various matters which could be dealt with in the plan's written statement as considered appropriate by the local authority. Most notable here is the regard that had been taken of social policies and considerations during the preparation of the plan, and of the availability of resources for the implementation of the plan's proposals.[21] This suggests that local plans prepared under the 1974 regulations could legitimately have had regard to social and economic factors insofar as they affect the use of land, despite the legislative limitation to 'the development and other use of land', though subsequent central government advice discouraged this interpretation.

The initial progress with the preparation and adoption of statutory local plans was slow. The first local plan (Coventry City Council's Eden Street Action Area Plan) was not formally adopted until December 1975. This reflected the disruptive effects of local government reorganization, but was also due to slow progress on strategic plans. As a result the policy framework which local plans were required to conform with was only emerging slowly. Central government, proceeding on a county-by-county basis, had in any case only completed issuing structure plan commencement orders in 1974. In the interim, local authorities had been encouraged to use informal local plans. Despite this slow start, local authority intentions, as recorded in their development plan schemes, were ambitious, if not completely realistic (Mabey and Craig, 1976; Bruton, 1983). Partly in response to this emerging experience, the DoE began to give advice on more substantive areas of local plan preparation. This trend is evident in central government advice notes and circulars from 1976 on. Local Plans Note 1/76 *Development Plan Schemes and Local Plans*, for example, offered advice to local authorities on various detailed aspects of local plans, despite the opening acknowledgement that 'one of the main purposes of the *Town and Country Planning Act 1968* was to limit the Secretary of State's involvement in Development Plans to the strategic level' (DoE, 1976, para. 1). The overall tone, however, is directive. Thus, 'work on informal plans should be avoided', although a valid role is recognized for informal material which could be used as 'supplementary planning guidance' to amplify statutory plan policies and provide a more detailed guide for developers and development control (*ibid.*, para. 14). The Note is firmly in favour of statutory as opposed to informal plans (*ibid.*, para. 10). Notwithstanding this advice, it is clear that many authorities at this time intended to continue to prepare and use informal plans, this strategy reflecting their doubts about the relevance of statutory local plans to their particular circumstances, as well as scepticism about the value of participation (Williams, 1978). Local Plans Note 1/76 also strengthened the criteria for deciding whether or not a topic was suitable for subject plan treatment, following Mabey and Craig's conclusion (itself echoed in the Note) that many subject plans were being proposed in inappropriate circumstances. Henceforth, subject plans would only be legitimate where 'the subject has such limited interactions with other planning matters that neither these matters, nor the subject itself, will suffer from being planned in isolation' (*ibid.*, para. 25). This is clearly an extremely stringent criteria, and would certainly rule out those topic studies identified as proposed subject plans by Mabey and Craig on population, industry, shopping and housing. Additionally, Mabey and Craig felt that mineral workings, derelict land, the recreational use of canals, disused railways and river valleys, Green Belts and country parks were suitable cases for subject plan treatment in that those topics met the criteria of limited interaction, but this is surely arguable. Further development plan scheme experience suggested a tendency to propose a large number of district plans, often covering very small areas, presumably along the lines of PAG's town and village plans. The potential problems with this approach were clearly indicated (Mabey and Craig, 1976), and authorities have tended to follow this advice, realizing that larger area plans tend to minimize the impact of the statutory preparation procedures. Lastly, the Note emphasized that 'local plans are land use plans', and that although other policies could be included as part of the reasoned justification to support land use proposals, 'the local plan will be the executive instrument only in respect of the land use element' (*ibid.*, para. 15).

Further advice on proposals in local plans was given in DoE Circular 55/77, a consolidation of earlier development plan advice. Proposals should either define a specific site for a particular development or other use, or define the area in which particular policies will be applied; proposals should be clearly shown on the proposals map and clearly distinguished as such in the written statement. Moreover, proposals should be *realistic*; only developments expected to be started within a period of some ten years should be formulated as proposals (DoE, 1977e, paras. 3.29 to 3.31). This latter point sought to balance the need to safeguard land for development with the desirability of avoiding unnecessary blight. Moreover, proposals could distinguish between *uses* but not *users*, so that housing proposals should not differentiate between public and private development (*ibid.*, para. 3.30). A later Local Plans Note (1/78 *Form and Content of Local Plans*) emphasized this restriction: 'Land can only be allocated for residential use generally' (DoE, 1978a, para. 16), as well as the point that local plan proposals should be based on Planning Act powers alone. Proposals could, therefore, relate only to the development or other use of land, despite the requirement of the 1974 Regulations to have regard to 'social policies and considerations'. The Note recognized the ambitions of some authorities to pursue a comprehensive policy approach, but offered no indication as to how policies other than those on land use could legitimately be included in local plans as anything other than a reasoned justification (*ibid*., para. 25).

This position of the DoE is based on a concern that local plans should only include proposals sustainable under planning legislation in development control. It is thus grounded on planning case law relating to development control conditions and refusals, effectively restricting local plan proposals to matters allowable as planning considerations in determining applications for development (Healey, 1979a, 1983). Yet as Healey notes, it is unclear how far such precedents should apply to local plan proposals; Hamilton, for example, has argued that a distinction between users may be feasible in a local plan context, even though not acceptable as a valid planning consideration for development control conditions or refusals (Hamilton, 1977). The DoE now emphasizes that development control criteria specified in local plans 'should not attempt to go beyond what can be achieved by means of development control powers' (DoE, 1984g, para. 4.15). Where planning permission might be refused or conditions imposed by reference to such criteria in local plans, authorities are advised to ensure that such decisions will be both reasonable and valid.

At the same time as the DoE was defining the acceptable nature of local plan proposals, other central government policy concerns were beginning to impinge on local plans, both in terms of content and procedures. A national concern for industrial performance, for instance, led to Circular 71/77 *Local Government and the Industrial Strategy* which called on local authorities to give priority to local plans with a large industrial content, to ensure the identification in local plans of suitable sites, and to include full provision for required infrastructure (DoE, 1977f). Procedurally, the same circular identified the problems of delay which were arising from the statutory requirement for draft local plans to await structure plan approval before being placed on deposit. These problems were regarded as particularly serious in the inner cities, which were themselves a growing policy concern for central government. A temporary palliative was the promise that draft local plans would be regarded as a material consideration in development control (DoE, 1977f, para. 17), but the status of these documents nevertheless remained uncertain. A more lasting solution came in the provisions of the *Inner Urban Areas Act 1978*, which allowed designated inner-city authorities to adopt local plans in advance of structure plan approval.[22] Despite this procedural change, however, some inner-city authorities were beginning to question the fundamental relevance of statutory local plans to the problems they faced, e.g. Liverpool (Duerden, 1978; Hayes, 1981). These legislative changes were incorporated into a new development plans memorandum, Circular 4/79 (DoE, 1979a) replacing Circular 55/77.

The change of administration in 1979 heralded a further shift in the attitude of central government towards the land-use planning system in general, and local plans in particular. An initial step was the proposed reduction of bureaucratic controls over local government activities, so as to give local authorities 'more choice and flexibility' (DoE, 1979b, para. 1). This move, however, did not signal a wider abdication of concern with local plan production on the part of central government. Ministers emphasized their requirements for a planning system which was 'efficient, responsive and speedy'; local plans, where needed, 'must be brought into being quickly'.[23] Local plans were seen as primarily facilitating and co-ordinating devices, providing 'the blueprint for land use management, for ensuring adequate supplies of land for housing and industry'.[24] Subsequent legislative amendments were made in the *Local Government, Planning and Land Act 1980*. As far as local plans were concerned, the Act extended the facility for local plan adoption in advance of structure plan approval to all authorities (s.88), dropped the requirement for a public local inquiry provided that all objectors to a plan agreed to have their objections dealt with as written representations (s.89, Schedule 14), and amended the requirements for publicity during plan preparation so that this was no longer required at 'issues' stage; only one round of 'public participation', at draft plan stage, is now required (s.89, Schedule 14). In total, it is claimed that these amendments to the planning legislation have weakened the position of the counties and the structure plan as a strategic guidance framework, and reduced the role of the citizen, both as a participator in local plan preparation and as an objector (Healey, 1983). Circular 23/81 *Development Plans* dealt with the development plan provisions of the 1980 Act and formalized earlier ministerial statements. Local plans should only be prepared where there is a 'clear need'; the work required to take the plan to deposit should be capable of being carried out 'within about a year' (DoE, 1981c, para. 10). Plans should not be prepared where little or no pressure for development is expected, being typically required 'to provide locations for the future supply of land for housing and industry, to define the precise boundaries of areas of restraint, or to co-ordinate programmes for development' (*ibid*., para. 11). To complete these detailed changes to the statutory system, new regulations were issued in 1982 with amendments incorporating the provisions of the 1980 Act. Other adjustments were also made, e.g. to titling conventions, while the requirements of the 1974 Regulations to have regard to resource availability and social policies were deleted.[25]

These amendments to the system since 1979 have now been brought together into a new development plans memorandum issued with Circular 22/84, which presently constitutes the principal source of advice from central government as to how local authorities should operate the statutory development plan system (DoE, 1984g).

Finally, more fundamental changes to the development plan system in the metropolitan areas have been proposed as a consequence of the abolition of the metropolitan counties and the GLC on 1st April 1986. These proposals initially envisaged that the London boroughs and metropolitan district councils would be given separate powers to prepare structure and local plans, but following consultations which emphasized the duplication of effort that would result, a simplified 'unitary development plan' has been outlined (DoE, 1984b; Angell and Taylor, 1985; Bristow, 1985). Each unitary plan will cover the whole area of a borough or district and be in two parts: a section outlining general *policies*, and a section containing specific *proposals*, applying the general policies to land in the area and accompanied by a proposals map. This suggestion preserves PAG's distinction between levels of plan-making, although the rationale for treating these levels together in a single document remains unclear. The *benefits* of the unitary development plan cannot thus be readily identified. In particular, the fundamental problem raised for the land use planning system by the abolition of the GLC and the metropolitan counties remains, since the strategic planning of London and the metropolitan areas will still be fragmented among the boroughs and districts, despite the promise of guidance on these matters from central government. There is a real danger that this guidance will serve to determine, or at least delimit, acceptable strategic policies, so that each unitary plan merely collects together strategic policies set by central government. The enhanced role for central government involved in the preparation of unitary development plans has for this and other reasons been subject to much criticism by professional bodies such as the Royal Town Planning Institute and the National Housing and Town Planning Council, who have argued that strategic work should continue to remain the responsibility of democratically elected councils.[26] The Royal Town Planning Institute comments that as far as central government involvement is concerned, unitary development plans amount to a reversal of the planning system set up in 1968, which itself was a result of government recognition of its own over-involvement in development plans. As regards local autonomy and flexibility, the statutory provisions for development plans thus appear to be in danger of coming full circle, at least as far as the metropolitan districts and London boroughs are concerned.

What implications has the development of central government advice, on the way in which local plans should be used, had for local planning practice? The extent and manner in which local authorities have taken up the local plan as a vehicle for local planning policies and proposals is discussed in detail in Chapter 4. However, one of the general problems faced by local authorities in preparing and using local plans ultimately stems from the fact that the economic context of growth, which was anticipated by PAG in the early 1960s and which informed many of their proposals, has not materialized. Partly as a consequence, action area plans (to deal with comprehensive change and regarded as of prime importance by the Group) are much less in evidence today than district plans (intended to deal with piecemeal change). At the same time, the usefulness of local plans for local authorities seeking to deal as far as practicable with the economic causes and social consequences of decline and stagnation has been further reduced by the evolving central government view of a local plan's proper role, form and content. Early advice, in line with PAG's views on local flexibility, focused mainly on the legislative procedural requirements. But partly in response to emerging experience of the consequences of such flexibility in practice, the DoE has moved more and more into substantive areas of plan preparation. This reflects a concern with the public credibility of planning, and with the protection of individual property rights and land values, as well as a desire for legally substainable policies and proposals, whose implementation can be secured under town and country planning powers alone. From this advice, it is clear that central government currently sees local plans as somewhat limited devices for land allocation and development co-ordination, a view which seems to assume that the economic context is still one of growth and which denies local authority attempts to integrate their socio-economic and land use policies in local plans and other policy vehicles.

Planning practitioners have not reacted favourably to these restrictions. The feeling that central government has unduly constrained the scope of local plans has been one of the main sources of local planning authority dissatisfaction with the statutory system, and thus one of the main reasons for the continued preparation and use of non-statutory local planning instruments (see Chapter 6). Specifically, it is argued that to be effective local plans must have a wide remit covering a range of policy areas, including social and economic issues as well as those relating to land use. As early as 1974, for instance, Perry could note that 'most of us will agree that local planning must reach beyond the confines of the Development Plan Manual' (1974, p. 492). Hambleton, pointing to growing concern with the limitations and relevance of the official view of local planning, argues that to restrict local plans to matters of land use is 'to slip back into the old sectional approach and allow procedure to triumph over relevance' (1976, p. 179). More recently, research for the Department has shown that some authorities have found statutory plans 'too restrictive in content and [have] felt the need to use more flexible packages of policies to deal with a lack of development pressure (if not actual decline), the management of land use rather than development and to link in with local authority corporate policies on inner city type issues' (Fudge *et al.*, 1983, p. 3). For local authorities the divergence between local requirements and the legitimate scope of the statutory local plan questions its usefulness as an effective vehicle for a range of local development policies, involving socio-economic issues as well as those of land use. This problem has been expressed as 'a sharp contrast between what the planners wanted to do and what the legislation actually allowed them to do' (Field, 1983, p. 38). Indeed, given the restrictions on the extent to which local plans can be used to progress non-land use policies, it is hardly surprising that many authorities appear to be filling local needs through the use of non-statutory planning instruments.

3.4 Local plan preparation and adoption procedures

3.4.1 Statutory requirements for local plan preparation and adoption (DoE, 1984g)

The statutory requirements for the preparation and adoption of local plans are set out in the following Acts, regulations and circulars:

- *Town and Country Planning Act 1971* as amended by the *Town and Country Planning (Amendment) Act 1972*; the *Local Government Act 1972*; the *Local Government Planning and Land Act 1980* and the *Local Government (Miscellaneous Provisions) Act 1982*;
- *Town and Country Planning (Structure and Local Plans) Regulations 1982*, **or** the *Town and Country Planning (Local Plans for Greater London) Regulations 1983*, and the *Town and Country Planning (Structure and Local Plans) (Amendment) Regulations 1984*;
- Circular 22/84 and the accompanying *Memorandum on Structure and Local Plans*.

Briefly, the Acts and regulations establish the functions and format of local plans as follows:

Functions
1 To develop the policies and general proposals of the structure plan and relate them to precise areas of land.
2 To provide a detailed basis for development control.
3 To provide a detailed basis for co-ordinating and directing both public and private proposals for the development and other use of land.
4 To bring local planning issues before the public (DoE, 1984g, para. 3.1).

Format
A local plan should consist of a written statement and a proposals map, although it can be accompanied by other supporting documents which do not form part of the plan. The written statement should include the authority's proposals for the development and other use of land for the plan area, including measures for the improvement of the physical environment and traffic management.[27] Such proposals should define specific sites for particular developments or other uses, or define specific areas within which particular policies apply. The proposals put forward in the written statement should be clearly distinguishable from the remainder of the text, which should include reasoned justification for the proposals.[28] Diagrams may be used to illustrate the reasoned justification for the proposals put forward.[29]

The proposals map should define the area of the local plan and identify all the proposals in the written statement. The map should be on an up-to-date Ordnance Survey base, at a scale considered appropriate by the local planning authority. Insets of the proposals map at a larger scale are permitted.

The purpose of the proposals map is to provide a comprehensive index of the plan's proposals; to this end 'each proposal should thus be related to the specific area of land to which it applies and its boundaries should be shown' (DoE, 1984g, para. 3.2.6). The plan and any accompanying documents should have a title indicating clearly the area to which the plan relates. It should also include the words 'local plan' and carry the name(s) of the authority or authorities who prepared the plan.[30]

As explained earlier in this chapter, each county outside London has to produce a development plan scheme which identifies the local plans that will be prepared within the area, the local planning authority/authorities responsible for preparing these plans and the priority to be allocated to their production.[31] Advice from central government makes the following clear:

1   That local plans are not needed in all areas, e.g. where there is little pressure for development (DoE, 1984g, para. 3.11).
2   Local plans covering the whole or greater part of a Borough or District and based on a comprehensive consideration of matters affecting the development of an area are generally likely to be more cost effective than a number of plans covering smaller areas or dealing with one or two particular aspects of development or other uses of land (DoE, 1984g, para. 3.14).
3   In certain circumstances a subject local plan, which is based on a particular type of development or use of land, may be appropriate. Such subject plans should concentrate on one or two issues across what might be an extensive area, e.g. intensive livestock farming units, Humberside CC; minerals, Bucks. CC.[32]
4   Action area plans are appropriate for an area where comprehensive development, redevelopment or improvement is proposed.[33] Action area plans no longer have to be designated in the appropriate structure plan. Since November 1980 local planning authorities may prepare action area local plans provided they are allocated responsibility for them in the development plan scheme. In exceptional circumstances, the Secretary of State can direct a local planning authority to prepare a local plan.[34]

3.4.2   Procedures to be followed in preparing and adopting a local plan
Figure 3.2 illustrates the procedures to be followed in preparing and adopting a statutory local plan. Briefly, the local planning authority made responsible for the preparation of a local plan in the development plan scheme or by the Secretary of State:

1   can utilize the survey information produced in connection with the relevant structure plan and/or undertake a new survey of the issues relevant to the proposed local plan;[35]

2   shall (*i*) give adequate publicity in the area of the plan to such matters they propose to include in the plan; (*ii*) publicize the fact that persons wishing to make objections or representations are able to do so; (*iii*) ensure that such persons are given adequate opportunity to make representations; the prescribed period for making such objections or representations is six weeks; (*iv*) consult with the district or county planning authority and other interested bodies as appropriate, with respect to the content of the plan; give them the opportunity of expressing their views and take those views into consideration;[36]

3   shall secure a certificate of conformity from the county planning authority confirming that the local plan conforms generally with the structure plan as approved by the Secretary of State;[37] the county planning authority should issue the certificate within one month; if the county are of the view that the local plan does not 'conform' in any way and the authority preparing the plan are disinclined to amend it so that it does conform, they should refer the matter to the Secretary of State;

4   shall notify the Secretary of State of their intention to place the local plan on deposit and send him two copies of the plan and a statement of the process of participation undertaken at the same time as advertising that the local plan is on deposit and setting out the procedure by which objections and/or representations can be made.[38]

If the Secretary of State is satisfied that the processes of consultation and participation meet the requirements of the regulations, then the process of adoption is allowed to proceed. If the programme of participation in any way fails to meet the requirements of the regulations then the matter is referred back to the authority producing the local plan to rectify the omissions.

The period allowed for objections to be made to the deposited plan is six weeks. Objections made in accordance with the regulations may cause a public local inquiry into the plan to be heard.[39] A public local inquiry need not be held into objections if all those persons who have made an objection indicate in writing that they do not wish to appear.[40] Where a public local inquiry is held, the local planning authority who prepared the plan shall (*i*) advertise at least six weeks prior to the inquiry that such an inquiry is to take place, and (*ii*) notify objectors that it is to take place.[41]

A person, usually an Inspector, is appointed by the Secretary of State (*i*) to hear the objections to the local plan at the Inquiry, and (*ii*) to prepare a report with recommendations on his/her consideration of the objections for the local planning authority preparing the local plan.[42] In exceptional circumstances a hearing, as opposed to an inquiry, can be held. This is conducted by a person, usually an Inspector, appointed by the Secretary of State in the same way as for a local plan public inquiry, although there is no requirement for notice by public advertisement to be given.[43] The local planning authority shall make the Inspector's report and their response to it

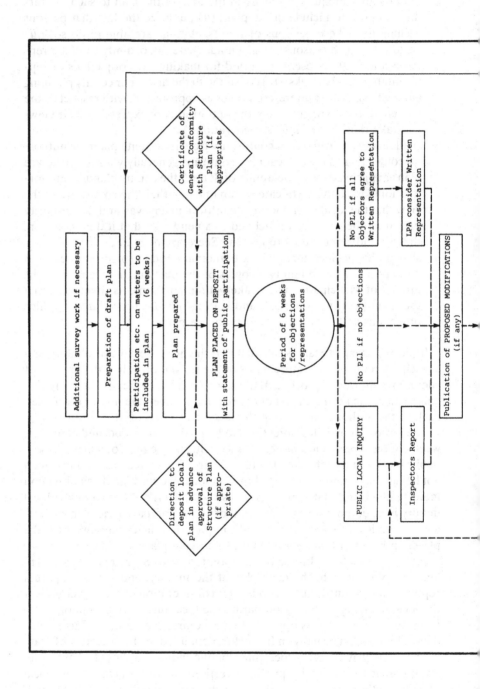

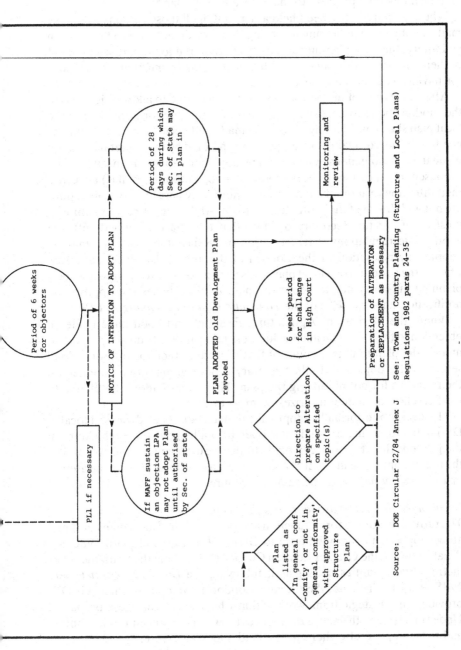

Figure 3.2 *Local plan adoption procedure in England and Wales*
Source: DoE Circular 22/84 (1984), *Memorandum on Structure and Local Plans*, HMSO, London, Annex J.

available for public inspection. At the same time, they should advertise the modifications they propose to make to the local plan.[44]

Where no inquiry has been held as a result of all those persons who have made an objection indicating in writing that they do not wish to take part in an inquiry, the local planning authority are required to prepare a statement of their decision with respect to each objection made, including the reasons for having reached those decisions.[45]

After the expiry of the period of six weeks required to allow objections to the modifications to be made, the local planning authority preparing the local plan shall consider any objections made and decide if a further inquiry needs to be heard, again before a person appointed by the Secretary of State to hear such objections. On resolving objections to the plan and/or the proposed modifications to it, the local planning authority shall (*i*) advertise their intention to adopt the plan, and (*ii*) notify any person whose objections are not withdrawn of their intention so to do. At the same time the authority must certify to the Secretary of State that they have complied with the appropriate regulations. After a period of 28 days the local plan may be adopted by resolution of the Council of the appropriate local authority. Prior to the adoption of a local plan the Secretary of State can direct the local planning authority not to adopt the plan until he issues the appropriate notification under s.14(3) of the *Town and Country Planning Act 1971*.[46]

Where the Ministry of Agriculture, Fisheries and Food have made an objection to the local plan and the local planning authority do not propose to modify the plan to take account of that objection, then the authority shall send MAFF's objections to the Secretary of State along with a statement of their reasons for not modifying the plan. They cannot adopt the plan until the Secretary of State authorizes them so to do.[47]

The decision to formally adopt a local plan should be advertised locally. During the following six weeks it is open to challenge in the High Court on the grounds that it fails to comply with the requirements of the Act and/or the regulations made under it.[48] As appropriate, the local plan can be altered or replaced by following the same procedure.

Local plans in Greater London and Scotland

The provisions for the form, function and preparation of local plans outlined above applied in a modified form to Greater London until April 1986 and the abolition of the Greater London Council (GLC), together with the other metropolitan counties in England, following the *Local Government Act 1985*. Prior to this date, the Greater London Development Plan (GLDP) provided the strategic framework within which local plans were prepared. Under this system, the major differences between the procedures in London and those applied elsewhere in England and Wales were:

1   there was no requirement to produce a development plan scheme for London;

2   the GLC could produce 'GLC action area plans'; otherwise, all local plans were the responsibility of the district planning authorities (the London boroughs); and

3   the provision that outside London the structure planning authority must certify, before it is adopted, that a local plan conforms generally to the structure plan was never extended to London; the principle was incorporated into the *Local Government, Planning and Land Act 1980* but the necessary commencement orders were never issued by the Secretary of State.

For a transitional period after abolition of the GLC and the other metropolitan county councils, existing structure and local plans are to remain in force pending the introduction of the new unitary development plan system, e.g. the approved GLDP will remain until replaced by successive unitary plans prepared by the London boroughs. Powers for the preparation of unitary plans will be brought into operation by commencement orders issued by the Secretary of State for each London borough and metropolitan district council. Each unitary plan when first prepared will incorporate in the second part any existing adopted or approved local plans, together with any proposed alterations, repeals and replacements, and any new proposals considered appropriate at the time.

In Scotland, the procedures for preparing and adopting local plans are similar to those in England and Wales, although in Scotland the preparation of statutory local plans is obligatory. The procedures followed are illustrated in Figure 3.3. However, the purpose and function of local plans have a slightly different emphasis as the following quotation illustrates:

The purpose of local planning is to guide development and changes in land use so that the physical environment can best serve the community. In helping to achieve this the most important functions of a local plan are:

   to stimulate and encourage development where appropriate;

   to indicate land where there are opportunities for change;

   to apply national and regional policies;

   to give a clear locational reference to policies for the development, change of use or conservation of land, and to proposals for development;

   to show how those who have an interest in the area, e.g. the authority, private owners, residents, commerce, industry, developers and investors could contribute to the implementation of the plan;

   to provide an adequate basis for development control;

   to indicate the intended future pattern of land use and development in the area by showing how existing development and the policies and proposals of the plan fit together (Scottish Development Department, 1984).

The procedures for the production of statutory local plans are now well established and in recent years have only been subjected to minor 'fine-

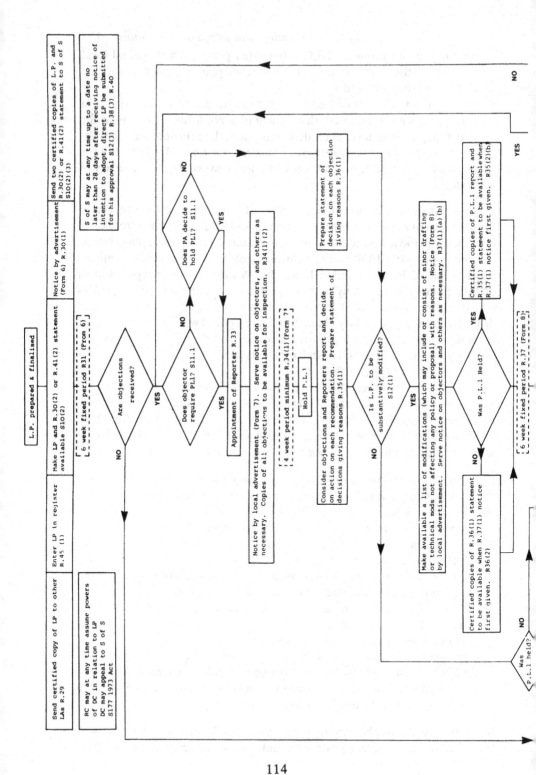

L.P. prepared & finalised

| Send certified copy of LP to other LAs R.29 | Enter LP in register R.45 (1) | Make LP and R.30(2) or R.41(2) statement available S10(2) | Notice by advertisement (Form 6) R.30(1) | Send two certified copies of L.P. and R.30(2) or R.41(2) statement to S of S S10(2)(3) |

RC may at any time assume powers of DC in relation to LP
DC may appeal to S of S
S177 1973 Act

S of S may at any time up to a date no later than 28 days after receiving notice of intention to adopt, direct LP be submitted for his approval S12(3) R.38(3) R.40

[6 week fixed period R31 [From 6]]

Are objections received?

NO

YES

Does objector require PLI? S11.1

NO

YES

Does PA decide to hold PLI? S11.1

NO

YES

Appointment of Reporter R.33

Notice by local advertisement (Form 7). Serve notice on objectors, and others as necessary. Copies of all objections to be available for inspection. R.34(1)(2)

[4 week period minimum R.34(1) (Form 7)]

Hold P.L.¹

Consider objections and Reporters report and decide on action on each recommendation. Prepare statement of decisions giving reasons R.35(1)

Prepare statement of decision on each objection giving reasons R.36(1)

Is L.P. to be substantively modified? S12(1)

NO

YES

Make available a list of modifications (which may include or consist of minor drafting or technical mods not affecting any policy or proposal) with reasons. Notice (Form 8) by local advertisement. Serve notice on objectors and others as necessary. R37(1)(a)(b)

Certified copies of R.36(1) statement to be available when R.37(1) notice first given. R36(2)

NO

Was P.L.1 Held?

YES

Certified copies of P.L.1 report and R.35(1) statement to be available when R.37(1) notice first given. R35(2)(b)

[6 week fixed period R.37 (Form 8)]

NO

Was P.L.1 held?

YES

114

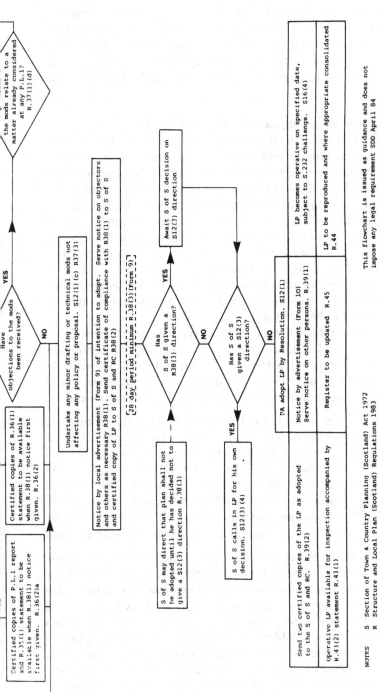

Figure 3.3 *Local plan adoption procedure in Scotland*
Source: *Planning Advice Note No. 30 (1984), Local Planning*, Scottish Development
Department, Edinburgh.
Note: This diagram was correct at December 1985.

115

tuning' changes. However, they are criticized on a number of counts the most significant of which are (*a*) the length of time needed to conform with the procedures prescribed in the regulations, and (*b*) the questionable legitimacy of local plans, where the local planning authority producing them invariably sits in final judgement on any objections made prior to formal adoption.

3.4.3   Time taken in preparing and adopting local plans

There is little hard evidence to support the generally held view that the length of time taken to produce and adopt a local plan is inordinately long. This is so notwithstanding specific instances of delay e.g. Humberside CC's Intensive Livestock Units Subject Plan, prepared between 1976 and 1980, where an objection sustained by MAFF following a public local inquiry in 1980 was not resolved until late 1983, finally allowing adoption of the plan in April 1984. MAFF required only minor amendments to be made to the plan.[49] Despite such cases, the available evidence indicates that it takes on average three years to complete the plan production process up to 'deposit' stage, with a further 16 months to negotiate the PLI, subsequent modifications and adoption stages, making a total of 52 months on average (Bruton, Crispin and Fidler, 1982).

Amongst the plans analysed to produce these figures, where there were no deposited objections and hence no inquiry, the post-deposit period was reduced by some 10 months on average, to 6 months (i.e. the need to proceed to an inquiry adds 10 months on average to the process). These figures are confirmed as being of the right magnitude by figures produced by:

1   the DoE (quoted in Healey, 1983, p. 123) which indicate that the average post-deposit period without any inquiry is 6.7 months. With an inquiry, it is 18.7 months. However, such averages should be treated with caution as they obscure the range of times. In the case of the 38 plans in the DoE analysis which proceeded to inquiry, the range of post-deposit times was between 7 and 29 months; and

2   Perry *et al.* (1985) which show that the post inquiry process from inquiry to adoption is on average 11.5 months, with a range of 4–25 months.

A review of the timing associated with the 10 stages of plan preparation and adoption (Table 3.3) suggests at first sight that there is some scope for reducing the length of time taken in stages 2 and 3 (preparation of draft plan and participation) and between stages 6 and 7, i.e. between deposit and inquiry. In the other stages, relatively little time is spent and any savings which might be achieved are unlikely to be significant. Indeed, the practice of the City of Cardiff Planning Department, where stages 2 and 3 are collapsed into one 12-month stage (approximately), thereby saving some 16–18 months overall on these two stages, could be followed by other authorities with advantage. This approach eliminates the period of public participation associated with the report of survey and puts emphasis on public participation for the draft local plan. To date, local plans produced by the City using this approach have not aroused adverse criticism from the public. However, despite the time savings achieved by Cardiff in this respect, a greater than average time has been spent on the other stages in the process, as a result of slippage in meeting committee cycles, pressure of other work and staff reductions, so that the overall average time taken in producing local plans is not significantly different from the average figures given in Table 3.3.

The period of 7 months between the end of the 'deposit' stage and the holding of the inquiry seems excessive until it is appreciated that (*a*) the number of Inspectors available to hear the inquiries is limited, and they have other work which may be programmed for some 4–6 months ahead, although the appointment by the Inspectorate of a special additional local plans panel of Inspectors may improve this situation; and (*b*) the planning authority producing the plan has to set out its response to all objections laid, has to fit the inquiry around other on-going work and has to find suitable accommodation in which to hold the inquiry. In these circumstances, it is unlikely that any significant systematic savings in time can be achieved on this stage of the process. Thus depending on the impact of the local plans panel of Inspectors it would seem that the shortest possible time for preparing and adopting a local plan where an inquiry is involved is 34–36 months – a far cry from the system of planning envisaged by PAG which was to be

Table 3.3 *Stages and timing in the preparation of an average local plan*

Stage in typical plan preparation	Timing (months)	Running total
1 Preparation and publicising of local plan brief	1.5	1.5
2 Preparation of Report of Survey and associated public participation	18.0	19.5
3 Preparation of draft local plan and associated public participation	12.0	31.5
4 Production of 'final version of local plan	2.0	33.5
5 Certification	1.0	34.5
6 Deposit	1.5	36.0
7 PLI	7.0	43.0
8 Inspector's report received	2.5	45.5
9 Modifications on deposit	4.0	49.5
10 Adoption of local plan	2.5	52.0
Total	52.0	

Source: Bruton, Crispin and Fidler, 1982, p. 18.

capable of responding rapidly to changing circumstances. But is it realistic to expect anything other than a long-drawn out procedure in producing and adopting a local plan? Given the following circumstances, it is inevitable that the production of a local plan will lead to the articulation of conflicts of interest:

1   Planning problems are 'wicked problems'.
2   Conflicts of interest are inherent in planning decision-making (Pahl, 1970; Simmie, 1974; Goldsmith, 1980).
3   Town planners are concerned to guide and control the spatial distribution of scarce resources.
4   The preparation of plans and the making of decisions associated with those plans exerts considerable influence over the way in which public and private investment into physical infrastructure is carried out.
5   Some people 'gain', whilst others 'lose' as a result of such investment decisions.

Indeed, given that one of the major functions of a local plan is 'to develop the policies and general proposals of the structure plan and to relate them to precise areas of land' (DoE, 1984g, para. 3.1) local plans will tend to generate specific and intense conflicts of interests. In the words of Goldsmith, '. . . planning, since it is primarily concerned with the use of physical space, is an activity central to the conflict about the allocation of space' (Goldsmith, 1980, p. 126). The problem is how to resolve these conflicts in a way which allows the local authority to fulfil its statutory obligations; to retain the general support of its electorate for what it is doing and to maintain its credibility:

> If the major conflicting interests associated with a particular local plan have not been considered and resolved in a way which the public regards as fair and reasonable, then the local authority can expect any attempt to implement those aspects of the plan to be opposed on every possible occasion (Bruton, Crispin and Fidler, 1982, p. 278).

The problems for the local planning authority of resolving such conflicts of interest at the same time as fulfilling its statutory obligations and maintaining the support of the public are complex in the extreme. The public and other interested bodies must be consulted through the participation process in a way which ensures that they are capable of influencing the final decisions. At the same time, the local authority must have the ability to break any 'dead-lock' situations which might arise, through compensation in the form of voluntary or compulsory acquisitions, and/or arbitration through the public local inquiry. The bargaining which is integral to this process of conflict resolution is time-consuming, and often iterative and convoluted. In these circumstances, it is not surprising that the time taken to prepare and adopt a local plan extends over a number of years. Indeed, the opportunity for local authorities to move to arbitration through a public local inquiry probably shortens what would otherwise be an open-ended process of bargaining, although the legitimacy of the local authority sitting in judgement on its own plan undermines some of the credibility associated with the arbitration role of the local plans public local inquiry.

Wilson (1977b) argues persuasively that one of the major reasons for preparing a local plan is the need to resolve a significant issue around which there are considerable conflicts of interest, and where the resolution of that issue is unlikely to be quick. Practice, by its selection of issues for treatment in statutory local plans and the time taken in securing the preparation and adoption of those plans tends to confirm Wilson in his judgement. At the same time, there is the feeling that in many instances four years is a small price to pay for establishing a statutorily approved basis for resolving major conflicts of interest in local planning matters.

The District Planning Officers' Society has recently addressed the issue of the length of time taken to produce statutory local plans and the move towards the production and use of non-statutory or informal local plans. It claimed that the main reason for the length of time taken with the statutory procedures was the provision for public involvement in the process and the protection of individual interests. A discussion paper produced by the Society suggested that the only practical way of reducing the time taken was to remove the requirement that an inquiry be held into objections laid against the plan, and to reduce public involvement in the process to consultation and comment on the draft plan (District Planning Officers' Society, 1982). Such a move would undoubtedly shorten the period of time taken to adopt a local plan by an average of 10 months. However, as Marsh (1983, p. 162) comments, arguments that the inquiry imposes unjustifiable delays on the local plan preparation process lack conviction:

. . . if the workload is heavy, it is usually because the inquiry highlights substantive problems which have to be resolved if the plan is to secure the political and public commitment essential for its long term prospects of successful implementation.

Indeed, the general reaction of many practitioners to the discussion paper was that to remove the inquiry into objections to a local plan would seriously undermine the already suspect legitimacy of local plans.[50]

3.4.4 The legitimation of statutory local plans
The *Town and Country Planning (Structure and Local Plans) Regulations 1982* (para. 29) require that:

Where, for the purpose of considering objections made to a local plan, a local inquiry or other hearing is held, the local planning authority who prepared the plan shall, as part of the consideration of those objections, consider the report of the person appointed to hold the inquiry or other hearing and decide whether or not to take any action as respects the plan in the light of the report and each recommendation, . . . and the authority shall prepare a statement of their decisions, giving their reasons therefore.

Thus local planning authorities are given the power of preparing local plans, and following consideration of objections, are also given the .power of approving or adopting those plans. By implication they are given very broad discretion as to the extent to which they accept or reject public opinion as reflected in objections to the plan. In practice, this discretion is restricted by the report of the Inspector who hears objections at the inquiry or hearing (when/if one is held) and the Secretary of State's powers of 'call-in'. Nevertheless the introduction of this system has caused some concern since, in effect, it makes the consideration of objections part of the process of plan preparation. This is a departure from the more generally accepted role of public local inquiries held in connection with appeals against the determination of planning applications where the Inspector fills an independent quasi-judicial or arbitration role. In the words of McAuslan (1975, p. 237), commenting on the *Town and Country Planning (Structure and Local Plans) Regulations 1974*, 'No part of the reformed planning machinery has been subjected to such strong criticism as this . . . which [is] regarded as infringing the principle of *nemo iudex in sua causa*'. Analysis of the outcome of a number of public local inquiries held into local plans by Bruton, Crispin and Fidler (1982, p. 285) found '. . . that the public at present does not trust the local authority to sit in judgement on its own plan proposals; the public almost unanimously feels that the final decision on local plan policies should be taken by the Secretary of State . . . .'. They concluded that 'The allocation of an administrative function for the local plans PLI could further undermine the credibility of the town and country planning system in this country' and point to the need for local planning authorities to be seen to be responding responsibly to objections made to local plans and the resultant Inspector's recommendations, '. . . so that the public can see for itself that an independent view is being sought and taken into account' (*ibid.*, p. 285). Thus the legitimacy of the local plan process can be considered at this time (1985) to depend on the outcome of the process of adoption following an inquiry and prior to the formal decision to adopt.

There are three important issues relating to the local plan adoption process after an inquiry which could affect the public's view of the credibility or otherwise of the local plan process. They are:

1   the extent to which local authorities accept the Inspector's recommendations on local plans;
2   the number of modifications proposed to take account of 'other material considerations' as provided for under the *Local Government, Planning and Land Act 1980* (s.14(1), *Town and Country Planning Act 1971*);
3   the extent to which objections are made to proposed modifications and the occurrence of second inquiries.

These issues were examined as part of the first stage of a DoE funded research programme and are reported by Perry *et al.* (1985). Based on an analysis of 76 local plans adopted between August 1981 and May 1983 after one or more inquiries, the findings show that less than 10 per cent (62) of all the 767 recommendations proposed by the Inspectors were not accepted by the local planning authorities. In addition, a further 3 per cent (21) of all recommendations were only partially accepted. Thus, 'in terms of numbers of recommendations, the incidence of non-acceptance is numerically not of great significance' (*ibid.*, p. 523). A detailed analysis of the profile of cases where recommendations were not acted upon by the local authority shows that:

1   over 60 per cent (50) relate to site-specific issues which are likely to arouse local controversy but which do not necessarily affect the wider policies of the plan;
2   the remaining 40 per cent (33) relate to a range of general or policy issues which the Inspectors recommended should be deleted (e.g. the prefacing of policy statements by reference to the uncertain financial resources available to implement policies – Brighton Central Area District Plan); or incorporated (e.g. the inclusion of a policy to improve cultural and entertainment facilities in the plan area – Hertsmere District Plan);
3   in most cases (54), the reason given by local authorities for not accepting the Inspector's recommendations relate to differences in interpretation of what constitutes acceptable planning standards (e.g. the recommendations that the extent and distribution of new housing land should be changed to bring it within a noise contour which was higher than that put forward by the local authority in the plan. The authority rejected this recommendation on the grounds that its interpretation of acceptable noise levels in the area was likely to relate better to locally acceptable standards than the Inspectors – Crawley District Plan).

Only a small number of recommendations (8) were rejected on the grounds that the Inspector's judgement was considered to be ill-informed, in that the recommendations could not be translated into practical policies, or were ambiguous or that the Inspector had failed to understand the policies (see Table 3.4).

In addition to local authorities not accepting all the Inspector's recommendations a number of local plans (19) were modified for 'other material considerations' under s.14(1) of the *Town and Country Planning Act 1971*, i.e. they were modified during the process of adoption to take account of considerations other than the Inspector's report, e.g. Rossendale BC amended the boundary of an Industrial Improvement Area in the Rossendale District Plan; Croydon LB revised the Croydon District Plan to take account of the Secretary of State's decision no longer to safeguard the line of a proposed M23 extension. Whilst such modifications allow an authority to respond rapidly to new circumstances, nevertheless the way in which these powers are used and the nature of the modifications introduced in this way, could reinforce the public's scepticism towards the whole process.

The third issue in the post-inquiry procedures, which could affect the legitimation or credibility of a local plan, concerns the way in which proposed modifications to the plan and objections to those modifications are handled. In the 76 plans forming the subject of their analysis Perry *et al.* (1985) found that:

1　objections were made to proposed modifications in 20 plans;
2　in 15 of the plans no second inquiry was held as the objections were judged by the local authorities to be of little significance or related to matters discussed at the first inquiry;
3　in one case, the objection resulted in a minor wording change, whilst
4　in four cases a second inquiry was held.

The evidence of this research project shows that overall local authorities do tend to accept the recommendations of the Inspector in processing local

Table 3.4　*Reasons given by local authorities for rejection/partial rejection of Inspector's recommendations*

	Type of recommendation	
Reason for rejection by LA	*Site-specific issue*	*Broader policy issue*
Planning need	36	5
Interpretation of environmental standards	7	6
LA propose to undertake additional study of the disputed issue	2	6
Financial uncertainties prevent firmer commitment	1	0
Issue is/is not of relevance to the plan	0	6
Issue covered by other statutory/ non-statutory responsibilities	0	3
Interpretation of county council policy differs	3	0
LA unable to translate recommendations into practical policies	0	2
Inspector's recommendation ambiguous	1	0
Inspector has failed to understand the policy	0	5
Total	50	33

Source:　Perry *et al.*, 1985.

plans to adoption. Where recommendations are rejected they tend to be concerned either with site-specific or minor matters. This indicates that the Inspector's independent status is seen as being important in the eyes of the local authorities. In addition, Marsh (1983, p. 147) suggests that Inspectors attempt to avoid making recommendations which could be interpreted by the local planning authority as encroaching unnecessarily into areas of decision-making which are the legitimate preserve of local politicians; in this way, the Inspector eases the way for the planning authority to concur with his findings. However, the tendency of local authorities to consider objections to modifications without causing a second inquiry to be held and/or to make modifications from 'other material considerations' could reinforce the view of the public that the final decision relating to local plans should be removed from the local authority producing the plan and placed in the hands of the Secretary of State. Until such time as there is a body of experience which convinces the public that to all intents and purposes the Inspector at the local plans PLI is fulfilling a quasi-judicial or arbitration role, then there will always be disquiet expressed at local planning authorities sitting in judgement on their own plan proposals. The more responsible local authorities are in responding to Inspector's recommendations, and in handling modifications to the plan, the quicker the legitimation of local plans will be accepted. To eliminate the possibility of holding a public inquiry would fundamentally undermine the role of statutory local plans in conflict resolution. To reduce significantly the public participation process in the preparation of local plans could contribute to a further reduction in the credibility of local plans.

3.5 Conclusions

This chapter has identified the formal contexts influencing the production of statutory local plans, and thus provides the necessary background for an account of how these plans are being used by local authorities (Chapter 4). These contexts relate to strategic or structure plan policy; the underpinning development plan legislation and its subsequent interpretation by the DoE through circulars and advice; and the statutory procedural requirements for local plan preparation and adoption. Taken together, these factors have diverse influences on local plan production which illustrate the relevance of the alternative and complementary theoretical perspectives discussed in Chapter 2. Both development plan schemes and certification, for instance, can be seen as embodying the strategic planning principle that policies set in the level above should influence those developed in the level below. The formal structure of the development plan system closely mirrors that of the strategic planning hierarchy, at least at the lower levels. Another way of viewing these policy devices however is as 'bargaining counters' used in a process of distributional bargaining to establish the relative levels of responsibility for plan-making between tiers of local government. Their role in influencing local plan policy also emphasizes that both county policy on local plan production (via the development plan scheme) and more substantive county policies as set out in the structure plan (via certification) must be regarded as part of the range of contingent factors which local authorities seeking to produce statutory local plans will have to take into account.

A further and particularly influential contingent factor impinging on local plan preparation is national policy advice from the DoE. These factors it should be noted interact, e.g. both development plan schemes and certification can be expected to work to the benefit of the county, but have been established by statute and developed by DoE advice. The DoE has sought – despite the initial claims during the passage of the new development plan legislation through Parliament in 1968 that one of the main aims was to divest central government of its previous responsibility for local planning issues – to become involved in the detailed operation of the new system, partly in terms of local plan procedures (a legitimate central government concern) but also in terms of form and content. This advice has typically emphasized the legal restrictions setting the boundaries to legitimate local plan policies, e.g. as to how far it is possible to distinguish between users of land as opposed to uses. It is clear that central government is sceptical about the value of local plans, encouraging local authorities to consider in development plan schemes the need for proposed plans and emphasizing a role for local plans as somewhat limited devices for land allocation and development co-ordination. In any case the provision of detailed guidance on local plan form and content differs from the original view of the PAG (1965, p. 35) that such matters should be 'left largely to local option'; a general conformity is desirable, but 'there is no need for rigid uniformity.' It can be argued that here the Group had in mind the operation of a range of local contingent factors, which would act to alter the context for local planning between authorities, and so restrict the relevance of detailed *material* advice. As PAG noted, flexibility as to form and content is desirable, for then 'the plans can be designed to meet the needs of the case and the kind of planning job required' (*ibid*., p. 35). Central government advice can perhaps best be seen as one factor among many that local authorities will have regard to in devising local planning strategies relevant to their needs. Certainly, planning practice has not reacted favourably to the evolving DoE view of local plans. The indications are that restrictions on local plan content have been one of the main reasons for the continued use of non-statutory documents, particularly where land use issues concern decline or management rather than development. This trend is itself against DoE advice, indicating that local authorities do not always give overriding weight to the provisions of circulars.

At the same time it is also argued (for example, by the District Planning Officers' Society) that the excessive duration of the statutory local plan adoption process is a further disincentive for authorities seeking to produce formal plans. Non-statutory approaches to local planning are being pursued instead. However, it is doubtful if the extreme diversity of approaches to local planning can be readily explained by a single factor such as difficulties with the statutory preparation procedures or restrictions on the legitimate form and content of statutory local plans. Rather these issues (which undoubtedly pose real difficulties for authorities attempting to prepare formal local plans) should be seen as simply part of the range of contingent factors influencing the 'choice' of local planning strategies. In this respect, other factors acting to dictate the need for a *statutory* plan (e.g. as a firm basis for restraint policies) may outweigh such associated drawbacks as the length of the preparation process. Given that the production of a local plan will almost inevitably lead to conflicts of interest which can only be resolved through a complex process of bargaining, it is in any case hardly surprising that the timescale of local plan preparation is invariably numbered in years. In a sense this expenditure of time and effort has to be seen as necessary if an effective statutory basis for local land use planning is to be secured. The public local inquiry, in providing a forum for arbitration on remaining objections to the plan, represents a key stage in this process of conflict resolution. This is so notwithstanding the fact that the credibility of the inquiry – and hence its function of local plan legitimation – is threatened by local planning authorities effectively sitting in judgement on their own local plan proposals.

Notes

1 The *Local Government Act 1972* created in England and Wales 53 county councils (6 metropolitan and 47 shire) and 369 districts (36 metropolitan, 333 shire). This structure paralleled that set up for London in 1965 by the *London Government Act 1963*, which created the Greater London Council and 33 London boroughs. The two-tier approach to the organization of local government has now been modified in London and the metropolitan areas by the abolition of the Greater London Council and the metropolitan county councils on 1st April 1986 following the *Local Government Act 1985*.

2 *Town and Country Planning Act 1971*, s.10C(1). Section 10C was inserted by the *Local Government Act 1972*, s.183(2).

3 *Ibid*.

4 Circular 23/81 for instance suggested local plans should only be proposed where the work required 'to enable it to be placed on deposit for objection is realistic, justified and can be carried out within about a year' (DoE, 1981c, para. 10).

5 By March 1982, 79 local plans had been adopted outside Greater London; 57 of these were district plans, 15 action area and 7 subject.

6 'Draft plans advice sold down Tyne', *Planning*, No. 491, 22 October 1981, p. 1; 'Row over Avon plan scheme', *Planning*, No. 472, 11 June 1982, p. 1; 'North-avon plan backed', *Planning*, No. 481, 13 August 1982, p. 1; 'Avon considers DoE local plan request', *Planning*, No. 484, 3 September 1982, p. 1.

7   That is, those prepared in response to the request for review made in DoE Circular 23/81.

8   See Sections 12(2), 14(2) and (5) of the *Town and Country Planning Act 1971* as amended by paragraphs 2(3), 3(1) and 3(2) of Schedule 16 of the *Local Government Act 1972*.

9   *Town and Country Planning Act 1971*, s.11(9) and s.14(2).

10   County planning authorities do not have to certify that a local plan that they themselves have prepared conforms generally to the approved structure plan, but the requirement that it should so conform remains (*Town and Country Planning Act 1971*, s.14(2)).

11   See 'Clean hands with mild, meek non-intervention', *Planning*, No. 464, 16 April 1982, p. 5.

12   See 'Row waits for inquiry', *Planning*, No. 516, 29 April 1983, p. 16.

13   See 'Heseltine picks up tab for Newcastle Shopping Change', *Planning*, No. 470, 28 May 1982, p. 3; 'Newcastle plan backed by Heseltine', *Planning*, No. 480, 6 August 1982, p. 1; 'Fog on the Tyne', *Planning*, No. 481, 13 August 1982, pp. 8–9.

14   See 'Certification wait on Oxford plans', *Planning*, No. 530, 5 August 1983, p. 1; 'Latest certificate battle resolved in district's favour', *Planning*, No. 533, 26 August 1983, p. 3.

15   See 'Leeds City forced to revise local plan', *Planning*, No. 471, 4 June 1982, p. 1.

16   See DoE Circulars 102/72 and 122/73, both entitled 'Land Availability for Housing'.

17   The comments of the Minister for Housing and Local Government are reported in full by Heap (1982, pp. 65–69); also see *Hansard*, February 29 1968, Standing Committee G, Cols. 190 *et seq.*

18   *Town and Country Planning Act 1971*, s.11(3)(a).

19   *Town and Country Planning (Structure and Local Plans) Regulations 1971* (SI 1971, No. 1109).

20   *Town and Country Planning (Structure and Local Plans) Regulations 1974* (SI 1974, No. 1486).

21   *Ibid.*, Regulations 16 and 17, and Schedule 2.

22   *Inner Urban Areas Act 1978*, s.12. This power was extended to all authorities by the *Local Government, Planning and Land Act 1980*, s.88.

23   Secretary of State's Address to the Town and Country Planning Summer School, 1979.

24   *Ibid.*

25   *Town and Country Planning (Structure and Local Plans) Regulations 1982* (SI 1982, No. 555). However, paras 4.8 and 4.9 of Circular 22/84 advise that local plans should be realistic, make the best use of all available resources, and include in the reasoned justification an indication of the assumptions made about the resources likely to be available for carrying out the policies and proposals formulated (Department of the Environment, 1984g).

26   Detailed criticisms of the unitary development plan proposals by the Royal Town Planning Institute are reviewed in 'Pinpointing problems in the unitary structure', *Planning*, No. 584, 31 August 1984, pp. 8–9. Those by the National Housing and Town Planning Council are given in *Housing and Planning Review*, Vol. 39, No. 4, August/September 1984, pp. 12–13.

27 *Town and Country Planning Act 1971*, s. 11(3).
28 *Town and Country Planning (Structure and Local Plans) Regulations 1982* (SI 1982, No. 555), Regulation 12.
29 *Town and Country Planning Act 1971*, s.11(5).
30 *Town and Country Planning (Structure and Local Plans) Regulations 1982* (SI 1982, No. 555), Regulations 10, 11 and 13.
31 *Town and Country Planning Act 1971*, s.10C.
32 *Town and Country Planning (Structure and Local Plans) Regulations 1982* (SI 1982, No. 555), Regulation 11(b).
33 *Ibid*. Regulation 11(a); *Town and Country Planning Act 1971*, s.11(4A).
34 *Town and Country Planning Act 1971*, s.11(7).
35 *Ibid*., s.6 and s.11(9A).
36 *Ibid*., s.12; *Town and Country Planning (Structure and Local Plans) Regulations 1982* (SI 1982, No. 555), Regulation 5.
37 *Town and Country Planning Act 1971*, s.14(2) and s.14(5).
38 *Ibid*., s.12; *Town and Country Planning (Structure and Local Plans) Regulations 1982* (SI 1982, No. 555), Regulations 24 and 35.
39 *Town and Country Planning Act 1971*, s.13; *Town and Country Planning (Structure and Local Plans) Regulations 1982* (SI 1982, No. 555), Regulation 27.
40 *Town and Country Planning Act 1971*, s.13(3) and s.15(4).
41 *Town and Country Planning (Structure and Local Plans) Regulations 1982* (SI 1982, No. 555), Regulation 28.
42 *Ibid*., Regulation 29(1).
43 *Ibid*., Regulation 28; *Town and Country Planning Act 1971*, s.13(1).
44 *Town and Country Planning (Structure and Local Plans) Regulations 1982* (SI 1982, No. 555), Regulation 31.
45 *Ibid*., Regulation 30.
46 *Ibid*., Regulation 32.
47 *Town and Country Planning Act 1971*, s.14(1A).
48 *Ibid*., s.244(1). See Bucks. County Council *v*. Hall Aggregates (Thames Valley) Ltd. and Sand and Gravel Association Ltd. (Court of Appeal, 14 November 1984), reported in *Journal of Planning and Environment Law*, September 1985, pp. 634–646; Great Portland Estates plc *v*. City of Westminster (House of Lords, 31 October 1984), reported in *Journal of Planning and Environment Law*, February 1985, pp. 107–115; Fourth Investments Ltd. *v*. Bury Metropolitan Borough Council (Queen's Bench Division, 23 July 1984), reported in *Journal of Planning and Environment Law*, March 1985, pp. 185–188 for recent examples of the attempts of property development, landowning and industrial interests to have adopted local plan policies quashed by making applications to the courts under s.244.
49 See 'Intensive effort fails to write the last word', *Planning*, No. 639, 11 October 1985, p. 5.
50 See, for example, 'Inquiries slow plan process?' *Planning*, No. 478, 23 July 1982, p. 12; 'Few marks for district planners', *Planning*, No. 479, 30 July 1982, p. 7; 'Inquiry report criticized', *Planning*, No. 482, 20 August 1982, p. 1; 'Forward planning stands to lose', *Planning*, No. 483, 27 August 1982, pp. 6–7; and 'Plan inquiries and credibility', *Planning*, No. 484, 3 September 1982, p. 5.

4 The statutory local land use planning system in Britain
II: Local plans in practice

4.1 Introduction

Despite the emphatic DoE advice on both substantive and procedural matters outlined in Chapter 3, local planning authorities have considerable discretion as to whether or not they should embark on statutory local plan preparation. This in part reflects the 1968 aim of central government to divest itself of any substantial involvement in the preparation of detailed local (as opposed to strategic) development plans. One way of interpreting the subsequent growth of DoE advice on local plans is to see it as an attempt to re-establish some influence in this area. However, without major legislative change these initiatives can only be framed as exhortatory policy statements; nevertheless they can be expected to be extremely influential. At the same time, local planning authorities themselves have begun to appreciate the difficulties associated with statutory local plan preparation, particularly with regard to the procedures involved and the various restrictions on form and content. It was suggested in Chapter 3 that these perceived limitations have led many authorities to move away from the statutory plan to a wide range of non-statutory instruments.

However, these issues cannot be taken to imply that local authorities have completely rejected the statutory local plan as a vehicle for local planning policies and proposals. Accordingly this chapter seeks to document the extent of local plan deposit and adoption as at March 1985, and from this to explore some of the variations in local plan practice and content. The discussion is structured around the four key local plan roles: the development of structure plan policies and proposals; the provision of a detailed basis for development control; the provision of a detailed basis for the co-ordination of the development and other use of land; and the bringing of local planning issues before the public. Together these roles encompass the function of local plans within the hierarchical model of policy control. An examination of how plans have sought to fulfil each of these roles therefore allows an evaluation of their effectiveness within the context of this model.

4.2   Local plan deposit and adoption

As noted above, local planning authorities were relatively slow in acting on the provisions for local plans set out in the *Town and Country Planning Acts* of 1968 and 1971, with the first plan being deposited in July 1975 and adopted the following December. Since then, however, progress on local plan deposit and adoption has been steady, with the rate of deposit increasing substantially in 1980 and 1981, and the rate of adoption following suit in 1981 and 1982 (Figure 4.1). This upsurge is likely to be due more to the substantial completion of the structure plan framework by the early 1980s than to the provisions of the *Local Government, Planning and Land Act 1980* allowing local plan adoption in advance of structure plan approval. By March 1985, 358 plans had been adopted, with a further 229 on deposit (Table 4.1).[1] Thus it is reasonable to assume that local plans have gained substantial acceptance as a vehicle for land use planning policies. This acceptance, however, has not necessarily been in the form envisaged by the Planning Advisory Group (PAG) in 1965. Most striking is the fact that only 34 action area plans have been adopted to date, with a further five on deposit; together these plans represent seven per cent of all adopted/deposited plans. However significant each of these plans is in the local context, the overall role is surely far short of that anticipated by PAG, for whom action area plans were 'the most important category of local plans' (PAG, 1965, p. 32). This role now appears to have been taken by the district plan, which accounts for 82 per cent of adopted plans and for 86 per cent of those on deposit (83 per cent of all adopted/deposited plans). It is difficult to infer trends from these figures, but it seems that the district plan is becoming at least numerically more significant at the expense of the action area plan, since subject plans account for a similar proportion of deposited and adopted plans. This is to be expected given that the anticipated growth in development associated with a growing economy and requiring action area treatment is not occurring. Moreover, prior to the *Local Government, Planning and Land Act 1980* action areas were identified in structure plans,

Table 4.1   *Deposited and adopted local plans by type of plan, England, March 1985*

Type of Plan	Deposited		Adopted		Total	
	No.	*(%)*	*No.*	*(%)*	*No.*	*(%)*
District	196	(86)	292	(82)	488	(87)
Action area	5	(2)	34	(9)	39	(7)
Subject	28	(12)	32	(9)	60	(10)
Total	229		358		587	

Source: Department of the Environment.

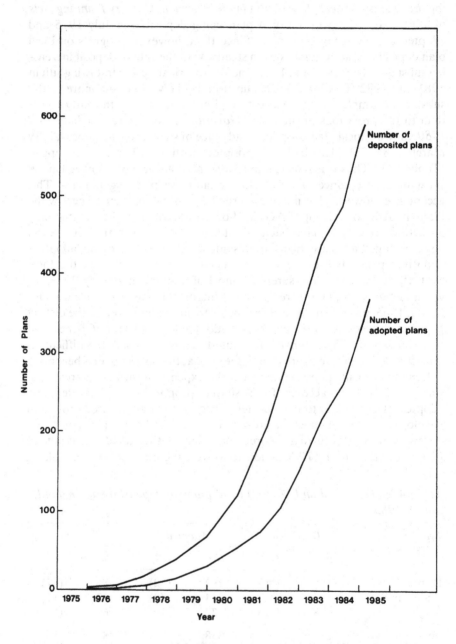

Figure 4.1   *Local plan deposit and adoption in England, 1975–March 1985*

and the proposed development had to commence within ten years from the date of the submission of the structure plan to the Secretary of State for approval; hence, local plans for such areas had to be produced quickly.[2] Finally, lower tier (district) authorities will tend to favour the *district* plan in particular because it provides more detailed guidance for development control than is available from the structure plan, thereby establishing a degree of local political autonomy within the bounds of structure plan conformity.

Table 4.2 indicates that in line with the presumption of the *Local Government Act 1972*, the majority of local plans have been prepared by the district councils and the London boroughs. These authorities have produced 89 per cent of adopted and deposited plans, the bulk consisting of 472 district plans (80 per cent of all adopted and deposited plans). The counties have produced 11 per cent of all plans, and have concentrated their efforts on subject plans, which out of a county-prepared total of 68 plans represent 74 per cent (50 plans). County-prepared subject plans in fact account for 83 per cent of all plans of this type. The majority deal with minerals, Green Belts, and countryside issues (Table 4.3). Topic studies on 'urban' issues such as employment or shopping do not feature, reflecting the influence of the now cancelled central government advice in Local Plans Note 1/76 that topics are only suitable for subject plan treatment if capable of being considered in isolation (DoE, 1976).

Table 4.2 actually gives a somewhat misleading picture of the extent to which the various types of local authorities in England have taken up the local plan, since the *number* of authorities of each type varies considerably. For example, there are 296 shire districts in England compared to 33 London boroughs and 39 shire counties. Table 4.4 takes this variation into account by giving average rates of local plan deposit and adoption. The

Table 4.2 *Deposited and adopted local plans by type of plan and authority, England, March 1985*

Type of Authority	District		Action area		Subject		Totals	
	No.	(%)	No.	(%)	No.	(%)	No.	(%)
Metropolitan counties	2	(1)	1	(3)	11	(18)	14	(2)
Shire counties*	14	(3)	1	(3)	39	(65)	54	(9)
Metropolitan districts	86	(17)	13	(33)	4	(7)	103	(18)
Shire districts	347	(71)	13	(33)	6	(10)	366	(62)
London boroughs	39	(8)	11	(28)	0	(–)	50	(9)
Total	488		39		60		587	

Source: Department of the Environment.
*Includes the Bakewell District Plan, prepared and adopted by the Peak Park Planning Board.

metropolitan districts are revealed as being on average the most prolific producers of local plans, especially district plans. The metropolitan counties and the London boroughs have also made relatively substantial use of local plans with averages of 2.0 and 1.5 plans, respectively. The shire districts have on average made least use of local plans, completing marginally fewer plans than the shire counties (1.2 to 1.4). Local plans (predominantly district) thus seem to have been taken up most readily in London and the metropolitan areas, with fewer coming forward on average elsewhere. This is not surprising given the complex nature of planning problems in these areas, with associated pressures for development and change. It could also reflect the political perspective of the controlling party in these authorities. Approved and adopted local plans in London and the metropolitan areas will (in conjunction with strategic guidance issued by the Secretary of State)

Table 4.3   *Topics of adopted and deposited subject plans, England, March 1985*

Topic	County		District		Total	
	No.	(%)	No.	(%)	No.	(%)
Minerals	14	(28)	2	(20)	16	(27)
Green Belt	14	(28)	3	(30)	17	(28)
Recreation/amenity	7	(14)	3	(30)	10	(17)
Countryside*	13	(26)	1	(10)	14	(23)
Other**	2	(4)	1	(10)	3	(5)
Total	50		10		60	

Source: Department of the Environment.
*Includes landscape, coastal and river valley issues.
**Intensive livestock units, waste disposal, airfield.

Table 4.4   *Average numbers of adopted and deposited local plans by type of authority, England, March 1985*

Type of Authority	Type of plan			
	District	Action area	Subject	All plans*
Metropolitan counties	0.3	0.15	1.6	2.0
Shire counties	0.4	0.02	1.0	1.4
Metropolitan districts	2.4	0.4	0.1	2.9
Shire districts	1.2	0.04	0.02	1.2
London boroughs	1.2	0.3	–	1.5
All authorities	1.2	0.1	0.1	1.4

Source: Department of the Environment.
*Discrepancies due to rounding.

provide an initial basis for the preparation of unitary development plans, the new type of development plan provided for the London boroughs and the metropolitan districts following the abolition of the Greater London Council and the six metropolitan county councils on 1st April 1986.

Table 4.5 identifies how many local authorities of each type have either placed on deposit or adopted at least one plan. Overall, 72 per cent of authorities have taken local plans to deposit stage or beyond, while 57 per cent have adopted one or more plans. There is considerable variation in these figures between different type of authorities, e.g. all but three of the London boroughs (91 per cent) have placed local plans on deposit and the majority (70 per cent) have at least one adopted plan, while the equivalent figures for the shire districts are much lower (68 and 54 per cent, respectively). However, perhaps the most striking feature of Table 4.5 is that over a quarter of authorities (28 per cent) have not yet deposited a single plan, the principal 'under-achievers' in this respect again being the shire districts. In some cases this is understandable, e.g. rural authorities with staff/resource limitations where the structure plan, perhaps in conjunction with informal local plans or policies of some kind, is thought to provide adequate guidance for development control. In other cases, however, the absence of deposited or adopted local plans is more surprising, e.g. Bristol City Council, Liverpool City Council, Middlesbrough BC. Some of these authorities may simply have been slow to proceed to deposit, particularly where the structure plan has only recently been approved, but others have deliberately embarked on wholly non-statutory local planning strategies in the belief that the formal local plan has little to offer (see Chapter 6).

Table 4.5 *Number of authorities with local plans on deposit or adopted, England, March 1985*

	Number of Authorities	Number of Authorities with one or more adopted local plan(s)		Number of Authorities with one or more local plan(s) on deposit		Number of Authorities with one or more local plan(s) adopted and/or on deposit	
		No.	(%)	No.	(%)	No.	(%)
Metropolitan counties	7	4	(57)	5	(71)	7	(100)
Shire counties	39	21	(54)	16	(41)	28	(72)
Metropolitan districts	36	25	(69)	24	(66)	31	(86)
Shire districts	296	160	(54)	112	(38)	202	(68)
London boroughs	33	23	(70)	13	(39)	30	(91)
Total	411	233	(57)	170	(41)	298	(72)

Source: Department of the Environment.

For those local authorities who have progressed plans to deposit and/or adoption, Table 4.5 conceals significant variation in terms of the *number* of plans prepared by individual authorities. For the district authorities and the London boroughs, who together account for 89 per cent of adopted and deposited plans, Table 4.6 gives the number of plans per authority. This again shows that a substantial number of authorities have not yet taken a plan to deposit, while at the other extreme a small number of authorities have each completed several local plans, e.g. North West Leicestershire DC, Warrington BC (seven plans), and Leeds City Council (eight plans). Only Salford City Council, however, has so far completed ten plans (six adopted, four on deposit). These figures, of course, relate only to *numbers* of plans and not to local plan *coverage*; some authorities have preferred to develop a single district-wide plan rather than a series of smaller area plans. This is certainly the case with many of the London boroughs (Field, 1983). The number of completed local plans is thus an indication of choice of strategy rather than of commitment to the statutory system.

Finally, what proportion of the total number of local plans that authorities intend to produce do those local plans completed to date (i.e. adopted or on deposit and not yet adopted) represent? Table 4.7 attempts an assessment of local plan progress. The figures for total local plan intentions represent an estimate of the number of plans that will have been produced once the present preparation programme is completed. These totals are based on numbers of plans *proposed* in development plan schemes at February 1982 (see Table 3.2, p. 90), together with the number of local plans *adopted* at this date, and include an estimate of local plans *proposed* by the London boroughs derived from Field (1983).[3] Overall, plans deposited or adopted to date represent just over a third (38 per cent) of total intentions. District councils have made below average progress on their proposed subject plans, and have completed only a third of intended action area plans, despite the urgency implied initially by the designation of action areas in structure plans,

Table 4.6 *Number of district councils and London boroughs by number of local plans adopted and deposited, England, March 1985*

Number of plans per district	Number of districts	(%)
0	102	(28)
1	127	(35)
2	75	(20)
3	30	(8)
4	17	(5)
5	9	(2)
6+	9	(2)
Total	369	

Source: Department of the Environment.

and now by their inclusion in the development plan scheme. However, the districts have managed the far more substantial task of completing over a third of their proposed district plans. This indicates where district priorities in plan production lie. The counties have completed the bulk of their proposed district plans, although only 18 plans are involved here in total, while maintaining average progress on their subject plans. Remaining commitments (all authorities) are as follows: 766 district plans, 71 action area plans, and 101 subject plans, a total of 938 plans. If the rate of deposit since 1981 (approximately 110 plans per annum; see Figure 4.1, p. 129) is maintained, these remaining proposed plans will all have been deposited in 8.5 years, i.e. by the end of 1993. The number of adopted plans, however, has and will lag behind the numbers on deposit for obvious procedural reasons. By March 1985, 1167 proposed plans remained to be *adopted* (229 of these, of course, had already proceeded to deposit). The average rate of plan adoption over the period 1982–1984 is some 80 plans per annum. Assuming this rate is maintained, all the remaining proposed plans will have been adopted in 14.6 years, i.e. by the year 1999. It is possible that these forecasts are unduly pessimistic, since work on many of the remaining proposed plans is no doubt already underway; future years may see an increase in the rate of deposit and, thence, of adoption. Nevertheless, there is still a substantial burden of local plan preparation work facing local authorities if current proposals are to be achieved, while many adopted plans will require review in due course. Finally, the programme of local plan preparation in London and the metropolitan areas will not be completed as a consequence of the switch to unitary development plans following the abolition of the Greater London Council and the metropolitan county councils, although for a transitional period the London boroughs and the metropolitan districts will be able to continue existing local plans work.

Table 4.7   *Local plan intentions and achievements, England, March 1985*

	County			District			Total		
Type of plan	No. total intentions*	No. Com-pleted**	(%)***	No. total intentions*	No. Com-pleted**	(%)***	No. total intentions*	No. Com-pleted**	(%)***
District	18	16	(89)	1236	472	(38)	1254	488	(39)
Action area	5	2	(40)	105	37	(35)	110	39	(35)
Subject	128	50	(39)	33	10	(30)	161	60	(37)
Total	151	68	(45)	1374	519	(37)	1525	587	(38)

Sources: Bruton, 1983a, Table 2; Field, 1983, Table 2; Department of the Environment.
* Total intentions for county and district authorities based on (*a*) respective numbers of local plans *proposed* in development plan schemes at February 1982 (Bruton, 1983a), together with an estimate of local plans *proposed* by the London boroughs derived from Field (1983), plus (*b*) respective numbers of local plans *adopted* at this date.
** Adopted and deposited plans.
*** Represents percentage of intentions; deposited/adopted.

4.3 Statutory local plans in practice

Local plans produced to date show that, within the confines of statute and regulations, a wide variety of approaches have been taken by local authorities to the question of local plan *content*. The legislative requirements are simply that a statutory local plan must consist of a map and written statement, formulating the authority's proposals for the development and other use of land in the area covered by the plan.[4] Subsequent DoE advice attempting to define more closely the nature of the local plan as a land use policy vehicle has apparently been interpreted freely by local authorities when applying it to their particular problems. For example, Thompson (1977b, p. 145) has written of the early development of statutory local plans that 'not surprisingly, local authorities are taking widely varying approaches ... reaction to the DPM (Development Plans Manual) identikit is partly inspired by a more sensitive response to local circumstances. Indeed, a local plan must be shaped by its operating context'. The result has been that 'local plan documents come in considerable variety' (Healey, 1983, p. 161). Local plans range from the substantial district-wide district plans prepared by the London Boroughs (e.g. the Royal Borough of Kensington and Chelsea District Plan, running to a total of 323 pages) to the single-sheet plan format devised by Hamilton DC (McGilp, 1981). There are, however, some common conventions, especially as to the treatment of policies and proposals following DoE advice in Circulars such as 55/77, 4/79 and latterly 22/84. The variety of approaches to statutory local land use planning can also be seen in terms of the *area* covered by the plan and the *issues* that are addressed. Each of the ten districts in Hertfordshire, for instance, is working towards comprehensive coverage to be provided by a single district plan, a strategy 'justified within the operational context of planning for restraint in the county' (Griffiths, 1978, p. 18). More common is the preparation of local plans for selected areas within a district which are judged to require the statutory treatment, e.g. Warwick DC's Lapworth District Plan, prepared to define Green Belt boundaries and to determine the level of local housing need in the area to accord with structure plan policy for rural settlements. There are more detailed variations in local plan content, however, than over issues or areas covered, and it is with a review of these variations that the remainder of this chapter is concerned.

Following Barnard (1980), the discussion is organized around the four roles envisaged for local plans by PAG and since developed by the DoE, viz.,

a. to develop the policies and general proposals of the structure plan and to relate them to precise areas of land defined on the proposals map;
b. to provide a detailed basis for development control;
c. to provide a detailed basis for co-ordinating and directing development and other use of land – both public and private; and
d. to bring local planning issues before the public.   (DoE, 1984g, para. 3.1)

The review is based on a number of adopted and deposited local plans, which together cover each of the three plan types.[5] This documentary material is supplemented where appropriate by practice reviews and published case studies of local plan experience. The review considers the way in which plans have sought to carry out or fulfil each of the local plan roles identified above. An assessment of how far plans have actually achieved these roles is then developed in the context of the hierarchical model of strategic planning discussed in Chapter 2.

4.4 The development of structure plan policies and proposals

Structure plan policies and proposals are not site-specific; it is the role of the local plan to develop what are hitherto only strategic policies and general proposals in order to show how they will apply to specific sites and areas. There are two aspects to the detailing of structure plan policies and proposals in local plans: the discussion of the strategic context, and the development of that context into specific local plan proposals.

The structure plan policy context which underlies and informs the local plan is usually referred to explicitly, e.g. action area plans prepared for areas specified in the structure plan as requiring comprehensive treatment within ten years typically indicate this as one of the rationales for plan preparation (e.g. Harrow Town Centre Action Area Plan, Harrow LB). Structure plan policies relevant to the local plan area are commonly gathered together in an appendix (e.g. Leicester City Council's Abbey District Plan), and/or referred to in the text of the written statement. As such, they may be cited as an introduction to relevant chapters, referred to in the reasoned justification in support of local plan policies and proposals, or quoted as part of local plan policy itself.

It should be noted in passing that one of the problems involved in referring explicitly to structure plan policies in local plans is that objections to the local plan, either at public participation or 'deposit' stages, may well raise strategic issues already determined by structure plan approval. The non-statutory Code of Practice on the conduct of local plan inquiries notes that while 'the local plan cannot alter the provisions of the structure plan . . . an objection to a local plan may relate to the way in which the local plan proposals seek to implement the structure plan' (DoE, 1984h, para. 2.6), but this attempt to limit the degree to which objections to the local plan raise strategic issues will not always be easy to apply in practice, nor will it always be accepted by objectors.

Having established the nature of the strategic context, local plans develop and apply the policies therein in two principal ways: by the definition of areas and boundaries using the proposals map, and by the closer specification of how structure plan policies will be applied in the local context, e.g. for the purposes of development control. The first of these roles involves the translation of broad structure plan allocations and policies into detailed local site- and area-specific proposals. Where the structure plan envisages population and employment growth, for instance, the role of the local plan is to allocate that growth to defined areas of land for future residential and industrial use. The Gloucester Local Plan thus defines a substantial area of land on the eastern fringe of the City for residential development, in order that the substantial growth in this area envisaged in the structure plan can be accommodated. The Abbey District Plan (Leicester City Council) has a similar function, although the plan also provided an opportunity to review earlier 'master plan' approaches (Lambert and Underwood, 1983). Barnard (1980, p. 45) comments that 'local plans have generally been effective in bringing forward housing land in accordance with the structure plan'.

In contrast to the allocation of areas for development, other local plans have been concerned to develop structure plan policies of development restraint or environmental protection through the definition of Green Belt boundaries, village envelopes, or other special landscape areas (e.g. the Western Wiltshire Green Belt Local Plan, Wiltshire CC; Cornwall CC's Countryside Local Plan). Elsewhere, local plans have applied structure plan restraint policies in respect of specific uses such as offices. The Royal Borough of Kensington and Chelsea District Plan, for instance, accepts the GLDP policy of restraint of office development outside the central areas of Kensington and goes on to specify that these locations will be in the 'immediate vicinity' of railway stations in the Borough. Proposals for office development within acceptable locations will still be subject to other plan policies, so that the locational requirement is a necessary but not sufficient condition for planning permission for office development.

The second way in which local plans apply strategic proposals is through the closer specification of structure plan policies for development control purposes. In some cases, this is seen to be necessary where structure plan policies are judged incomplete in themselves. Slack (1983), for instance, notes that one of the roles of Ashford BC's Rural Settlement Plan will be to determine the location and level of local needs for employment and housing, since these questions remain unanswered in the Kent structure plan. The Birmingham Central Area District Plan details structure plan policy to encourage office development within the city centre by defining a 'primary office area' and establishing additional development control criteria for office development elsewhere. Finally, Barnard (1980) notes that local plans have also translated general proposals into development control policies for specific areas pending positive action to implement the intended developments. This has occurred, for example, where the structure plan has included a long-term road proposal, carried forward in the local plan by the safeguarding of a corridor of land.

4.5 The provision of a detailed basis for development control

4.5.1 Introduction

The hierarchical model of policy control underpinning the development plan system is such that just as structure plans provide a context for local plans, so local plans are intended and expected to provide a policy context for development control. The link between the development plan (including local plans) and the operation of development control is specified by s.29 of the 1971 Act, which requires local authorities to 'have regard to the provisions of the development plan, so far as material to the application, and to any other material considerations' in determining planning applications. The courts have clarified the weight to be given to the development plan in development control; it is clear that in reaching their decisions on planning applications local planning authorities should consider all material considerations, of which the development plan is one (Chapter 7). This interpretation belies the significance of the development control process for the implementation of local plan policies and proposals. No additional powers or resources result from the adoption of a local plan, and it is only under the provisions of s.29 that local plan policies and proposals have to be taken into account by an implementing agency (in this case, by the local planning authority in reaching decisions on planning applications). The main role for local plans thus 'lies in the way local planning authorities conduct their regulative powers in relation to private sector development initiatives' (Healey, 1983, p. 98). In turn, this emphasizes the importance of central government support for local plan policies through the appeals system as a key factor in deciding how significant and effective a local plan will be as a vehicle for land use policy. The development control/appeals system is thus very important for local plans, since it is only when making decisions on planning applications that the local plan *has* to be taken into account.

As far as local plan content is concerned, this suggests that plans will give a good deal of attention to providing a suitably detailed basis for development control. The legislation allows authorities to formulate in such detail as they think appropriate proposals for the development and other use of land,[6] and so embodies considerable discretion. Circulars make it clear that local plan proposals may either define a specific site for a particular development or other use, or define the specific area within which particular policies will be applied (DoE, 1984g, para. 3.22), so that 'the basis for development control can consist of any combination of proposals for the development of specific sites and policies which can be applied to specific areas' (Barnard, 1980, p. 50). Local plans have in fact formulated their policies and proposals in three main ways: on a *plan-wide* basis; within a defined *area*; and for specific *sites*. Most local plans express their policies and proposals through a combination of some or all of these methods. The proposals map is then used as an integrating device, not only defining the sites for particular developments or uses and the areas to which particular policies apply, but also indicating where those policies which apply to the whole of the plan area are formulated in the written statement.

4.5.2 Plan-wide policies and proposals

Healey (1983) distinguishes two forms of plan-wide policies, the *general principles* which will guide consideration of development proposals, and more detailed *performance criteria* to be applied to such proposals. Most plans include, often under the appropriate topic heading, general statements of principle which establish the philosophy and strategy of the plan. For instance, the District Plan prepared by Hammersmith and Fulham LB sets out a series of objectives which, when applied to the key planning issues within the Borough, lead to an overall strategy for land use and development providing the basis for plan policies. Similarly, the Gloucester Local Plan defines a series of 'core' policies which are used to set the tone for each topic chapter and are also gathered together at the end of the plan. The core policies are supplemented by 'implementation' policies of a more specific nature, e.g. the core policy seeking to promote a five years supply of industrial land is supported by implementation policies identifying suitable sites (also defined on the proposals map). Other plans contain more restrictive general proposals in the form of presumptions against particular forms of development. For example, the Westminster District Plan has the following policy to prevent the loss of existing housing throughout the City: 'planning permission for the change of use of residential accommodation to other uses will not be permitted except in special circumstances' (para. 5.15).

Performance criteria can be applied to all forms of development or restricted to specified uses. A typical example is the criteria established by Humberside CC's Coastal Caravans and Camping Subject Plan that 'extensions to existing sites will not be permitted when they amount to more than 1 hectare, or 25% of the size of the original site area . . . whichever is the greater . . . .' (Policy 4.3.3). Similarly, Humberside CC's Intensive Livestock Units Local Plan contains a number of policies referring to 'protected areas' where the policy concerned applies, e.g. 'new . . . units . . . will be allowed only if sited outside certain protected areas' (Policy 1.2). These areas are subsequently defined in terms of proximity to existing housing or land allocated for housing in development plans. A common concern for shopping policies is to restrict the spread of non-retail activities in shopping streets. Warwick DC's Leamington Town Centre Local Plan provides a quantified version of such criteria, e.g. such changes of use are not to be permitted when this would result in 20 per cent or more of the frontage length between any two road junctions being used for non-shopping purposes. Finally, the Gloucester Plan provides a statement of the criteria to be used in judging applications for 'starter homes' development, e.g. on sites accommodating less than 150 dwellings not more than 30 starter homes will normally be permitted. Local plans can also include criteria which relate to more detailed aspects of the development rather than to whether or not proposals will be acceptable in principle. Typical here are standards relating to car-parking provision or form and design. The Haringey District Plan includes guidelines to be used when considering proposals for house alterations and extensions, which are intended to prevent overdevelopment in terms of bulk or density. Comparable policies are also established on retail development (design of shopfronts, advertisement control), and industrial and warehousing proposals (design, materials, access), which are in addition to general environmental and design policies applicable to all development (high buildings, materials, landscaping). Lastly, many local plans set controls over the physical intensity and bulk of buildings by the use of density provisions in the case of housing, and plot ratio standards for more intensive uses such as office developments.

One general issue which is raised here concerns the level of detail that local plans should adopt in specifying development control policies. Some plans refer to performance criteria published separately as supplementary planning guidance; this is a practice which has its advantages in terms of ease of amendment, but is not without its drawbacks (see Section 4.5.4).

4.5.3 Area-specific policies and proposals
Many local plans define areas inside the plan boundary within or outside of which special policies apply. This is particularly common among wider area plans or those for complex areas such as city centres, where a form of policy expression midway between plan-wide policies and site-specific proposals is needed to deal with the variety of issues involved. The Camden Borough Plan for instance defines a series of 'special policy areas', whose 'complexity or special character have justified more detailed policies either to protect or enhance their function and character in the face of considerable pressure for change and development or because they offer significant opportunities for major land use change' (para. 11.2). In addition, Camden LB has prepared 'local area plans' for the priority areas of Fitzrovia and West Hampstead, non-statutory documents designed to show how Borough Plan policies apply in each area while emphasizing local community needs. On the basis of this experience, initiatives for other priority areas in Camden are being developed within the framework of the Borough Plan, with the approach in each case dependent on particular local problems and characteristics, i.e. contingent factors (Newby, 1985). This strategy reflects the methodology of 'mixed scanning' adopted by Camden, whereby an initial broad scan (the preparation of a borough-wide local plan) is supplemented by a finer scan concentrating on those areas identified as needing urgent attention (Thompson, 1977a,b; Fudge, 1984a).

In general, area zonings act to establish policy distinctions which are additional to the policies applying through the plan area, but area-specific policies can also be used to alter the way in which plan-wide policies such as performance criteria are applied. For example, the Haringey District Plan lists a series of 'defined industrial areas', inside of which there is a presumption in favour of industry, and where industrial development will be encouraged even though the authority's industrial development standards cannot be fully achieved. Outside these defined areas, industrial development 'may' be permitted – provided that the relevant standards can be fully met. This policy thus does not seek to *prevent* industrial development outside the specified areas, but rather to steer such development with the promise of relaxed performance criteria. More usual are area-specific policies where the aim is to restrict certain developments to within the area, or to exclude them from the area. The prime example of such a latter area of restraint is the Green Belt, within which policies of severe restriction on all new developments apply. For example, the Greater Manchester Green Belt Local Plan defines the boundaries of the Green Belt in the county on a consistent basis, rationalizing previous policies whereby areas of open land had been variously defined as approved, submitted, draft and provisional Green Belt. Area-specific policies can also be linked to specified uses, such as the 'areas unsuitable for diplomatic uses' defined by the Kensington and Chelsea District Plan. The Birmingham Central Area District Plan defines a 'primary office area', within which office development will be favourably considered, and a further policy area (the 'primary shopping area') where new office developments will only be considered favourably if a retail use is provided at ground level. This plan, therefore, has two distinct area policies for office development. A further refinement has been introduced by the City of Manchester City Centre Plan, which has taken an innovative approach in its use of small area policy frameworks. The plan begins with a statement of aims, objectives and themes, which lead to policies and proposals at two levels. The first of these comprises a series of 15 general policies applying throughout the plan area and organized on a topic basis, e.g. tourism, shopping, housing, the location and amount of office development. These policies provide general guiding principles, but they do not in themselves constitute adequate guidance for development control. Accordingly, they are supplemented by a series of small area proposals applying to 27 small areas which together comprise the plan area. These small area frameworks form the core of the plan; each area is considered in turn and appropriate proposals made. Together, the general policies and small area proposals undoubtedly provide, as is claimed, a general framework for managing change rather than a shopping list of proposals which may date quickly. But the emphasis on robustness and flexibility means that as Healey comments 'it is not possible to determine exactly what the local authority would find acceptable on a particular site', and thus raises the question of whether the plan gives a sufficient basis for those with an interest in a particular site to determine the local authority's policies and proposals (Healey, 1984, pp. 118–119). In practice, such specific matters of detail could be resolved through consultations with the local authority and/or by submission of a planning application. Nevertheless, the danger here is that the plan's policy flexibility could be exploited as ambiguity, e.g. in the context of an appeal against refusal of planning permission (McNamara, 1986). The issue of the amount of detail that should properly be included in statutory local plans has recently been examined by the courts in a case involving the City of Westminster District Plan (see Section 4.5.4).

A second type of area policy specified in local plans relates to various area-based initiatives derived from national legislation. These include National Parks, Areas of Outstanding Natural Beauty, conservation areas, sites of importance for nature conservation, General Improvement and Housing Action Areas (see Chapter 5). Both structure and local plans, for example, should include the land use and development control policies to be applied in National Parks and Areas of Outstanding Natural Beauty, defining the boundaries to which these policies are to apply on the key diagram and proposals map (DoE, 1984g, paras. 4.14 and 4.35). Analogous local derivations also feature, such as Haringey's 'Housing Investment Areas', wherein housing renewal activity (rehabilitation and selective clearance) will be concentrated. In effect, this is a statement of intent made on behalf of other council departments rather than a policy capable of implementation via the planning legislation.

4.5.4 Site-specific policies and proposals

These identify particular sites for which development is proposed. In terms of development control, they indicate a strong presumption in *favour* of the proposed use and *against* other uses for the site in question. Departmental advice is that 'local plans should normally include only proposals for development which may reasonably be expected to start within about 10 years . . . where there are firm proposals for development, proposals to safeguard land may extend over a longer period' (1984g, para. 4.5). This stems from a concern to avoid the dangers of 'planning blight'. Plans have varied in terms of the extent to which site-specific proposals have been made; a key factor here seems to be the nature of the problems and opportunities in the plan area. Both the amount of vacant land that already exists in the area and the scale of development that is envisaged will be particularly important. Some local plans have continued the tradition of 1947 Act plans in showing proposed land uses over the whole or most of the plan area. The Beckton District Plan (Newham LB), for example, has a proposals map whose site-specific proposals (for industry, housing and open space) cover the bulk of the plan area. This reflects the plan's stated role of co-ordination of substantial development activity and the presence of considerable amounts of vacant land within the area. In contrast, other plans have taken a more selective approach to site proposals, often reflecting the obvious constraints imposed by existing development. Thus, much of the area covered by the Leamington Town Centre Local Plan comprises an outstanding conservation area, so that the plan contains few proposals for substantial new development, the main emphasis being on the management and maintenance of the existing urban fabric. The bulk of new development is envisaged to fall onto a series of 'key sites'; many of the specific proposals for such uses as offices, shops and housing are linked to these sites, although there are also more general policies relating to these and other activities and issues. Thus, the absence of site-specific proposals over much of the plan area does not mean that these are true 'blank areas', i.e. where, for development control purposes, there is no presumption either for or against any particular development. The Leamington Plan covers such areas with a series of plan-wide and area-specific policies, e.g. an area policy which defines an outer limit for the commercial core and allocates the remainder of the plan area primarily for residential use. The Central Leicester District Plan (Leicester City Council) does incorporate 'white areas' where a variety of uses may be permitted, but these have caused problems resulting from the uncertainty involved (Lambert and Underwood, 1983; Green, 1982).

The use of local plans to provide a detailed basis for development control has in practice raised questions concerning the proper *scope* of such policies and the degree of *detail* that needs to be incorporated within the statutory plan itself. The first of these issues concerns whether or not local plan policies can distinguish between *users* as well as land *uses*. For example, a degree of controversy surrounds the use of local needs policies, which attempt to allow local planning authorities operating general planning policies of restraint on new development to grant permission for certain types of development on the ground that the development will cater for local needs, e.g. employment, housing. Such policies raise complex issues concerning policy definition, expression and effectiveness, as well as the legality of the development control strategies through which they are pursued (Loughlin, 1984). As far as local plans are concerned, DoE advice has stressed the legislative requirement that proposals should relate to the 'development and other use of land' by stipulating that proposals distinguishing between users are invalid. Thus, 'proposals for housing should include, without distinguishing between them, proposals for both public and private development' (DoE, 1979a, para. 3.35). Local plans which have not followed this advice have attracted objections at PLI stage, the argument being that distinctions between users are outside town and country planning powers. The deposited Camden District Plan, for instance, included proposals to restrict the expansion of independent education facilities. These proposals incurred objections on the grounds that they restricted the activities of lawful institutions and personal freedom of choice, that they were not based on planning powers and were hence *ultra vires*, that secondary education was a matter for Parliament and not for a local authority, that there was a demand for independent education in the Borough which should be met, and that there was no adequate reasoned justification. The Inspector commented that the Camden proposals were unreasonable, illogical and discriminatory, and recommended deletion; the plan was modified accordingly (cited by Barnard, 1980).

The Courts have also recently considered the view that local plan policies should not differentiate between users. A case involving the City of Westminster District Plan illustrates the arguments involved. Following the adoption of the Plan in April 1982, a property company active in the City challenged two sections of the Plan dealing with (*a*) office development, and (*b*) industry. The application, made under s.244 of the 1971 Act, failed in the High Court, but succeeded in the Court of Appeal. The Appeal Court decision resulted in sections of the plan containing the disputed office and industry policies being quashed. An appeal by Westminster City Council to the House of Lords was only partially successful, in that only the industrial policies were reinstated, the office policies remaining quashed.[7] The question of the validity of distinguishing between users arose in relation to the industrial policies. Here, the plan contained a policy designed to protect existing industrial activities within the City. Specifically, the plan stipulated that planning permission for major rehabilitation or the redevelopment of industrial premises containing industrial use would not normally be granted where such development could be to the disadvantage of existing or potential industrial activities. Counsel for the property company argued that this policy was not within the powers of the planning legislation, being concerned with particular users of land rather than with the development and use of land, and that in formulating the policy the Council had had regard to an irrelevant factor, namely the interests of individual occupiers of industrial premises within the City. The District Plan was thus unreasonable, since it precluded the carrying out of development or redevelopment which would be justified in land use terms, in its desire to protect the interests of existing occupiers. Although the High Court found nothing objectionable in the policy concerned, the Court of Appeal took the view that it amounted to the protection of existing individual occupiers and was, therefore, illegal; the policy was quashed. Reinstating the policy on appeal by Westminster City Council to the House of Lords, Lord Scarman held that the plan was here attempting to secure a genuine planning purpose, i.e. the continuation of industrial uses considered important to the diverse character, vitality and functioning of Westminster. This aim could be promoted by protecting from redevelopment sites of certain classes of industrial use – in this case long-established clothing, fur and leather, and paper, printing, and publishing firms. This would inevitably mean that existing occupiers would be protected, but as a consequence rather than a part of the planning purpose of the plan. The industrial policy was thus reinstated *not* because the House of Lords held that it was valid to distinguish between users in local plans, but because they felt that the particular policy under consideration did not set out to do so. This was despite the fact that an unavoidable consequence of the operation of the policy would be to protect existing *users* as well as established *uses*.

A further problem for local planning authorities is that user-oriented policies included in local plans, such as attempts to satisfy local needs, and which survive the PLI and possible challenge under s.244, may not be enforceable anyway through development control. At the PLI into the Camden plan, one objector sought to add an exception to the restrictive office policy to the effect that office uses providing for local employment opportunities would be exempt. The Inspector commented that under existing planning legislation it would be impossible to differentiate for development control purposes between offices providing local employment and other types of office use (Barnard, 1980). This problem of enforceability through development control will still remain even if local plan policies themselves are not bound by the limits of development control conditions and refusals (see Hamilton, 1977).

The second issue raised by the role of local plans in establishing development control policies concerns the appropriate level of detail that should be included in the plan itself. The DoE has long accepted that some detailed material will be inappropriate for inclusion in a local plan, and recommends that this should be progressed as informal (i.e. non-statutory) 'supplementary planning guidance'. The idea of supplementary guidance is a policy invention of the DoE and has no legislative foundation. The current development plans memorandum envisages 'a continuing role for planning guidance which supplements the policies and proposals contained in structure and local plans' (DoE, 1984g, para. 1.14), material which is best left out of statutory plans either because it is too detailed or liable to frequent change. Supplementary guidance may include practice notes for development control requirements, development briefs and detailed or sketch layouts for housing or open space developments. Such documents should:

1　be published separately from the policies and proposals of the statutory development plan for the area;
2　be kept publicly available; and
3　be consistent with the structure and any local plan for the area.

Proposals for the development and other use of land should be included in the statutory plan rather than in supplementary guidance. Such guidance may be taken into account by the Secretary of State as a material consideration in matters which come to him for decision. Moreover, the weight to be accorded to it will increase when it has been prepared in consultation with the public and made the subject of a Council resolution (DoE, 1984g, paras. 1.14 and 1.15). When policies contained in supplementary material which meets these requirements are used in the determination of planning applications, they should thus be taken into account by the Secretary of State in his consideration of any resultant appeals against refusal.

Typically, where supplementary planning guidance is to be used this intention is stated in either local plan policies or the reasoned justification.

Occasionally supplementary guidance is (contrary to the advice of the DoE) included within the plan covers as an appendix (e.g. Humberside CC, Intensive Livestock Units Local Plan; Kirklees MBC, Huddersfield Local Plan). The deposited Huddersfield Local Plan included a policy on sites for residential development that 'the council will issue guidelines for development of the sites . . . [which] will specify any areas within each site that are not to be developed and will comment on how such areas should be laid out and used' (Policy HD/D3). The guidelines themselves, included within the deposited plan as an appendix to the chapter on land for development, amounted to a series of individual outline briefs for the sites concerned. The reasoned justification to Policy HD/D3 stated that 'the Council reserves the right to amend or expand these guidelines during the course of the plan period'. This being the case, the inclusion of such statements within the plan itself raises the prospect of frequent changes to the published plan being found necessary as a consequence of the detailed information involved. Following the public local inquiry, the Inspector recommended that the status of the guidelines should be clarified, and that they should be treated as supplementary planning guidance as described in para. 1.14 of DoE Circular 22/84, i.e. should be issued as a separate appendix which may be readily updated. Policy HD/D3 was considered valid, since the guidelines represented legitimate supplementary detail rather than major land use planning policies.[8]

Notwithstanding the emphasis of the DoE on the use of supplementary guidance as a vehicle for development control material, the recent judgements of the Appeal Court and the House of Lords on the office policies of the City of Westminster District Plan sound a warning as to the legitimate division between matters of policy and of detail. Here, the issue was not that the plan was too detailed (cf. the Huddersfield plan) but that it was not detailed enough. In particular, the judgements emphasize that supplementary guidance should not attempt to introduce any new land use policies or proposals (Bruton and Nicholson, 1984d, 1985b). Briefly, the City of Westminster District Plan stipulates that planning permission for office development outside a defined 'Central Activities Zone' will not be granted except in special circumstances. The Plan envisages that these 'special circumstances' will be set out in non-statutory guidance prepared after consultation following adoption of the plan. In the High Court, Counsel for the applicants argued that, as a result, the Plan failed to include any, or any sufficient, formulation of its proposals for office development, and that it was unreasonable to require developers to look at a non-statutory document to find out the Council's intent; this should be included in the plan. Though these arguments were not accepted by the High Court, they were by the Appeal Court and subsequently by the House of Lords. As a result, the office policies of the plan, both inside and outside the Central Activities Zone, were quashed. The issue here was not whether the office policy was bad in law (as was the case with the industrial policy), but turned instead on the question of the division between statutory and non-statutory policy. The policy contained in the local plan was judged to be inadequate as an indication to the public of which planning applications were going to be found acceptable and which were not. The nature of this inadequacy turned on the fact that the acceptability or otherwise of particular locations outside the Zone for office development could not be gleaned from the Plan, which failed to include any such statements of policy. However, the non-statutory guidelines subsequently made some provision for office development outside the Zone, most notably at the specified locations of Paddington and Marylebone Stations. The guidelines identified particular locations where office development would not normally be refused (subject to other plan policies), and set out other criteria ('special circumstances') which, if met, would allow the development of small amounts of office floorspace elsewhere. The guidelines, therefore, acted to introduce land use policies on office *location* which were not in the Plan, particularly in respect of such development at Paddington and Marylebone Stations. Indeed, the District Plan had apparently ruled out office development at these (and other) locations outside the Zone, despite the fact that the strategic GLDP had identified both Paddington and Marylebone Stations as preferred locations for office development. As a result of this division of policy between statutory plan and informal guidance, applicants for planning permission for office development outside the Central Activities Zone could not determine from the Plan alone the policy of Westminster City Council on such schemes. Both the Appeal Court and the House of Lords found this state of affairs to be 'wholly unreasonable and improper'.[9] Had an alternative division of policy been adopted by the Council, e.g. the Plan containing land use allocations and other criteria on the 'special circumstances' pertaining to office development in the City, with the non-statutory guidelines comprising more detailed matters such as car-parking requirements or design issues, a different view might well have been taken. This judgement, therefore, emphasizes the point that supplementary guidance is effectively restricted to detailing or expanding on land use policies established elsewhere in statutory plans, and thus stresses the role of local plans in setting the policy framework for development control. At the same time there is the still unresolved question of whether there is an acceptable level of detail below which local plans need not proceed and supplementary guidance can take over; and if so, what this level is. The only certain way to answer this problem, in order to avoid the possibility of a challenge under s.244, seems to be that local plans must be as comprehensive and detailed as possible. This has some obvious implications for local plan form and content, and for the timescale of local plan preparation and adoption, and subsequent alteration and modification. Nonetheless, as Barnard (1980, p.59) comments, 'it would seem sensible that enough detail should be given in the local plan to indicate to the public and to developers the principles the local planning authority will apply in judging planning applications'.

4.6   The provision of a detailed basis for the co-ordination of the development and other use of land

For the Planning Advisory Group, local plans were intended to provide a guide to developers and a basis for the co-ordination of public and private action in development and redevelopment. The primary function of local plans would not be as a control mechanism but 'as a positive brief for developers, public and private, setting the standards and objectives for future development' (PAG, 1965, p. 31). Local plans were thus envisaged as devices to *facilitate* rather than *prevent* development, a role which would be carried out by the setting of a detailed site-specific basis serving to co-ordinate individual developments. By implication, local plan proposals should be feasible, particularly in terms of the availability of finance to the implementing agency. In fact, local plans have varied considerably in the extent to which they have considered the *resource implications* of their proposals, both for the local planning authority and for other public and private sector agencies, and the *degree of commitment* to plan proposals on the part of those public and private sector agencies who will be responsible for implementation. In any case, a prerequisite for the successful co-ordination of development is the provision by the local plan of a comprehensive and up-to-date set of proposals which are themselves co-ordinated. Together, these aspects of local plan implementation can be used to show how and to what extent local plans have attempted to achieve an integration of development proposals within their areas.

The main hurdle faced by local plans in achieving a co-ordinated set of policies and proposals is the common practice of organizing the written statement in a series of discrete topic chapters on such matters as shopping, housing, transport and community facilities. This practice presumably originated from the list of eleven topics given in the 1974 Regulations as 'matters to which proposals are required to relate';[10] the 1982 Regulations contain no such list. However much the topic approach assists local plan preparation (in terms of initial survey, policy formulation, and public participation), individual topics need to be inter-related so as to identify inter-dependencies between proposals which would otherwise remain unaddressed. This much was recognized by the 1974 Regulations, which stipulated that district and action area plans be based on a comprehensive consideration of matters affecting the development and other use of land.[11] All plans will achieve a degree of spatial integration of proposals through the proposals map, as required by the legislation, but this cannot satisfactorily show how proposals will relate over time (i.e. cannot deal with questions of phasing or programming). The proposals map need not be confined to a single sheet: the statutory 'proposals map' of the Greater Manchester Green Belt Local Plan for example comprises 76 map-sheets which collectively show the precise location of the Green Belt boundary. In addition to the proposals map, a variety of other methods and techniques of presentation have also been used, often in combination, in an attempt to show the relationships between topic policies. For instance, the proposals map may be supplemented by a summary list of policies or proposals (e.g. Bexley Heath Town Centre Action Area Plan, Bexley LB), or the plan may include diagrams emphasizing the key structural relationships between proposals in a manner analogous to the structure plan key diagram, e.g. the 'Borough Diagram' of the Haringey District Plan. Relationships can also be indicated in the reasoned justification to individual policies by cross-referencing other relevant policies and proposals in support. This technique serves to strengthen the individual policies concerned by demonstrating that they have not been formulated in isolation. Relationships on a wider and more general scale can be established by setting priorities, either between topics or between areas. For instance, the Kensington and Chelsea District Plan, whose main aim is to maintain and enhance the character of the Borough as a residential area, gives first priority to the protection of existing residential areas and their amenity, second to the protection of services and jobs for local residents, and third to the maintenance of the Borough's role as part of inner/central London. The Haringey District Plan establishes 'priority areas for investment' which cut across the plan's topic-ordered policies. Plan priorities can also be set implicitly through the general objectives and philosophy of the plan. A further technique is the use of chapters which focus on defined sectors of the plan area rather than on topics. The Leeds Central Business Area District Plan includes the usual range of topic chapters setting the policies of the plan, but also incorporates a series of chapters focusing on four sectors which taken together make up the plan area. These sectors do not form policy areas as such, but are simply areas drawn up on the basis of existing common characteristics so as to define the impact of the plan's policies and proposals on them.

A more sophisticated way of handling policy relationships is to incorporate within the plan some treatment of how these relationships are expected to vary over time, and which goes beyond the usual statement that a plan's policies and proposals have been designed to be implemented by the end of the plan period. Some statement on the phasing of proposals may be necessary, either as a result of their inter-dependency or in order to achieve plan objectives e.g. the Totton Town Centre Action Area Plan, New Forest DC; Buckinghamshire CC, Minerals Subject Plan. Information on the timing of proposals, either in terms of their relative phasing or as a programme of when all the plan's proposals are expected to start and finish, is usually contained within a separate 'implementation' chapter, which also often incorporates some material on the costs of proposals, resource availability, and the agencies who will be responsible for specific projects, e.g. the Kensington and Chelsea District Plan; the Harrow Town Centre Action Area Plan, Harrow LB. These factors (timing, agency, resources) are often particularly well integrated where substantial public landownership and development is involved, e.g. in action areas such as Coventry City Council's Eagle Street and Eden Street plans; Leicester City Council's Abbey District Plan; the Beckton District Plan (Newham LB).

The inclusion of phasing and programming information within local plans raises the more general issue of how local plans and the policies and proposals they define can be kept up-to-date. Circular 22/84 suggests that formal plan alteration, repeal or replacement will be necessary where the local plan is no longer in general conformity to the structure plan; where the plan needs updating to take account of a proposal for a major development with substantial implications for the plan area or of the cumulative effects of departures where these are tending to distort the plan; or where the Secretary of State directs the planning authority to prepare proposals for local plan alteration, repeal or replacement (DoE, 1984g, para. 3.90). The procedures for local plan alteration, repeal or replacement are the same as for the adoption of the original plan. There is at present only limited practical experience with the formal review of local plans, although some proposals for alteration have been made, e.g. by Camden LB and Watford BC to their Borough Plans. Nonetheless, local plan review can be expected to become of increasing concern to local planning authorities as the context in which adopted plans operate changes. Pressures for plan review will stem from experience of the use of adopted policies in development control and the effect of appeal decisions; from progress in implementing plan proposals; from changes in the locational pattern of private investment; and from changes in the priorities and programmes of public agencies (Fudge *et al.*, 1983). Structure plan reviews will alter the strategic policy context and may require adjustments at the local level. Fudge *et al.* suggest that many authorities, having gone through the statutory adoption procedures, are unwilling to face the same procedures for altering the plan, while there may also be difficulties in mobilizing and justifying resources for plan review where plans are seen as once and for all statements. One alternative to the formal review of local plans is the process of informal annual reviews undertaken by Gloucester City Council in connection with the Gloucester Local Plan (Stuart and Beaumont, 1984). Each annual review considers progress in implementing the objectives of the adopted plan, whilst also updating policies, programmes and expenditure. The following advantages are claimed for this process: ability to monitor policies on a regular basis; revision of information and inclusion of new trends; amendment of policies and proposals where necessary; provision of a comprehensive base for budgeting; provision of a vehicle for bringing forward major new proposals within an overall context. The reviews are presumably intended as non-statutory supplementary planning guidance and Stuart and Beaumont report that the DoE Regional Office has supported the Gloucester approach. However, the reviews appear to go beyond the limits of such guidance as envisaged by the DoE in Circular 22/84 and by the Court of Appeal and the House of Lords in the Westminster case, particularly with regard to the inclusion of new land use proposals.

A comprehensive, integrated and up-to-date set of policies and proposals is only a starting point so far as their implementation is concerned. In considering exactly how a plan strategy is to be carried forward, plans have to have regard to the linked questions of resource availability and the attitudes of implementing agencies. One way of seeing the problems that these questions have raised is as the difficulties posed for the effective and continuous management of change within the relatively fixed and static policy framework provided by statutory local plan provisions. These difficulties have arisen notwithstanding the fact that the 1974 Regulations, for example, required local authorities to include in the written statement 'such indications as the local planning authority preparing the plan may think appropriate' of the regard paid to the resources likely to be available for carrying out the proposals formulated in the plan, as well as to the extent and nature of the relationship between the proposals.[12] The 1982 Regulations make no mention of these factors, although Circular 22/84 does emphasize the need to address resource questions in local plan preparation (DoE, 1984g, paras. 4.8 and 4.9). Some plans avoid a direct treatment of resources by claiming that all their proposals can be implemented within the timescale of the plan, omitting estimates of the availability of finance or of the cost of proposals (e.g. the Bromsgrove Local Plan, Bromsgrove DC; Leicester City Council's Soar Valley South and Aylestone Local Plan). Most plans, however, provide some treatment of the financial aspects of their proposals, while a number take a wider view of 'resources' and consider such factors as land availability, the capacity of the local building industry, labour supply, or even the plan's implications for the staffing of the development control function.

As far as coverage of financial resources is concerned, Barnard (1980, p. 63) comments that this 'has often been superficial with little quantitative analysis', e.g. some plans give a description of the sources of finance, but fail to show how these are expected to relate to plan proposals, which remain uncosted. Many local plans acknowledge that the implementation of their policies and proposals will require a combination of public sector expenditure and private investment. The former is commonly divided into expenditure incurred by the local authority and that incurred by other public agencies such as statutory undertakers, so that three categories of implementing agency are recognized. Of these, the treatment of private investment is usually the most simplistic, often being limited to a statement to the effect that resource availability is here largely outside the influence of the local planning authority, being determined instead by national and local economic forces. To be realistic, therefore, plan proposals dependent upon the private sector for implementation must be flexible in terms of the timing of development. This must be balanced against the need to provide certainty in terms of which private development proposals will be acceptable to the local planning authority over the plan period. Thus, the Birmingham Central Area District Plan recognizes the difficulties of predicting or controlling private sector investment, but points out that a key factor in private development initiatives is the degree of confidence in the future that is felt by investors, and proposes to encourage private enterprise by, for example, the limited use of public sector resources for 'pump-priming' activities, and stability in land use plans and other public sector programmes.

More detail is usually given on anticipated expenditure by the local authority preparing the plan and by other public agencies, but a full comparison of the likely availability of resources and costings of proposals is rare. The treatment of local authority expenditure more usually comprises a broad descriptive account of local authority finance on the one hand, and, on the other, schedules of costings normally limited to development already programmed (i.e. committed) by the service departments of the authority in the early years of the plan period. The links between the two remain unspecific, e.g. 'having regard to the financial resources available the Council will put forward further proposals from time to time in accordance with the general policies of the plan' (Haringey District Plan, para. 10.7). Where general accounts of local authority finance are provided, these typically emphasize the constraints and uncertainties contingent upon central government restrictions on local authority spending, as well as underlining the point that plan proposals are subject to the usual council financial appraisals and budgetary procedures (e.g. Gloucester Local Plan; Westminster District Plan).

As far as expenditure by other public agencies is concerned, detailed costings (and timings) are rarely available for inclusion in local plans except where already programmed. Again, some plans avoid this problem by only including development projects as plan proposals where the agency involved has confirmed that the financial resources will be available and that work will begin within the 10-year plan period (City of Westminster District Plan). One problem here, however, concerns the level of commitment which public agencies can be expected to give to local plan projects and proposals, particularly where capital expenditure is involved. In the context of the preparation of the Beckton Local Plan, this difficulty has been characterized by Byrne (1978, pp. 185–6) as follows:

Local plans are statutory documents, putting forward policies and proposals for a ten-year time horizon . . . . However, outside the Planning Department a different system of planning is gathering force. This is a system of 'sectoral planning' which contrasts with the 'areal' approach of the Local Plan . . . Generally speaking, these plans consist of general strategy statements accompanied by detailed short-term programmes of action, which are regularly reviewed and rolled-forward . . . . service 'planners' are bidding for and allocating resources on a different time-scale [from the statutory local plan] and possibly without any 'areal' appreciation of their activities. The crucial issue is that increasingly, the finance for public sector development will emerge from the sectoral policy and programme systems . . . [the local plan and sectoral policy] systems could develop along parallel, separate paths, with little relationship established between proposals on the one hand and resources on the other.

To reconcile this potential conflict, the Beckton plan sought to consider programming and implementation in some detail. However, two problems emerged (Byrne, 1978, p. 186):

Firstly, the programming information was immediately 'bled off' into Service Plans as soon as it appeared in the draft version of the Local Plan. As a result, commitments have been made well in advance of the Local Plan being adopted. Secondly, the programming information dated quickly. To be of use, it should be annually updated as a whole. In practice, the only chance of updating was between the draft version of the Plan and the approved Plan for 'deposit'. Now that the Statutory Local Plan is locked on the conveyor belt of deposit–public enquiry–adoption–modification, it becomes increasingly less useful as an input into the individual sectoral Service Plans.

The activites of service departments and other spending agencies may thus appear to be initially plan-led, either because the plan draws together pre-existing commitments or because the process of plan preparation does actually influence resource commitments. But given the timescale of modifying local plans and the primacy of sectoral policy systems in allocating resources, the relevance of the adopted plan as an influence on the spending activities of other public agencies can be expected to steadily diminish. Healey's verdict (1983, p. 142) is harsh but accurate: for most 'government departments and public agencies, local plan considerations are at best peripheral to the primary concerns and, at worst, an irritating infringement of their own autonomy to determine the principles and priorities for allocating resources'. Faced with these uncertaintities (over the national economic context, central government allocations, and the commitment of other agencies), it is hardly surprising that local plans have tended to limit themselves to specifying already-programmed development, and otherwise simply indicating the general constraints on resource availability. However, there would seem to be no reason why programming information could not be progressed and updated as supplementary planning guidance prepared within the context of adopted local plan(s), since such statements by themselves would not act to introduce new land use policies and proposals. This begs the wider question of the adequacy of local plans as guiding summaries of the intentions of other public and private sector agencies. Indeed, the difficulty of assessing the availability of resources to implement local plan proposals is actually one specific example of the problems that face local planning authorities in attempting to use local plans to secure the co-ordination of physical developments by public and private sectors at the local level. Local plans by themselves commit no one (not even the local authority's own service departments such as housing) to carrying out the developments which they propose, nor do they make the resources available for such developments. In attempting to manage change in the environment, local planning authorities have thus had to fall back onto the other means available to them of restricting or regulating the use of land – primarily development control, in which the provisions of the local plan will be only one of a number of considerations, but also including various forms of 'planning by agreement' (Grant, 1982). Such agreements may be based on contract or proprietary relationships (where the authority has freehold ownership or the developer is willing to enter into a sale and leaseback arrangement with the authority), or can be made under s.52 of the 1971 Act, which allows authorities to enter into formal agreements for planning purposes. None of these methods for the restriction or regulation of land use *need* to be specified in a local plan before they can be concluded, but all can be used in the implementation of local plan proposals. However, these various methods provide only a limited degree of effective control over the timing or occurrence of development, and hence only allow a limited extent of co-ordination to be achieved by the local planning authority.

As far as development proposals to be implemented by the private sector are concerned, for instance, all these means of controlling land use are initially dependent, to a greater or lesser degree, upon private development initiatives actually coming forward. Where local plans address this problem, it is usually recognized that this is an area over which they have little control (e.g. Sutton Coldfield District Plan, City of Birmingham). One way in which authorities have sought to convert their largely responsive role towards private development proposals is by the 'bringing forward' of land for development by the use of supplementary planning guidance in the form of planning and development briefs. The Leamington Town Centre Local Plan (Warwick DC), for example, indicates that policy statements and planning briefs will be issued for various sites to explain the plan's policies and to set out what is required of the developer. One of the key proposals of the plan, for shopping development, has been progressed in this way, using two briefs and a competition to select a developer, the whole process resting not only on development control powers but also on substantial local authority ownership of the site in question (Bruton and Nicholson, 1984b). The use of briefs in this promotional way, of course, still requires a modicum of developer interest to be successful.

As far as development by public sector bodies such as government departments and statutory undertakers is concerned, from the viewpoint of the local planning authority the situation is no more satisfactory. Each of these agencies will have its own service plans incorporating proposals for actual development or with land use implications, and it will seldom be possible for the local planning authority to determine the content of such programmes. Indeed, the planning department may not be able to influence the proposals of other local authority departments, let alone those of agencies with a much larger remit than a single local authority area. As far as bodies outside the local authority are concerned, local planning authorities preparing a local plan are required to consult the other planning authority or authorities for the area and to take their views into consideration, and they are also asked to consult government departments or other public bodies carrying out responsibilities which are likely to be affected by the plan's proposals.[13] However, the following quotation from the development plans memorandum (Circular 22/84) hints that local plans will largely act to bring together and summarize already established proposals, rather than actually having any substantial impact on the direction of these programmes: public authorities and bodies 'should be consulted on the matters which concern them to ensure that where relevant the policies of these other bodies are accurately represented in the local plan and that any potential conflict is resolved' (DoE, 1984g, para. 3.40). Byrne (1978, p. 186) comments, again in the context of the Beckton Local Plan, that while the Borough Council's 'house building programme in Beckton will be a major component of its overall housing programme . . . for other agencies, housing development in Beckton will be only a relatively small part of their activities, very susceptible to external adjustments'. To realize local plan proposals 'an inter-agency programme is required, setting out "who is doing what, when and how", and this needs to be regularly reviewed and rolled-forward. The statutory Local Plan cannot perform this task, since its primary concern is with policies and proposals, not programming . . . . This suggests the need for an annual inter-agency investment programme' (Byrne, 1978, pp. 186, 189). Such a programme would offer a means of co-ordinating public sector investments in the environment, and of managing this aspect of environmental change in accordance with local planning policies. Not only do local planning authorities have no power in local plans to determine the when and how of development proposed by other public agencies, but the development control process – the main tool available to local authorities in implementing local plan proposals – is of only restricted relevance when public sector development proposals do come forward. The Crown, for instance, is not bound by planning legislation, so that there is no legal obligation for government departments (including health authorities) to obtain planning permission for their developments. However, a formal requirement exists whereby departments consult local planning authorities before embarking on development which would, but for the exemption, require permission (DoE, 1984f). The operation of development control is also circumscribed, but to a lesser extent, when development proposals stem from the local authority itself (e.g. from a service department such as housing and education) or from the statutory undertakers. For instance, many of the development demands of the statutory undertakers arising from the carrying out of their function of service provision have deemed permission under the 1977 General Development Order.[14] The relative independence of statutory undertakers from local plan proposals creates a twofold problem for plans, since infrastructure provision by such undertakers as the water, gas and electric authorities is likely to be a prerequisite for any substantial new development, while these and other agencies such as British Rail may have large-scale landholdings whose release is needed to secure development. Yet the local planning authorities' powers over these bodies is limited, and in this context plans often restrict themselves to hopeful statements, e.g. 'the British Waterways Board and British Rail are expected to play a key role in securing and promoting development. Some of their substantial land ownerships in the Riverside will need to be released if all the development proposals are to proceed' (Leeds Central Business Area District Plan, para. 9.16).

4.7   The bringing of local planning issues before the public

4.7.1   Introduction
This final role for local plans differs from its predecessors in that it is primarily addressed during local plan production. The manner and degree in which local plans have carried out this role is thus linked to *procedures* rather than to *content*. Nevertheless, the way in which a plan is produced and presented will still be relevant. Much of the detailed DoE advice on presentation has this in mind, e.g. emphasizing that the plan should be so laid out as to enable the public to ascertain whether their property interests are likely to be affected. Factors here include scale and comprehensiveness of the proposals map (DoE, 1984g, paras. 3.26 and 3.29). Other aspects of presentation, such as clearly distinguishing proposals from the reasoned justification and ensuring that the plan is self-contained, will also help to clarify the likely effects of the plan to the public (Barnard, 1980). Advice and guidance on these issues has been a long-standing feature of the legislative and administrative framework for local plan preparation. For this reason, it is probably fair to assume that most recent local plans are reasonably comprehensible policy documents, although Healey (1983, p. 205) warns that 'the way plans present policies may obscure rather than reveal local planning policies except to those with a sophisticated understanding of the development system'. If this is the case, then the more active stages of public involvement established in the statutory provisions for local plan preparation (i.e. public participation and the public local inquiry) will become correspondingly more important.

4.7.2   Public participation
The statutory requirements for public participation are simply that the local

planning authority should give adequate publicity in the area to the matters proposed to be included in the plan, give those interested an adequate opportunity to make representations, and make them aware of this opportunity.[15] The six-week period for publicity and public participation is prescribed in the Regulations.[16] The DoE advises that while authorities may undertake further work, this 'will not normally be necessary to meet the statutory requirements and should only be undertaken where the authority are satisfied that the additional work and delay entailed is clearly justified' (DoE, 1984g, para. 3.32). This is to see the process of public participation as a discrete and technical stage in local plan preparation. However, Bruton and Lightbody (1980b) have argued that the view of participation as a technical exercise concerned primarily to improve communication between local authorities and the public, allied with a lack of concern with the wider theoretical underpinnings of participation (Damer and Hague, 1971; Hampton, 1977; Thornley, 1977), has led local planning authorities to conduct participation exercises within a framework which ignores the realities of power and political decision-making and is oblivious of the basic requirements for successful communication. As a result, such exercises have under-emphasized (*a*) the problems of reconciling public participation with representative democracy; (*b*) the fact that the introduction of public participation into planning changes the way in which decisions are made; (*c*) the fact that conflicts of interest will arise; and (*d*) the fact that the main aims of many of the participants is to influence the decision-makers (Bruton and Lightbody, 1980b, p. 58).

Bruton and Lightbody (1980b, pp. 57–66) suggest that many of the problems encountered in practice could have been anticipated through a consideration of the theoretical political, communication and social issues which are implicit in participation. In brief, these theoretical issues and their implications for programmes of participation are as follows:

1 *Political theory* identifies different forms of democracy ranging from little or no participation (elitist) to participatory democracy, which is characterized by wide discussion and consultation. These different concepts of democracy embody different levels of power transfer from the elite decision-makers to the public. Arnstein's (1971) 'ladder of participation' draws on these notions, setting total control by the public at the top of the ladder and manipulation of the public by the decision-makers at the bottom. Political theory identifies the fundamental question facing elected members in our system of representative democracy as the extent to which power should be transferred to the public in a participation programme.

2 *Communication theory* (e.g. Berlo, 1960) offers basic guidelines to aid the planner in drawing up programmes of participation: there should be two-way communication between the planner and the public; more than

one channel of communication should be used; and the channels of communication should have multiple entry points into the communication system, in order that 'message blockages' caused by the failure of intermediaries to pass messages on may be bypassed. The planner should also ensure that the 'code' in which his message is transmitted to the 'receiver' is known or understandable, and that the content of the message is meaningful.

3 *Social theory* identifies three broad perspectives on the ways in which social interaction takes place; these perspectives are neither mutually exclusive nor mutually exhaustive. Each perspective envisages a different role for public participation. These perspectives are:

(a) *Consensus* (e.g. Parsons and Schills, 1951) whereby society is a stable system characterized by a common acceptance of culture, values and organization and planning is a technical apolitical exercise conducted in the public interest; social and other problems are wholly attributable to breakdowns of communication between decision-makers and the public; hence participation can be used to overcome such communication problems, thereby leading to consensus.

(b) *Pluralist* whereby society is seen to consist of diverse groups with differing interests and values and planning is essentially a political activity where politicians are the final arbiters; social problems arise from the under-representation of certain interests in the democratic system; participation offers a channel for these groups to influence decision-makers, with the planner acting as an advocate (e.g. Davidoff, 1965).

(c) *Conflict* whereby the existence of incompatible interests in society and a continuing demand for scarce resources ensures that conflict is endemic; social problems arise out of fundamental conflicts between groups; public participation is an overtly political activity where participants are concerned to gain a redistribution of power and resources.

The views expressed by government when public participation was first introduced into the town planning system in the late 1960s show an implicit acceptance of the consensus perspective. Thus the then Chief Planner at the Ministry of Housing and Local Government stated that one of the main aims of the new planning system proposed by PAG was seen as '... increasing public understanding of the system ....' and 'simplifying administration ....' (James, 1965, p. 22), while Skeffington stated unequivocally '... we see the process of giving information and opportunities for participation as one which leads to greater understanding and co-operation rather than to a crescendo of dispute' (Ministry of Housing and Local Government, 1969, p. 5). However, critics of the consensus view such as Simmie (1974) argue that it provides an inadequate perspective on the nature of society and of change therein. Society is clearly not made up of a collection of groups who all ascribe to the same values. This fact implies that public participation, far from leading to agreement and consensus, is more likely to provide a framework for the articulation and resolution of conflicts of interest. Here the pluralist and conflict perspectives suggest that the values held by different groups will influence their attitudes to different issues; that different groups will view the same issue in different ways, and that different groups will perceive the same problem differently. These perspectives thus lead to an alternative view of the implications of introducing participation into the town and country planning process; rather than leading to consensus, participation 'inevitably leads to conflicts of interest and results in a change in the balance of power insofar as the taking of decisions on local issues is concerned' (Bruton and Lightbody, 1980b, p. 155). Moreover,

The formal introduction of public participation in town planning would therefore seem to provide a framework within which these conflicts [of interest] can be clearly and formally articulated, rather than a device to achieve consensus . . . experience of the outcome of public participation exercises in planning in the recent past would seem to suggest that motivation and an ability to bargain are . . . important in resolving the conflicts inherent in the distribution of resources through local planning activities' (Bruton and Lightbody, 1980b, pp. 156–157).

In effect, public participation is an integral part of the process of policy formulation and of the management of change in the environment.

Local plans intervene in private rights of development at a level of detail at which specific sites (and therefore specific interests in land) are identified. Public participation provides a framework within which conflicts of interest arising out of this intervention can be discussed and negotiated with those concerned. As Bruton and Lightbody (1980b) show this is likely to entail a lengthy process of bargaining which may only be curtailed through arbitration at the public local inquiry. An additional role for participation is to generate support for local plan policies, especially where there are considerable conflicts of interest over land use and development issues. Indeed, 'the fact of engaging in an apparently public debate on policy alternatives helps to legitimate the policies . . . the process may also co-opt some sections of the public into giving support to policies' (Healey, 1983, p. 139). However, the opportunity to debate 'alternatives' on a meaningful basis is likely to be limited where, as the DoE recommends, only limited participation is undertaken on the basis of a draft plan. The full benefits of legitimation may only be secured through more extensive participation than statutorily required. Fudge *et al*. (1983) identify various 'contextual features' underlying local authority participation strategies where the minimum requirements were exceeded:

1   The demand for participation in different areas will vary, usually being greater where there are pre-existing groups, such as amenity associations, which expect to be centrally involved in policy formulation. Organizations formed to oppose particular proposals will also require more time to be spent on negotiations.

2   Local authorities have a duty to ensure accountability in matters affecting private property stemming from the judicial role of local plans.

3   Councillor attitudes and 'open' styles of administration may lead to a tradition of continuous contact with the public and interest groups, e.g. Warwick District Council (Bruton and Lightbody, 1980b).

4   The time spent on public participation is also related to the controversial nature of the planning issues involved. For example, in Hampshire extensive public consultation was required on proposals for growth due to the degree of local opposition.

Local plan proposals for radical change can be expected to generate hostility, and public consultation provides local authorities with a means of smoothing their progress (Grant, 1982), especially in terms of reducing the number of objections subsequently laid against the deposited plan. However, the effectiveness of public participation is difficult to estimate in this respect. Bruton, Crispin and Fidler (1982) report that, although all of the local authorities included in their study considered that their participation programmes had gone further than the statutory requirements and given ample scope for the public to comment, approximately 50 per cent of objectors at the subsequent public local inquiries had not been involved in the respective participation programmes. A plan might attract few objectors because of genuine attempts by the local authority to elicit and meet the views of potential objectors, but conversely participation might only serve to stimulate greater public awareness of the plan and thus potentially *increase* the number of objections (Barnard, 1980). There is also evidence to suggest that some parties, such as landowning and development interests, deliberately wait until the formal inquiry stage before objecting to the plan, perhaps because participation is by its very nature open-ended and not necessarily conclusive. Moreover, however comprehensive the participation programme, the inherently distributional and political nature of local land use planning decisions, particularly where controversial proposals are involved, will usually mean that objections are inevitable. The possibility that a public local inquiry could be avoided through a comprehensive and lengthy programme of participation thus appears to be slim (Bruton, Crispin and Fidler, 1982).

The most common methods of securing public participation are through public meetings, questionnaire surveys, the distribution of leaflets and news-sheets, exhibitions and the use of local press and radio. These methods cover both the giving and gathering of information from the various public groups who are to be consulted, including local amenity and 'interest' groups, Chambers of Trade and local industrial and commercial firms; members of the public with residential or property interests in the plan area; and the wider public (Bruton and Lightbody, 1980b, pp. 66–68). The public response to these traditional 'mass' methods of participation has been characteristically low, but will usually be higher where controversial proposals are involved.

Barnard (1980) points to various innovations, designed to elicit a more considered response or to involve people missed by the traditional approaches, e.g. community panels, area discussion groups and community forums. The study of plan preparation by Fudge *et al.* (1983) identifies a number of difficulties with the way in which local authorities have organized their participation programmes:

1 Inadequate consideration of the timing and purpose of consultation, e.g. participation taking place before proposals are fully formulated.
2 Inadequate thought given to the aims of the exercise and a tendency to use standard methods, leading to a low or predictable response.
3 Staff availability/expertise can affect the duration of consultation exercises, particularly for small authorities. The geographical area covered by a plan (and hence the number of organizations wishing to be considered) is also a factor affecting the length of participation programmes, e.g. for one authority participation on the draft district-wide plan took four months.
4 Both reports of survey and draft plans are commonly too technical and/or generalized to promote public discussion.
5 It is difficult to confine the response to matters dealt with in the plan, e.g. the public might raise strategic issues.
6 Handling the response can itself be time-consuming.

In most cases practice is responding to these problems by developing more sensitive approaches to participation in terms of timing, method, and staff resources. Fudge *et al.* confirm that public consultation does not guarantee that all issues are raised or that all objections are made before the plan is placed on deposit. Commenting that this may be due to inadequate consultation, to insufficient appreciation of the strength of public feeling, or to the practice among property interests (landowners and developers) of reserving their comments to final objection stage, they suggest that whatever the reason 'if there is a real concern with accountability and the local authority wish to ensure that those wanting to object know of their rights then publicity at this stage (deposit) would seem to be as important as that at draft stage' (Fudge *et al.*, 1983, p. 65). To this end, South Staffordshire DC carries out limited publicity at deposit stage; this is also seen as a way of minimizing late objections.

Finally, it is interesting to note a quotation from the Skeffington Report

(Ministry of Housing and Local Government, 1969, p. 3) which implies that, for pragmatic reasons, public participation in planning was introduced in part as a response to the complex and 'wicked' nature of planning problems and the need to attempt to take account of contingent factors in dealing with those problems:

Life . . . is becoming more and more complex, and one cannot leave all the problems to one's representatives. They need some help in reaching the right decision, and opportunity should be provided for discussions with all those involved . . . Planning is a prime example of the need for this participation, for it affects everyone . . . This becomes all the more vital where the demands of a complex society occasion massive changes; changes which in some areas may completely alter the character of a town, a neighbourhood or a rural area. The pace, intensity and scale of change will inevitably bring bewilderment and frustration if people affected think it is to be imposed without respect for their views.

Public participation in local plan preparation: a case study of Warwick District Council [17]

In recent years, bargaining has gradually come to be seen as an integral part of the planning process (Bruton, 1980; Fudge *et al.*, 1983; Healey, 1983), and in Chapter 2 reference has been made to the principles of distributional bargaining. If planning is about conflicts of interest, and if conflict involves distributional bargaining, then it is reasonable to assume that in any formal process of local plan preparation, some or all of the features of distributional bargaining will be in evidence. Thus it should be possible to identify:

1   which parties in the exercise adopt a commitment, and how this commitment is communicated to the other parties in the exercise;
2   whether or not bargaining agents are established and used;
3   whether intersecting negotiations and/or restrictive agenda are used;
4   the extent to which compensation and/or arbitration are used to reach agreement.

With these issues in mind, the process followed by Warwick District Council in preparing two local plans was analysed as part of a five-year research project undertaken for the DoE on public participation in local planning (Bruton and Lightbody, 1980b). The plans concerned are the Southtown Plan (an informal action area type of plan produced for an area in physical and economic decline immediately adjacent to Leamington town centre) and the Lapworth District Plan (a plan for an area of Green Belt on the edge of the West Midlands conurbation under pressure for residential and recreational development). The major protagonists in both plans were the District Council, the County Council and local groups who opposed certain policies in the plans.

Evidence from the project shows that pronounced conflicts of interest developed during the process of plan preparation, whilst in both cases some form of distributional bargaining was undertaken. All parties attempted to win relative to their value systems, e.g. the District Council by producing plans for areas with perceived problems and attempting to establish with the County Council its right to produce local plans for its area; local resident opposition groups by attempting to secure the rejection of policies which harmed their interests. The 'opposing' groups adopted clear commitments against policies put forward in the plan and communicated their commitment to the District Council, e.g. through the press; protest meetings; and correspondence. They also established resident groups which were used as bargaining agents. By contrast, the District Council did not publicly adopt a counter-commitment to the opposition groups – presumably because it was, at the time, in the process of consulting the public about draft local plan proposals and had to be seen to be keeping an open mind on the matter. However, the dealings between the County and District Councils over the two plans make it quite clear that the District was committed to demonstrating its right and ability to prepare local plans in its area, whilst the County was committed to ensuring that any local plans produced were in accordance with policies contained in the structure plan. In the words of McAuslan and Bevan (1977, p. 10) writing about the Warwick experience with Southtown, '. . . it is essential to realize that as much as being about substance, each plan is staking a claim for the power and influence of the authority who prepare the plan, the Southtown exercise being no exception'.

Throughout the negotiations, all parties attempted to avoid mutually-damaging behaviour, e.g. the District Council made concessions over traffic-management proposals in the Southtown plan and residential development in the Lapworth Plan, partly to avoid the risk of 'damage' at the then forthcoming local elections. Similarly, opponents of policies in the plans appeared to be guided by their expectation of the likely attitude of the District Council towards conflict and pressure, e.g. similar traffic-management proposals had been successfully opposed in the past in the Southtown area and the opposing groups were convinced that they would be successful in connection with the proposals in the Southtown plan.

The District Council resorted to the use of an arbitrator to resolve conflicts which reached deadlock, e.g. the DoE was involved to resolve the issue of county/district responsibilities over the Southtown plan, while a public local inquiry was used to resolve the deadlock between speculative builders and the District Council over the extent of proposed residential development in the Lapworth Plan. No attempt was made to use compensation, or intersecting negotiations.

Although the statutory bodies appeared to have the greatest scope for exercising power through their statutory powers and financial and other resources, the informal opposition groups exerted considerable influence during the plan preparation period. Indeed, this influence was such that it fully achieved the objectives of the groups – the abandonment (albeit temporarily) of the proposals for Southtown and the acceptance of only limited residential infill development in Lapworth. This can be explained on two counts – firstly, the informal associations were highly motivated to oppose the plan proposals and, secondly, they were most proficient in conducting their bargaining to achieve their objectives.

One noticeable feature of the two case studies was the limited involvement of what might be termed the 'general public' in the distributional bargaining aspects of the participation process. The main participants were the statutory bodies, pressure groups, and large commercial undertakings – the major and minor elites. This, again, is to be expected if distributional bargaining is one of the main themes underpinning local plan preparation. No place will be found at the bargaining table for those without either statutory powers or a strong commitment forcefully argued. If the situation which emerged at Warwick District Council is typical, the implications for planning decision-making are considerable. Until such times as it is accepted that plan-making is about conflict, and that consensus is derived only through conflict, the non-politically accountable major elite groups in society will exert a dominant influence on planning decisions – an influence which is concerned to further what they as a group value.

To redress this imbalance, a marked change in attitude towards the role of plan preparation and public participation is needed on the part of both politicians and professional officers. Public participation needs to be seen as part of the decision-taking process rather than as a process which will eventually lead to agreed and acceptable decisions. It needs to be seen as an aid to informing the political decision-takers of the attitudes of certain elite groups to planning proposals. It should be seen as a process which can lead to distributional bargaining; to pressure; to trade-offs, where the role of the elected member is to balance these pressures against the wider public interest.

To enable this arbitrating role to be performed successfully, the politicians need to be centrally involved in evaluating the response of the public to local plan proposals. It would seem equally desirable for the politicians not to be involved in the details of plan preparation until the public participation programme is operational. In this way the adoption of a commitment to particular aspects of a plan is avoided and the politician is less biased in evaluating the response of the public and balancing the wishes of pressure groups against the wider public interest.

Such an approach to public participation would strengthen the role of the politicians in the planning decision-taking process. The introduction of public participation in town planning may well have led to a better understanding by the public of planning and the planning system. It has not speeded up the process of preparation and approval. Rather it has substituted a form of conflict resolution which enables the major and minor elites to achieve their objectives more easily, at the expense of the public interest, however defined.

4.7.3 The local plan public local inquiry

The stages in the local plan preparation process from deposit to adoption are often seen by local planning authorities as lengthy and problematic. These stages centre on the public local inquiry (PLI), held to hear objections against the plan made during the deposit period. The procedures laid down in the 1982 Regulations for local plan adoption are specific and detailed, largely because of the fact that local plans are statutory documents affecting individual rights in property. The average length of time taken for a local plan to reach the deposit stage has been put at 36 months, with a further 16 months required to reach the adoption stage if an inquiry is held; in general terms, the need to proceed to a PLI can add in the region of 10 months on average to the process (Bruton, Crispin and Fidler, 1982, p. 282). This feature of the statutory requirements is unlikely to encourage local planning authorities to proceed their plans to deposit and adoption unless the support of a statutory plan is considered necessary, e.g. in the face of complex conflicting interests. Instead, such plans may be used on an informal basis. Notwithstanding this point, Fudge *et al.* (1983) report that authorities which have progressed plans through to statutory status have found the procedures less difficult than expected, while they also value the increased status of a plan which has been subject to thorough examination in an open public forum. In general, arguments against the PLI which focus on its time and resource costs (see for example District Planning Officers' Society, 1982) tend to give insufficient emphasis to its role in overall plan preparation and legitimation. The statutory procedures which allow for objections to be made to a plan and for the holding of a public local inquiry have several functions in practice (Healey, 1983, p. 158). These are:

1 Identification and final resolution of issues still outstanding after plan preparation and participation.
2 Continuation of the negotiative processes between the local authority and interested parties.
3 Provision of an arena for continuing public debate about local authority proposals, especially where there is conflict with particular groups.
4 Provision of an opportunity for property interests to identify and object to adverse consequences of the plan as it affects their interests (even though they may also appeal at the level of planning applications or purchase proposals).
5 Provision of an opportunity to review the competence of the plan as drafted, thus providing a final external check on both local authority *ability* and *discretion*. Here, the DoE may be involved as an objector on policy matters.

Although clearly the degree to which each of these roles can be distinguished in individual inquiries will vary from case to case, together they emphasize the links between the PLI as the principal forum for plan scrutiny and the preparation process in general.

As we have seen (Chapter 3), local plan PLIs differ from other inquiries within the statutory land use planning process, such as those held to determine appeals against development control decisions, by virtue of the fact that the final decision on whether or not to modify the plan rests with the local authority rather than the Inspector. The local plan PLI must thus be seen as 'primarily an administrative device to assist the local planning authority in reaching decisions on the most appropriate set of solutions for the problems addressed by the plan' (Bruton, Crispin and Fidler, 1982, p. 277). A further distinctive feature of the local plan PLI is that the inquiry takes place within a plan preparation process which embodies statutory provisions for public participation, so that the role and purpose of the PLI has to be viewed in the context of this process in general and of public participation in particular. In contrast, other planning inquiries are discrete events at which those excluded from the policy-making process can challenge the resultant decisions on a definitive basis. Finally, there are statutory regulations governing other types of local inquiries, a factor which encourages a formal and structured approach to their conduct; local plan PLIs however are guided instead by a non-statutory Code of Conduct, which allows the Inspector scope and flexibility to adapt procedures as he or she considers appropriate (DoE, 1984h). These distinctive features of the local plan PLI have led to uncertainty among participants as to whether the proceedings are judicial or administrative in nature (Bruton, Crispin and Fidler, 1980). Inquiries in general are typically and traditionally judicial or quasi-judicial events in an administrative process of decision-making, but the local plan PLI emphasizes the latter rather than the former, primarily as a result of the decision-taking responsibility of the local authority. Bruton, Crispin and Fidler (1982) report that local authority officers typically view the PLI as a device whereby policies and proposals in the plan, and objections to them, are explored in depth, the role of the Inspector being to provide independent advice on the most appropriate policies to adopt. Many objectors in contrast view the PLI as a judicial instrument capable of providing a definitive ruling from the Inspector on unresolved conflicts of interest. They therefore see the PLI not as the final stage in public participation but as a discrete event outside of this process. However, this interpretation of the PLI is founded on a misconception of the formal decision-taking relationships between Inspector and local authority. Bruton, Crispin and Fidler (1982) report substantial confusion on this issue among objectors, with a large proportion apparently assuming that the Inspector takes the final decision. Subsequent clarification of the actual role of the Inspector demonstrated considerable dissatisfaction among objectors with a situation where the local authority is responsible for sitting in judgement on proposals which it has itself put forward. This is a problem which threatens the credibility of both the PLI and of local plans, and is likely to remain despite the usual local authority practice of accepting the bulk of Inspectors' recommendations (Perry *et al*, 1985).

The conduct of local plan PLIs will differ from inquiry to inquiry due to the fact that the proceedings are governed by a non-statutory Code of Practice rather than a set of statutory rules. This allows the Inspector to adapt the procedures to suit local circumstances, setting the tone of the PLI accordingly, e.g. as to the degree of formality appropriate. There is no possibility of an appeal to the courts if the Code is not followed. Experience with the Code indicates that the conduct of PLIs has overall been fair, orderly and impartial, with 'substantial agreement amongst the officers and objectors . . . that the Inspectors have handled the proceedings with considerable skill, clear impartiality and according to the circumstances treating, for example, unrepresented objectors rather different from counsel' (Bruton, Crispin and Fidler, 1983, p. 283). Local authority satisfaction with the proceedings is likely to be a factor in determining the way in which the subsequent recommendations of the Inspector are treated by the authority. The most common complaint of unrepresented objectors concerns the extensive use of legal representation by local authorities. This creates a stark contrast between the formal proceedings at the inquiry (possibly including cross-examination) and the pre-inquiry informal contacts that the objector may have had with local authority officers. However, Bruton, Crispin and Fidler (1983) found no evidence that the unrepresented objector has had his case weakened by lack of professional advocacy, largely because Inspectors have demonstrated the utmost consideration, carefulness and fairness towards such objectors (see also Marsh, 1983). Nevertheless the widespread use of advocates by local authorities and some objectors at PLIs potentially generates an atmosphere which is essentially hostile and adversarial, and this is at odds with the place of the PLI within a planning process which actively encourages the public to participate in plan preparation.

The content of local plan PLIs can also be expected to vary, primarily because different local plans address different planning issues. Most objectors appear to find out about their right to make an objection via the local authority, e.g. deposit may be publicised extensively or awareness generated by the earlier period of public consultation. Those interests who have deliberately chosen not to become involved in participation will also be awaiting deposit so as to make their objections. The Code of Practice provides information on the rights and method of objection to deposited local plans, but as many as one-third of all objectors may go through the inquiry process without seeing a copy (Bruton, Crispin and Fidler, 1983). The DoE (1984h) now advises local authorities to make copies of the Code available at as early a stage as possible. Knowledge of rights of objection is obviously fundamental to the success of the PLI and of the local plan itself; in particular, objections to the plan are best dealt with at PLI stage rather than via appeals against refusals of planning permission based on the adopted plan. Objections should be made in accordance with the 1982 Regulations, which in practice means that 'all objections made to the plan within the prescribed six-week period are duly made and must be considered by the planning authority' (DoE, 1984g, para. 3.60), provided they are made in writing and state the matters to which they relate and the grounds on which they are made. The statutory provisions are for objections to be made to the local plan as deposited, and this includes both the plan's policies and proposals and the reasoned justification. The Code of Practice emphasizes that 'objectors should be as specific as possible in setting out both the matter to which they object and the changes which they are seeking' (DoE, 1984h, para. 2.5). Objections made on the basis of the omission of policies and proposals from the plan are admissible; objectors can put forward *alternative* proposals or solutions in order to sustain an objection, and suggest *additional* proposals if they consider the plan is deficient in some way. Supporting representations may also be made. The Code notes that objections to deposited local plans should not challenge previously approved structure plan policies, but can relate to the way in which the local plan proposals seek to implement the structure plan. However, counties which only certify local plans under pressure from the DoE may well expect to raise more fundamental objections. Authorities may also consider objections made outside the deposit period; late objections are usually accepted, and even those lodged during the PLI itself may be heard.

During the pre-inquiry stage, the authority has to consider the relevance of objections made to the plan, with the Inspectorate being consulted on ambiguous cases. Objectors may also be involved in these pre-inquiry discussions, with the authority seeking to clarify the nature of an objection, consider ways in which the plan might be changed, or explore the extent of common ground. Through processes of negotiation and compromise objections may well be resolved at this stage. While the final decision as to relevance rests with the local planning authority, the Inspectorate advise that doubtful cases should be admitted and heard at the inquiry, and Inspectors have sometimes admitted late objections without consultation with the local authority (Bruton, Crispin and Fidler, 1983). A wide view of relevance has usually been taken, with challenges to structure plan policy or objections which raise non-land use issues having been accepted and heard at the PLI. Healey (1983) suggests that this receptiveness on the part of both local authorities and the Inspectorate may reflect uncertainties about local plan form and content, the continuation of a consultative style forged during public participation, or a more practical concern to ensure that no interested parties can subsequently claim that their views have been suppressed. As long as an objection is heard and considered there exists no recourse to the courts except that provided by s.244 of the *Town and Country Planning Act 1971*; here, persons aggrieved by a local plan may apply to the High Court within six weeks of adoption to quash the plan in whole or in part, but such an application can only be made on the grounds that the plan is not within the powers conferred by the Act or that the procedural requirements in relation to adoption have not been complied with.

PLIs vary widely in terms of the number of objections involved, the range of issues to which objections relate, and also in terms of who objects. In the sample of 76 local plans adopted after one or more PLI studied by Bruton, Crispin, Fidler and Perry (1984) the average number of objections heard was 105, but the range was from one (Belper–Kilburn Local Plan, Amber Valley DC; Walton Park Subject Plan, Warrington BC) to 1980 (Minerals Subject Plan, Buckinghamshire CC); Healey (1983) reports over 8000 objections having been made to Greater Manchester CC's Green Belt Subject Plan. A wide range of issues are raised by objections at PLIs, which extend from those questioning the philosophy and principles of the plan to those based on the individual grievances of land and property owners. Objections at PLIs can be classified as to whether they concern specific land and property proposals or broader policy issues and aims. The range of objections thus includes procedural objections; substantive objections concerned with allocations of housing and industrial land, Green Belt matters, transportation proposals, and recreational, environmental, conservation and community facility proposals; objections on the grounds of implementation, resources and development control policies and proposals; and those concerning the overall strategy of the plan (Bruton, *et al*., 1982; Marsh, 1983). An extreme example of the latter cited by Bruton, Crispin and Fidler (1983) relates to the Haringey Central Area Action Area plan, where 44 (out of a total of 47) objections were lodged by Friends of the Earth to the central philosophy of the plan, which was seen as concentrating limited resources in one part of the plan area. There is also a wide range of objectors, from residents living within the plan area to major multi-national companies with a financial interest in the area. A distinction can be drawn between those objectors who have a 'title' interest in land and property likely to be affected by proposals, and those objectors who have no direct property interest, i.e. third parties. The range of objectors may include other local authorities, e.g. neighbouring district councils, county councils (perhaps introducing objections concerned with the conformity of the local plan proposals to the structure plan), and parish and town councils; statutory undertakers and central government departments whose land, property or interests are affected; national amenity societies, such as the Civic Trust or the CPRE; local amenity, ratepayers and residents groups, perhaps set up specifically to fight plan proposals; various pressure groups lobbying on specific issues, e.g. housing (House Builders Federation), farming (National Farmers Union) or the environment (Friends of the Earth); and individual firms from the small local to the multi-national (Bruton, Crispin and Fidler, 1983).

Have local plans succeeded in bringing local planning issues before the public through public participation and PLIs? This is not an easy question to answer, partly because generalizations are difficult; the 'success' of local plans in this respect will depend on the scale and nature of the participation programme, the degree of controversy surrounding the issues involved, and the responsiveness of the public. However, the general problems facing authorities attempting to use local plans as vehicles for exercises in 'participatory democracy' include: the minimum statutory requirements (six weeks) for publicity, which the DoE regards as adequate; the difficulty of securing a reasonable and representative response to participation exercises; the constraints set by the need to maintain 'general conformity' to the structure plan; and the dilemma of whether to consult only on the basis of a draft plan (thereby limiting the discussion of alternatives) or to precede this stage with consultation on issues (and risk an equivocal response). Local planning authorities can be expected to meet these problems as they think appropriate, e.g. by extending participation beyond the six-week statutory minimum. In doing so, they are likely to have in mind not so much a disinterested identification of a notional public interest as a concern with the legitimation of the plan as a formal basis for local land use planning.

4.8   An evaluation of the statutory local plan[18]

This review of local plans in practice shows that authorities have taken a wide variety of approaches to the four local plan roles as given by the DoE, reflecting the range of local circumstances and contingent factors which local planning authorities preparing local plans have had to deal with. However, this account of current practice says little about how *effective* local plans have been (or can be expected to be) in meeting the requirements of the hierarchical model of policy control, either as a link between structure plans and development control, or as a device intended to integrate public and private sector development proposals. Accordingly, an evaluation of the statutory local plan is now taken up, which considers the local plan as used in practice against the theoretical framework of strategic planning developed in Chapter 2 and itself implicit in the British development plan system.

In terms of the hierarchical model of strategic planning set out in Chapter 2, the statutory local plan can be firmly located at Level 3, which deals with the local implications of the regional strategies set out in Level 2 above. The theoretical requirements of Level 3, if the link between national socio-economic policies and development on the ground is to be sustained, are that planning documents here should

1   translate general policies and broad proposals into detailed plans where appropriate;

2 establish a basis for the co-ordination of public and private investment
   into development and re-development; and
3 work out in detail the timing of proposed changes and the resources
   required.

These theoretical requirements mirror closely the DoE's own view of the
functions of local plans, emphasizing the hierarchical model of policy control
underpinning the present development plan legislation. The various components of the British planning system may thus be usefully viewed against
the hierarchical model (Figure 4.2) so as to assess strengths and weaknesses.
In evaluating the role of the statutory local plan, therefore, the key question
is how successful have local plans been as a component of such a policy
model? i.e. in translating higher level policies into detailed spatial strategies,
together with a statement of the availability of resources and phasing needed
for implementation.

As far as translating higher level policies into more detailed local
strategies is concerned, the relevant policy vehicle that the local plan is
supposed to supplement is the structure plan. On the technical point of
furnishing detailed site proposals to meet structure plan allocations, local
plans appear to have performed satisfactorily (Barnard, 1980), aided no
doubt by the machinery of development plan schemes, certification and local
plan briefs. In terms of the hierarchical model, these instruments occupy a
position *between* Levels 2 and 3, seeking to control or aid the development
and expression of the local implications of higher level strategies. Both
development plan schemes and certification, formal devices concerned to
ensure consistency in plan preparation and policies between tiers, are a
practical manifestation of the theoretical principle that policies set in the
level above should constrain and guide those in the level below. In this sense,
although they have developed in the context of a county/district split of
planning functions, similar instruments could also be used to effect within a
unitary development plan authority to ensure coherent vertical strategy and
policy formulation. Whatever the value of these linking devices, however,
local plans suffer from one particularly serious handicap in seeking to carry
forward the regional strategies set out in Level 2. This relates to the way in
which socio-economic policies, as opposed to their spatial development and
land use implications, can legitimately find expression at the local level. The
local statements of Level 3 are supposedly based on both the socio-
economic and spatial policies evolved at the level above that setting regional
strategies. In the British system, however, both *national* and *regional* socio-
economic policy is itself often only poorly expressed, leaving structure plans
(which like local plans are restricted to establishing policies and proposals
for the development and other use of land) with something of a dilemma. In
these circumstances, 'the role of the structure plan becomes uncertain –
should it, in the absence of a socio-economic rationale for land use develop-

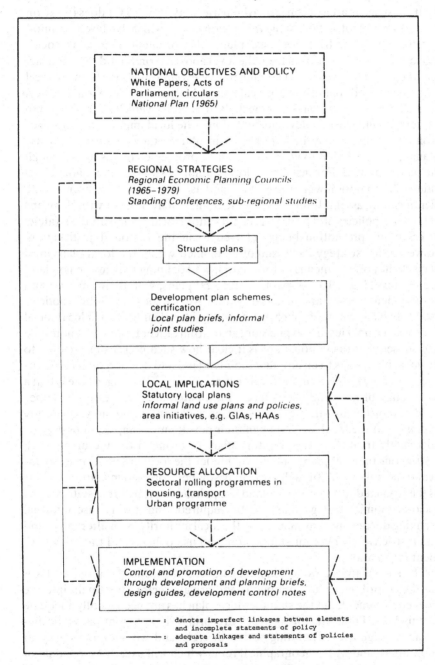

Figure 4.2 *Statutory and non-statutory elements of the British planning system*
Source: Bruton, M.J. and Nicholson, D.J. (1985), 'Strategic land use planning and the British development plan system', *Town Planning Review*, Vol. 56, No. 1. Figure 2.

ments, restrict its area of concern to land use matters only? Or does it take on board the task of establishing this rationale? – which by law it cannot' (Bruton, 1984, p. 87). For local plans, this problem of a weak socio-economic strategy inherited from the level above is compounded by the fact that the legislative basis and administrative evolution of the statutory local plan has stressed spatial and physical issues above social or economic policy. As a result, the effectiveness of local plans as policy vehicles is restricted to matters of land use and development alone; the local implications of socio-economic policies may be referred to insofar as they relate to land use change, but the extent to which they can be progressed themselves through the plan is limited. For instance, a local authority strategy of economic and industrial promotion will have certain land use ramifications in terms of site identification, assembly and development; these may be dealt with by means of land use policies and proposals in a statutory local plan, with the strategy of economic promotion being referred to in the reasoned justification. However, the strategy itself cannot be defined within the local plan; local plans do not offer a means of formulating and defining policies on non land use matters (DoE, 1984g, para. 4.2). Local plans, with their land use and development remit, are necessarily reactive to socio-economic policies established elsewhere. In such a situation, it is unclear how such local social or economic policies find expression, apart from officer reports to committee and subsequent resolutions; nor is it clear how such policies are related to the local plan, so as to ensure that their land use and development consequences are fully reflected in the local spatial strategy. The danger here is that a wide range of non land use policies, formulated and progressed by either local authorities or other public agencies, will not be fully and coherently expressed at the local level, though their development implications *may* be adequately treated by the statutory plan. In terms of the hierarchy, such restrictions on local plan content will hinder the local articulation of socio-economic policies. Although these may still be expressed in statutory plans as the reasoned justification for land use policies, a more direct statement of socio-economic strategy may still be required, and in this case informal planning documents may be used by the local authority as an alternative and less restrictive channel for stating non land use policies and their development implications.

The spatial strategies developed at Level 3 theoretically provide a basis for the co-ordination of both public and private sector investment into the built environment, but the statutory local plan has not been wholly adequate to this task. Public sector agencies are primarily autonomous vertically-organized agencies, operating within discrete and well-defined policy or topic areas, and with a geographical remit which will usually be regional or national rather than local. The decisions of such agencies that loom large in the context of the planning issues of an individual local plan may actually be relatively insignificant when seen against the background of the agencies' total operations. Although the local planning authority will be most concerned with the *local* implications of the activities of public corporations and the statutory undertakers, these effects are unlikely to be significant for the agencies themselves, whose prime concern will be the efficient operation of their particular industry or policy area. Local plans do nothing to provide local planning authorities with an effective means of *controlling* (as opposed to seeking to *influence*) the activities of such public bodies, even though their activities will often have significant land use implications. Despite consultation, most public agencies outside the local authority, and perhaps some municipal service departments, are likely to see the adopted local plan as at best an irrelevance, and at worst a hindrance to their local-level activities. For private sector development activity, the local plan does at least have some teeth via the development control process. However, despite the use of such initiatives as development briefs, the development control process is primarily reactive, and the local plan, as one component of the development plan, forms only one of a range of considerations material to any given application. Moreover, the relationship between local plans and development control is complex, and the two elements are not necessarily always fully integrated (Pountney and Kingsbury, 1983a). The statutory local plan *can* potentially provide an integrated spatial base for public- and private-sector development activity; the key problems likely to prevent this integration from working effectively are the relative independence of other public agencies and the difficulties of implementing the plan's proposals through the development control process.

Lastly, the hierarchical policy model suggests that local plans should establish the phasing of proposals and consider the availability of resources needed to implement their proposals. But as we have seen, local plans have tended to limit themselves to a review of already-programmed development, otherwise simply indicating the general constraints on resource availability. This problem is a corollary of the first, for just as socio-economic policies remain outside of the plan, so do the resources needed to implement the development implications of such policies. Instead of being directed by the local plan, these funds are largely under the control of either local authority spending departments or of other public agencies, where they are being increasingly handled by service programmes, again 'vertically' organized by policy topic, e.g. those under the urban programme, TPPs, HIPs. The failure to link land use policies and proposals and available resources means that the theoretically close relationship between Levels 3 and 4 is, in actuality, weak. Thus, programmes in Level 4 may proceed independently both of each other and of the statutory local plan, whose policies and proposals they are supposedly implementing. A number of attempts have been made by different authorities to overcome these weaknesses. In Liverpool, for instance, the rejection of the statutory planning approach has led to 'a new sort of local planning . . . based on the need to service and implement programmes and to relate planning activity – and particularly the identification of local objectives and problem solving – to the wider resource base' (Hayes, 1981, p. 77). Proceeding in the context of Partnership, local planning in the city now takes place within an enhanced corporate environment where the main role is to serve topic-based programmes and to direct resources to particular problems at the local level. Similarly, in Sheffield the recognition that effective implementation is dependent upon both political and financial commitment has led to the identification of a number of spending programmes, including the five main blocks of capital allocation, the urban programme and EEC grants, whose allocation is linked to rolling programmes of projects included in both statutory and non-statutory local plans and studies (Bajaria, 1982). Finally, the three-tier system of local planning developed by Birmingham City Council incorporates, for the inner areas, a series of informal co-ordinating plans whose speed of preparation can meet both the needs of urban renewal and the annual budget cycle of the Partnership programmes (Middleton, 1982). These examples imply that the statutory policy output of Level 3 is not being reflected, in the main, by the short-term spending programmes making up the fourth level, resource allocation. They also suggest that the formally provided vehicles at this third level are inadequate to meet the variety of contingent factors which can arise at this level.

The practice of circumventing these difficulties by linking informal plans to spending programmes cannot be regarded as a best solution, for it raises questions as to the effectiveness of such approaches in dealing with the local implications of regional and strategic policies. In particular, a concern with resources and implementation which focuses on the fourth level in the hierarchy may lead to a more short-term and incremental approach to local planning than is suggested by the theory of strategic planning. In this situation, the links with structure plans remain far from clear. The danger is that the relatively independent nature of these non-statutory initiatives will lead to the development of policies for the areas concerned which are either not to be found in approved statutory plans or which actively conflict with such plans.

At the same time as various non-statutory approaches to local planning are being developed by local authorities in response to the various problems associated with the local plan, a number of additional formal initiatives have been introduced which have a bearing on land use and development e.g. plans for Housing Action Areas (HAAs), General Improvement Areas (GIAs), Industrial Improvement Areas (IIAs), Conservation Areas, National Park Plans for Management, Statements of Intent for Areas of Outstanding Natural Beauty (see Chapter 5). These plans generally complement the statutory and non-statutory land use plans, and because they are invariably resource based, they are important in securing development and change on the ground. More recently, the introduction of initiatives such as Enterprise Zones, Urban Development Corporations and Operation Groundwork has accentuated the trend of picking up isolated problems, or areas with severe problems, and giving them special treatment (Bruton, 1983c). Given the number of free-standing plans now existing at this level, confusion and uncertainty about the relative role and contribution of the respective plan vehicles is inevitable. The system now places enormous demands on the co-ordination of the planning activities of a number of bodies if the direct link between general policies for socio-economic change and development on the ground is to be provided.

In short, statutory local plans appear to be too limited a policy vehicle to fulfil the theoretical requirements of Level 3 by themselves. There are problems in expressing local socio-economic policies as a result of the restriction to matters of land use, difficulties in establishing a meaningful basis for the co-ordination of development given autonomous vertically-organized public agencies and the problems of implementing plan policies through development control, and limitations on the extent to which the necessary resources can be controlled or even influenced by the plan itself. These various difficuties may be seen as illustrations of the criticisms levelled at the hierarchical approach to organization by Barrett and Fudge (1981), namely that the assumptions that policy is formulated at the top (or centre) and that implementers act as agents for policy-makers are unrealistic. Barrett and Fudge point out for instance that policy may be a response to pressures and problems experienced on the ground, and that agencies will often be 'autonomous or semi-autonomous, with their own interests and priorities to pursue and their own policy-making role' (1981, p. 12) and this is indeed the case. Nevertheless a *formal* hierarchical structure of sorts is discernible in the land use planning system (see Figure 4.2), even if the ways in which this structure is operated in practice show considerable variation from the model. The analysis here has attempted to highlight these variations so as to illustrate some of the difficulties and issues involved with the use of statutory local plans. The problems that are posed for local plans and local planning (e.g. by the difficulty of securing effective inter-agency co-ordination) remain whether we view these as regrettable shortcomings in the context of the hierarchical approach or as central features of public policy only reconcilable through processes of negotiation and bargaining. This is to suggest that in evaluating local plans and planning a combination of theoretical perspectives is likely to be more fruitful than a single approach applied in isolation.

Faced with these problems with local plans, planning practice has responded in the way suggested by the contingency approach to planning and management by producing a wide variety of non-statutory instruments. Thus, to meet the limitations of the statutory local plan as a vehicle for socio-economic policies, a range of informal land use plans and other local policy statements have been developed by local planning authorities as substitutes or complements to the statutory plan. These may be tied to corporate or other programmes in a bid to relate development planning more closely to available resources. The wide variety of such instruments that have been produced is testimony to the wide variety of contingent factors that exist. Such approaches, by themselves or in combination with statutory plans, may then satisfy the requirements of this level in terms of the translation of general proposals into detailed plans. The various area-based initiatives noted above may also contribute to the formulation and implementation of local policies, despite the absence of any formal link either to the other plans developed at this level or to higher level strategic proposals. These points clearly suggest that to understand local *planning* as opposed to local *plans* the discussion must be widened to incorporate the other policy vehicles currently in use at this level of the strategic planning hierarchy. Accordingly, Chapter 5 reviews the various formal area-based initiatives with land use implications, while Chapter 6 considers the scale and variety of non-statutory local planning documents that local planning authorities have evolved.

4.9 Conclusions

Local plans are currently being deposited and adopted on a substantial scale following the virtual completion of the structure plan framework, while individual plans show markedly different approaches being taken to the four central local plan roles. This had led to variety in plan content (structure plan context given in detail or mentioned only in passing; plan-wide or area- or site-specific policies and proposals, usually in combination as appropriate; resource and implementation issues treated in detail or largely ignored) and to the manner in which plans have been produced (length and intensity of participation programmes, role of the PLI). The primary influences here have undoubtedly been the confines set by statute and regulations, together with formal and informal DoE advice. Within these limits however local authorities appear to have experimented with the statutory provisions, the secondary influences here being the local planning issues the plan must face and the operating environment in which it must be 'implemented'. For this reasons, 'local plans thus appear local in orientation and specific to particular areas or issues' (Healey, 1983, p. 164), so that local plan documents are characterized by variety as well as consistency, bearing out once again the importance of contingent factors.

The commitment to the statutory system that current progress on local plan deposit and adoption represents, however, cannot hide the basic weaknesses of the local plan as a land use policy vehicle when it is considered against the hierarchical model of policy control developed in Chapter 2. Local plans cannot be used directly to establish local socio-economic policies, although they may be used to deal with the land use implications of such policies. Nor does an adopted plan give the local authority any effective control over the local activities of public agencies, while there are a number of difficulties in implementing plan proposals through the development control process. The resources necessary to implement the development implications of local plan proposals are not directed by the plan, being instead under the control of local authority spending departments or of other public agencies. Many local planning authorities have developed non-statutory approaches to local planning in a bid to meet these problems. A further difficulty is that of successfully integrating local plans with the variety of area-based policy and planning vehicles which have recently been established. This, however, is a general problem of integration which cannot be wholly accounted for by the shortcomings of the statutory local plan alone.

Notes

1 Tables 4.1 to 4.7 are based on DoE schedules of local plan deposit and adoption in England up to the end of March 1985.
2 The *Local Government, Planning and Land Act 1980* removed the former requirement that action areas should first be identified in the structure plan. Structure plans, proposals for alteration, and replacement plans can no longer identify action areas but in appropriate areas local planning authorities may prepare action area local plans provided that they are allocated responsibility for them in the development plan scheme (DoE, 1984g, para. 3.18).
3 These latter figures are likely to be an under-estimate since Field's figures only refer to commitments to prepare plans of each type, and do not include data on numbers of intended plans.
4 *Town and Country Planning Act 1971*, s.11(3)(a).
5 The following 30 plans have been referred to in the review:

Action Area Local Plans
Bexley LB: Bexley Heath Town Centre Local (Action Area) Plan.
Coventry City Council: Eagle Street Action Area Plan.
Coventry City Council: Eden Street Action Area Plan.
Harrow LB: Harrow Town Centre Action Area Plan.
New Forest DC: Totton Town Centre Action Area Plan.

Subject Local Plans
Buckinghamshire CC: Minerals Subject Plan.
Cornwall CC: Countryside Local Plan.
Greater Manchester CC: Green Belt Local Plan.
Humberside CC: Coastal Caravans and Camping Plan.
Humberside CC: Intensive Livestock Units Local Plan.
Wiltshire CC: Western Wiltshire Green Belt Local Plan.

District Local Plans
Birmingham City Council: Birmingham Central Area District Plan.
Birmingham City Council: Sutton Coldfield District Plan.

Bromsgrove DC: Bromsgrove Local Plan.
Camden LB: A Plan for Camden; Borough Plan; Proposals for the Alteration of the District Plan.
Gloucester City Council: City of Gloucester Local Plan.
Hammersmith and Fulham LB: District Plan.
Haringey LB: District Plan for the London Borough of Haringey.
Kensington and Chelsea Royal Borough: District Plan.
Kirklees MBC: Huddersfield Local Plan; Town Centre Inset Plan; Moldgreen Inset Plan.
Leeds City Council: Central Business Area District Plan.
Leicester City Council: Abbey District Plan.
Leicester City Council: Central Leicester District Plan.
Leicester City Council: Soar Valley South and Aylestone Local Plan.
Manchester City Council: Manchester City Centre Local Plan.
Newham LB: Beckton District Plan.
Swale BC: Faversham District Plan.
Warwick DC: Lapworth District Plan.
Warwick DC: Leamington Town Centre Local Plan.
City of Westminster: District Plan.

6   *Town and Country Planning Act 1971*, s.11(3)(a).
7   See: (*a*) Great Portland Estates plc *v.* Westminster City Council (Queen's Bench Division, Woolf, J., February 25 1983); (*b*) Great Portland Estates plc *v.* Westminster City Council (Court of Appeal, Lawton, Purchas, Dillon, L.J.J., December 6 1983); (*c*) Great Portland Estates plc *v.* Westminster City Council (House of Lords: Lord Fraser of Tullybelton, Lord Wilberforce, Lord Scarman, Lord Roskill, Lord Bridge of Harwich, 1984).
8   Kirklees MBC, Huddersfield Local Plan, Report on the objections made at the public local inquiry and in writing, February 1985, p. 74.
9   Judgement of Lord Justice Dillon, Court of Appeal, December 6 1983.
10 *Town and Country Planning (Structure and Local Plans) Regulations 1974* (SI 1974, No. 1486), Schedule 2, Part 1.
11 *Ibid*., para. 15(1). Again, this requirement is absent from the 1982 Regulations.
12 *Ibid*., Schedule 2, Parts II(iv) and (vi).
13 *Town and Country Planning Act 1971*, s.12(1A); DoE, 1984g, para. 3.40, Annex C.
14 *Town and Country Planning General Development Order 1977* (SI 1977, No. 289), Classes XVIII–XX.
15 *Town and Country Planning Act 1971*, s.12(1).
16 *Town and Country Planning (Structure and Local Plans) Regulations 1982* (SI 1982, No. 555), Regulation 5.
17 The details of the bargaining process undertaken in connection with the Southtown and Lapworth local plans may be found in Bruton, M. J. (1980), 'Public participation, local planning and conflicts of interest', *Policy and Politics*, Vol. 8, No. 4, pp. 423–442. The outcome of the process is summarized here.
18 This section appeared in a modified form in Bruton, M. J. and Nicholson, D. J. (1985), 'Strategic land use planning and the British development plan system', *Town Planning Review*, Vol. 56, No. 1, pp. 21–41.

5 The role and content of other formal plans and programmes

5.1 Introduction

The statutory local plan, as a part of the overall development plan, is intended to provide local authorities with a definitive means of setting out policies and proposals for the development and other use of land within their areas. However, in addition to the statutory local plan there are a large number of other formal initiatives, plans and programmes which have been established in an explicit attempt to manage environmental change and which have direct or indirect land use and development implications. These cover a range of policy areas (e.g. housing, conservation, the environment, urban development) and take a variety of forms (e.g. grants for defined areas, public sector development agencies, topic-based investment programmes, areas of restrictive or liberal planning control). Many of these initiatives, like the statutory local plan itself, are of relatively recent origin, and have been introduced on an *ad hoc* basis rather than as part of a sustained attempt to ensure that the British planning system as a whole meets the requirements of a coherent hierarchical model of policy control.

Although these measures have few common characteristics apart from their formal nature, one trend which has emerged is the tendency to focus on a particular topic or policy issue in isolation, and/or to define geographical areas (often limited in extent) where the provision of the initiatives in question will apply. This 'area'-based approach persists despite much criticism that such spatial boundaries are largely irrelevant, particularly where questions of social policy are concerned (Hatch and Sherrott, 1973; Townsend, 1976). Yet there may be sound administrative and political reasons underlying the use of priority area social policies (Donnison, 1974). In any case, the difficulties of dealing with urban deprivation through area policies do not necessarily mean that the area-based approach as a whole has nothing to offer. Hambleton (1977), for instance, suggests that area approaches have considerable potential, e.g. in promoting new approaches and learning about the efficacy of current policies, and in assisting in the renovation of management and political processes at the local level by relating policies to areas which are meaningful to local residents.

However, the links between the policies, proposals and resources of these formal measures and the policies and proposals set out in statutory local plans are often unclear, to say the least. Pointing to a proliferation of different sorts of plan or policy instrument which are concerned to provide a framework for individual service programmes, Fudge (1976, p. 174) comments that 'all these plans and programmes raise questions about how different kinds of planning at the local level fit together and relate to strategic or corporate plans'. The danger is that the planning system will become increasingly fragmented, and that the local plan itself will become more and more irrelevant, both as a vehicle for, and as a statement of, policies and proposals for the development and other use of land. This chapter reviews the role, function and content of the principal formal initiatives, plans and programmes with a bearing on land use and development, and considers how these measures relate to the statutory local plan in the light of the theoretical strategic planning framework proposed in Chapter 2. For this purpose, the various policy measures involved are collected into four broad groups: those dealing with housing; conservation and the environment; urban development and regeneration; and investment programmes.

5.2   Housing

Housing renewal policy up to the late 1960s emphasized the clearance and redevelopment of older housing, but the *Housing Acts* of 1969 and 1974 shifted the emphasis to improvement and rehabilitation. These Acts introduced three area-based measures designed to foster housing improvement. *General Improvement Areas* (GIAs), established by the 1969 Act, embody the change from area clearance to area improvement, providing local authorities with a specific grant to spend on environmental works. *Housing Action Areas* (HAAs), introduced by the 1974 Act to extend the rehabilitation initiative, allow authorities to identify a further category of area between those designated as GIAs and those suitable only for clearance. Whereas GIAs were seen as areas of fundamentally sound and easily retrievable properties capable of providing good living conditions for many years to come and unlikely to be affected by redevelopment or major planning proposals, HAA designation was to cover the worst areas of housing stress where bad physical and social conditions interact. The HAA policy thus includes social as well as physical objectives. Early DoE advice on the size of these areas noted that successful GIAs had been of the order of 200–300 houses, and suggested that HAAs should also be confined to this range (DoE, 1975b, para. 15). Designation of GIAs and HAAs is by resolution of the local authority following consideration of a report on the area(s) involved. Finally, *Priority Neighbourhoods* (PNs), also established by the 1974 Act, were designed to prevent the housing position in or around stress areas from deteriorating further and to stop stress from 'rippling out' from areas such as GIAs or HAAs where action was being taken. In practice, very few PNs were designated, and the *Housing Act 1980* (s.109) terminated the PN provisions by repealing Part VI of the 1974 Act. Established PNs continue to run, subject to their designated life of five years. The principal effect of these area designations is to provide higher rates of grant for house renovation than available elsewhere, although certain other powers also accrue to the local authority, e.g. the ability to initiate compulsory improvement of some tenanted dwellings in GIAs and HAAs. No formal planning documents are provided by the legislation as a base for action in GIAs, HAAs or PNs, apart from the reports that must be submitted to local authorities prior to designation.

These area-based measures are thus devices for concentrating resources into small areas of older housing, so as to achieve housing and environmental improvement in pursuit of both physical and social objectives. How do these area measures relate to the policies and proposals set out in local plans? DoE advice in Circular 13/75 *Renewal Strategies* emphasized the independence of action taken under housing legislation from that proposed under planning legislation. Thus, authorities were advised to 'ensure that housing renewal proposals can be reflected, as necessary, in the statutory planning framework'; local plans were thus to *reflect*, rather than control, such proposals. Moreover, 'the declaration of special housing areas, public participation and implementation can – and, in many cases, should – proceed *independently* of the related development plan . . . where local conditions call for it, the declaration of a special housing area should not wait on any associated plan-making' (our emphasis) (DoE, 1975a, para. 8). Confusingly for local authorities, they were also asked to consider, later in the same circular (para. 28), how they could develop a corporate approach to renewal, and in particular how housing action could be associated with other complementary activities and services of the local authority and of other public bodies. The DoE clearly did not envisage local plans playing a key role in this corporate approach to renewal. Later development plan advice emphasizes the distinction between housing and planning legislation. Local Plans Note 1/78 (now cancelled by DoE Circular 22/84) for example distinguished between proposals for the development and other use of land and those matters arising from the non-planning functions and powers of the local authority and of other authorities and bodies, such as government departments and health authorities. These latter activities will often be closely linked with planning functions, and may also generate proposals in local plans. Designation of an HAA is cited as an example, whereby proposals will also be required for environmental improvement and traffic management. The non land use planning activities themselves, however, must not be distinguished as proposals, nor shown on the proposals map, although they may be referred to in the reasoned justification. This distinction is argued to be important for four reasons:

1 To clarify which parts of the plan have statutory force following adoption.
2 To ensure the plan is not unnecessarily affected by termination or alteration of the non-planning activities (e.g. if an HAA is terminated).
3 To help the local authority express full and substantiable proposals for land use and development.
4 To avoid confusion with procedures stemming from different statutes (DoE, 1978a, para. 17).

The development plans memorandum (DoE Circular 22/84) stresses that local plans should not seek to designate, through policies and proposals, areas such as GIAs, HAAs, or PNs where special grants will be available, although plans should indicate any land use proposals to be applied within these areas (DoE, 1984g, para. 4.2).

Central government has thus sought to maintain the distinction between the planning and housing powers available to local authorities *on the basis of the empowering legislation*. This emphasizes that the local authority acts as local housing authority and as local planning authority under different and distinct statues. But few would argue that physical renewal of the housing stock is not a legitimate local plan concern. Local plans can *reflect* GIA, HAA or PN designation, but cannot themselves establish such areas. In the context of the departmental organization of local government, this will do little towards achieving a comprehensive housing renewal policy and indeed could emphasize departmental separatism. Local plans prepared by the planning department can formulate policies on renewal, but cannot control the statutory tools available for carrying such policies out, which will usually be administered instead by a service department such as housing. This need not be a problem where authorities have developed a satisfactory corporate approach, but municipal housing departments are more usually relatively independent organizations by virtue of their strong and well-established legislative base, financial resources, and political significance. However, planning departments may still become involved in renewal initiatives, through either the provision of information, e.g. towards formulating the pre-designation report, or via the production of informal 'special area work' dealing with such aspects as design criteria (see Chapter 6). Finally, the relative independence of these housing area measures also has implications for the attempts of local planning authorities to ground their local plans on participation programmes in a bid to achieve public support for their policies. Unlike the legislation governing local plan preparation, the procedures leading to HAA designation for instance contain no provision for participation. Indeed, Circular 14/75 actually acknowledges that, bearing in mind the dangers of anticipatory blight, authorities may not wish to carry out public participation prior to designation, which instead may be 'replaced' by publicity and consultation afterwards (DoE, 1975b, para. 33). There is a danger here that the public will come to suspect that their participation in the process of local plan production is largely a hollow exercise, at least as far as policies for housing renewal are concerned.

5.3 Conservation and environment

There are a large number of formal and quasi-formal initiatives dealing with the protection, conservation and enhancement of both the rural and the urban environment. Individual measures deal with a variety of specific land use and built environment issues and problems, and make use of a number of means of implementation. Their links to the development plan system are equally diverse.

5.3.1 National Parks

Perhaps the most well known of these initiatives are the ten *National Parks*, designated in the 1950s by the then National Parks Commission (now the Countryside Commission following the *Countryside Act 1968*) under the *National Parks and Access to the Countryside Act 1949*. The Parks account for some nine per cent of the area of England and Wales. The statutory purposes of the National Park designations are to define 'extensive tracts of country . . . for the purpose of preserving and enhancing the natural beauty of the areas specified . . . and for the purpose of promoting their enjoyment by the public'.[1] The twin aims of conservation and recreation will not always be compatible, and in such situations it is accepted that landscape conservation takes precedence over recreational use.[2]

Following the *Local Government Act 1972*, the National Parks are administered by joint and special planning boards for the Peak District and Lake District, and by county council committees for the remaining eight Parks. Each of the boards or committees is provided with permanent staff headed by a National Parks Officer. These authorities have important functions under the *Town and Country Planning Acts*, e.g. development control. The two boards are also required to produce structure plans for their areas, while strategic planning policy for the other Parks is set within county-based structure plans. In this latter case the tendency is for county-wide rural settlement policies to be generally applicable to areas both inside and outside Park boundaries, with stricter development control conditions being imposed in areas seen as needing special protection. Where structure plans have been prepared by National Park authorities, more explicit attention has been given to National Park issues, with specific policy responses coming forward as a result (Cloke, 1983). A number of statutory local plans prepared wholly or in part by National Park authorities are now coming forward (Peak Park Planning Board, the Bakewell District Plan; North Yorkshire Moors National Park Committee and Ryedale DC, Helmsley Local Plan; Exmoor National Park Committee and West Somerset DC,

Dulverton Area Local Plan), while a number of informal local plans have also been produced.

As well as reconstituting National Park administration, the 1972 Act also required the new Park authorities to each produce a five-yearly management plan, the National Park Plan (NPP). The NPP is a statutory vehicle for the formulation of policy on the *management* of the Park and for the *exercise* of the Park authority's functions.[3] The requirements as to preparation procedures are simply that copies of the proposed plan should be sent to the Countryside Commission and to district councils in the Park, and their observations taken into account. This contrasts with the detailed statutory requirements for structure and local plan preparation, and has been justified on the grounds that NPPs are not instruments of control, and as expressions of intention will usually relate to those of the authority itself and to those of other authorities, with their agreement (Hookway, 1978). The DoE and the Countryside Commission however have emphasized that the Park authority will need to consult widely during the preparation of a NPP, so as to ensure that the policies in the plan command the general approval of local and other interests (DoE, 1974b). National Park authorities appear to have followed this advice, e.g. setting up working parties, including local representatives, of a wide range of interests which have helped in policy formulation on specialist topics (Dennier, 1978, 1980). No doubt this has ensured that local contingent factors have been addressed and taken into account. The view of planning as the management of environmental change (rather than as simply the production of plans) comes to the fore in NPP preparation, particularly insofar as local interests have been incorporated into the NPP preparation process.

Management policies in NPPs are usually derived from the statutory aims of the conservation and enhancement of the landscape, and the provision of opportunities for the public to enjoy the Park. National Park authorities, however, must also satisfy other aims for the Park area, and in particular the requirements imposed by s.37 of the *Countryside Act 1968* to have due regard to the needs of agriculture, forestry, and to the economic and social interests of the area. These broader aims are supposedly dealt with in full in the development plan, being included in the NPP insofar as they provide a context for management. The development plan thus deals with the planning of the Park and the NPP provides a framework for its management. The DoE and the Countryside Commission recognize however that in practice things may not be so clear cut: 'planning and management are not distinct activities . . . the processes are complementary and inter-dependent. Therefore, the content of each type of plan will influence the content of the other' (DoE, 1974b, para. 1.6). Hence, for instance, *new* planning policies may develop in part out of the work of preparing the NPP, though the statutory development plan remains the main means whereby such policies are formulated (*ibid.*, para. 1.10). This early 1974 advice is largely reiterated in the current development plans memorandum (Circular 22/84). National Park authorities have indeed found it difficult in practice effectively to separate out land use planning and management issues, particularly the point at which National Park objectives begin to determine policies on subjects which feature in both plans, e.g. historic buildings, minerals, public transport (Dennier, 1978). Cloke (1983, p. 291) points out that the resolution of the obvious conflicts between conservation, recreation and local needs is central to the planning task in National Parks, and that 'very often it is the local needs objective which has been the first to be sacrificed in the complex policy equations which delegate priority to landscape, townscape and recreation in these areas'. NPPs have thus tended, perhaps not surprisingly, to emphasize National Park objectives above socio-economic issues when faced with the tension between conservation and local needs. This is not to gainsay the attempts of some Park authorities to meet local needs, e.g. the controversial use of s.52 agreements by the Lake District Planning Board in an attempt to ensure that new dwellings are occupied solely by 'local' people (Shucksmith, 1981; Loughlin, 1984).[4] Nor does this imply that NPPs have necessarily proved to be successful devices for land use management. Although hailed by the Director of the Countryside Commission in 1978 as 'a milestone in the development of planning' (Hookway, 1978, p. 20), experience has shown that their management by agreement basis has often proved ineffective when faced with major development proposals (Blacksell and Gilg, 1981).

Mention should also be made of the unique *Lee Valley Regional Park Authority*, which under the *Lee Valley Regional Park Act 1966* is required to prepare a plan showing proposals for the future land use and development of the Park. In preparing this plan (and amendments to it) the authority is required to consult with the local planning authorities concerned and with appropriate statutory bodies; the Authority also consults local interest groups and the general public, although there is no statutory obligation to do so. The practice of consulting widely on draft proposals undoubtedly reflects a desire to take account of local contingent factors and interests and to ensure that the Plan provides an effective basis for the management of change in the Park area. The latest version of this Plan sets out a revised strategy for future developments; indicates areas where detailed management plans and policies are required (to be prepared within the framework of the Plan); includes land use policies and proposals, advice on planning controls, and a rolling programme for implementation; and establishes guidelines for monitoring the effectiveness of the Plan (Lee Valley Regional Park Authority, 1985). The Plan reflects the remit of the Park Authority to develop and manage the Park for leisure and recreational activities and uses, including the provision of nature reserves. Local planning authorities are required to include the proposals of the Park Plan within their development plans, with these plans carrying a presumption in favour of leisure development within the Park. The Authority also has certain development control powers, including the right to be consulted by local authorities on planning applications which are likely to affect any part of the Park, and the right to request the reference of such an application to the Secretary of State if the views of the Park Authority and a local planning authority on a particular application are in conflict. The Plan sets out policies and proposals in respect of the Authority's development control functions. However, the Park Authority is not a planning authority, and must apply for planning permission for its own proposed developments in the usual way.

5.3.2　Areas of Outstanding Natural Beauty

The *National Parks and Access to the Countryside Act 1949* also allows the Countryside Commission to designate *Areas of Outstanding Natural Beauty* (AONBs), again for the statutory purposes of the preservation and enhancement of natural beauty. AONBs are areas considered as unsuitable for designation as National Parks (e.g. due to limited size), but which nevertheless require special treatment by local and central planning authorities to safeguard their landscape qualities. Before designation the Commission must consult with the local authorities concerned, give public notice of their proposals and invite representations; in practice AONB proposals are subject to wide consultation. Designation by the Commission is subject to approval by the Secretary of State. By 1983, a total of 35 AONBs had been designated and confirmed, with one awaiting confirmation and a further six proposed (Countryside Commission, 1983). Confirmed designations amount to some 11 per cent of the area of England and Wales, with individual AONBs varying in size from 16 km^2 (Isles of Scilly) to 1738 km^2 (North Wessex Downs).

The legislation provides no special planning arrangements for AONBs as there are in National Parks, nor is there any provision for a formal management vehicle such as NPPs. Thus, the objective of 'preservation' has perforce been largely advanced through the usual local authority development control process, and 'enhancement' through the discontinuance of existing uses, derelict land reclamation, tree preservation and planting, and management agreements. Implementation of the objectives of AONB designation is thus closely linked via development control to the development plan system, and development control data relating to AONBs has been used to assess the effectiveness of planning control (Blacksell and Gilg, 1977; Anderson, 1981). Local planning authorities are in fact required to consult the Countryside Commission on development plan proposals affecting an AONB,[5] and in practice the AONB designation is usually reflected in appropriate structure and local plan policies. The Commission may also give advice to either the Secretary of State on proposals for development, or to the local planning authority on planning applications affecting an AONB, *if consulted*.[6] This does not amount to a power of veto. Indeed, the Commission has recently sought to strengthen the position of the AONB designation in the development plan system. This has been a response to problems of administration and to the difficulties of relying on the development control process to achieve the preservation of natural landscapes. Administrative difficulties arise because in most AONBs planning and management responsibilities are split between county and district councils, while 14 AONBs cross county boundaries. The Commission argues that AONBs should be treated as a single unit for planning and management purposes, and to this end encourages the establishment of Joint Advisory Committees for AONBs, comprising representatives of local authorities, amenity groups and land user interests (Countryside Commission, 1983). As far as the preservation of landscapes through development control is concerned, by 1978 the Commission had come to the view that 'negative policies of development control are no longer enough', and called for 'a new emphasis on relating public and private investment and land management practices to landscape conservation goals and social objectives' (Countryside Commission, 1978, p. 2), i.e. it was arguing for the establishment of a system to manage change in the environment. Commission policy on AONBs now not only calls for the adoption in structure and local plans of development control policies emphasizing the conservation of natural beauty, but also advocates two specific planning vehicles, statements of intent and management plans. Statements of intent were originally introduced in 1968 by the Commission as a pre-designation document for AONBs, intended to set out concisely the effects of designation and the policies to be pursued by the local authorities concerned. Henceforth such statements have been found to be useful working strategic policy documents in clarifying planning, management and development control policies. For example, a statement of intent for the Forest of Bowland AONB is now being progressed by Lancashire CC, having formerly been intended as a statutory subject plan on recreation in the area. The proposed statutory format was dropped due to administrative difficulties over joint working, the timescale likely to be involved, and the DoE's restrictive attitude to the preparation of subject plans. These can be seen as contingent factors influencing the local choice of an appropriate policy vehicle. Development control decisions will not be taken on the basis of the statement of intent alone, but in conjunction with the structure plan and other local plans. However, policies set out in the statement may feed into and affect those contained in forthcoming statutory plans. The Commission has continued to advise local authorities to prepare statements of intent, for whilst structure and local plans are vehicles for planning, management and development control policies, 'the Commission considers that all the policies relevant to an AONB should be brought together into one document' (Countryside Commission, 1983, p. 10). The strategic statements should be supplemented where required by more detailed management plans, with both documents linked to the statutory development plan by referring to them in structure and local plans covering AONBs. The DoE has partly accepted the Commission's arguments, recognizing that statements of intent are useful informal aids to the planning and management of AONBs, and confirming that they can be regarded as supplementary planning guidance for development control purposes. However, 'the statutory development plan remains the definitive means of setting out local policies for development and other use of land' (DoE, 1984g, para. 5.6).

A variety of formal and informal documents are thus available with which to establish management policies in National Parks and AONBs. These coexist, somewhat uneasily, with structure and local plans. Statements of intent for AONBs at least have a defined relationship as supplementary planning guidance to the statutory development plan system. NPPs on the other hand have no such clearly specified relationship, the only guidance being that although development plans and NPPs have different functions, where appropriate the content of one should influence that of the other. Some overlapping of policy issues therefore seems unavoidable, and could lead to confusion over the relative status of policies in NPPs and development plans.

5.3.3. Heritage Coasts

A number of other special area designations aimed at landscape and nature conservation have also been introduced. *Heritage Coasts*, established by the Countryside Commission in 1970, have the two main objectives of conserving outstanding coastal scenery and enhancing the public enjoyment of the coast. These objectives are clearly similar to those of AONBs, and in practice the areas designated as Heritage Coast often overlap with AONBs, leading to the charge that two labels are being attached to the same article, thereby adding 'another patch to a highly colourful quilt of designations, definitions and notations in Structure and Local Plans' (Himsworth, 1980, p. 62). There are no statutory designation procedures for Heritage Coasts, although the boundaries can be indicated in development plans along with the land use policies to be applied within them. Structure plans should include a policy describing general boundaries and these should be illustrated on the key diagram as a basis for their precise definition in local plans (DoE, 1984g, para. 4.35). The boundaries of Heritage Coasts are therefore established by the development plan system, unlike those of National Parks and AONBs. Heritage Coast management has been primarily via development control policies in structure and local plans, with in some cases the appointment of a Heritage Coast or project officer to co-ordinate practical experiments on the ground, together with the preparation of an informal management plan (Cullen, 1981). For example, Cornwall CC's Countryside Local Plan defines the inland boundaries of ten stretches of Heritage Coast within the county, and considers the priorities for the preparation of management plans and the implementation of coastal management measures. The plan also establishes a number of policies for the control of development and car parking in Heritage Coast areas, although with only minor exceptions all the Cornish Heritage Coasts lie within AONBs and so also benefit from relevant structure and local plan policies (Cornwall CC, 1983).

5.3.4 Nature Reserves and Sites of Special Scientific Interest

Nature Reserves

As regards a specific concern with nature conservation, the Nature Conservancy Council has since 1949 established a number of *National Nature Reserves* as sites of national or international importance. The statutory purposes of declaration are to facilitate research into and to preserve the natural habitats of the areas concerned.[7] By 1982, a total of 182 Reserves had been declared on 1390.8 km^2 throughout England, Scotland and Wales (Nature Conservancy Council, 1983). Reserves are either owned or leased by the Council (27 and 13 per cent, respectively, of total Reserve area), or are the subject of Nature Reserve Agreements with the landowners and occupiers (60 per cent). These arrangements should be supported with development plan policies where appropriate. Local planning authorities are encouraged to take full account of nature conservation factors both in formulating structure and local plans and in the consideration of planning applications (DoE, 1977g). Informal support may also be suitable, e.g. the Conservancy Council and Swansea City Council have collaborated on the development of a strategy for the designation of Crymlyn Bog as a Reserve, jointly producing a non-statutory 'Action Plan' which details a programme of reclamation and landfill and establishes a basis for a Reserve Agreement between the two authorities as respects the local authority land involved (1982). In addition to National Reserves, in consultation with the Conservancy Council local authorities can establish *Local Nature Reserves*. These can be protected and regulated through ownership or agreement, or through byelaws made by the local authority.[8]

Sites of Special Scientific Interest

The Nature Conservancy Council can also establish *Sites of Special Scientific Interest* (SSSIs) under the *National Parks and Access to the Countryside Act 1949* and latterly the *Wildlife and Countryside Act 1981* as areas of special interest by reason of their flora, fauna, geological or physiographical features. Designation is by statutory notification of the local planning authority, landowners and occupiers, and the Secretary of State; the Council must also consider representations or objections before making a formal declaration.[9] By 1982, 4026 SSSIs had been notified in Great Britain, representing 13,669 km^2 (Nature Conservancy Council, 1983). Management and site protection is through advice rather than by acquiring a legal interest, with voluntary management agreements in force in a limited number of cases.

Planning controls are limited, with local authorities simply being required by the 1977 General Development Order to consult the Conservancy Council before granting permission for development of land in an SSSI. The effective exclusion of agricultural and forestry activities from the planning system means that the value of a site can easily be destroyed without any planning application being required. The *Wildlife and Countryside Act 1981* however has introduced further protection, the Council for instance being able to notify owners and occupiers on designation of an SSSI of 'special interest' features and of operations likely to damage them, which may thereafter only be lawfully carried out after notification to the Council, who may either consent or offer to enter into a management agreement. However, if no agreement is reached within 4 months the Council has no further powers to prevent the work from proceeding. Nevertheless, the 1981 Act now makes it possible 'to convert the general policy designation into one having direct and specific effect in controlling a range of operations' (Grant, 1982, p. 313), although the Act does not strengthen planning control itself over operations affecting SSSIs.

5.3.5   Conservation areas

The foregoing initiatives all focus on rural landscape and site protection, but there are also conservation and environment measures set within an urban context. For example, local authorities may define areas of special architectural and historic interest the character of which it is desirable to preserve or enhance, and designate these as *conservation areas* under the *Town and Country Planning Act 1971*.[10] Conservation area designation usually has several objectives:

1   The safeguarding of listed and other buildings by statutory powers, e.g. the blanket control of building demolition, and the use of grants and loans for improvement, repair and maintenance.
2   Closer control over new development, particularly materials, colours, building lines and height.
3   Critical assessment of existing developments.
4   A greater attention to details such as street furniture, signs and lighting.
5   Encouragement of voluntary local initiatives (Woodruffe, 1976, citing a 1971 policy statement by East Suffolk CC).

By 1980, over 4800 conservation areas had been declared in England and Wales, with designation initially orientated towards historic towns rather than villages. However, over half of district councils in rural areas had designated villages as conservation areas by 1980 (Historic Buildings Council for England, 1981). In National Parks or AONBs conservation area designation can be used to strengthen policies and restrictions against development in specific locations (Cloke, 1983).

The 1971 Act requires local authorities to give special attention to the desirability of preserving or enhancing the character or appearance of a conservation area,[11] so that designation will thus be only a preliminary to action. DoE Circular 23/77 *Historic Buildings and Conservation Areas – Policy and Procedure* emphasizes that action to be taken must be considered in the context of structure and local plan policies. Specifically, the local plan should be used to set out detailed development control policies and the measures to be taken for maintaining and enhancing the area's characteristics, e.g. diversion of development pressures, traffic management (DoE, 1977b, para. 37). Conservation areas are thus established outside of the development plan system but in effect should be linked closely to it. Authorities also have a duty to formulate and publish their proposals for conservation area preservation and enhancement, and to submit these for consideration to a public meeting.[12] These proposals may be prepared on an informal or non-statutory basis (see Chapter 6), but the DoE prefers their inclusion 'where appropriate' in structure and local plans (DoE, 1984g, para. 4.36). Conservation area advisory committees may be established to promote the area designation and to act as consultees on planning applications affecting the area's character and appearance (DoE, 1977b, para. 44). As with the incorporation of local interests in the preparation of National Park Plans, the activity of planning is here explicitly seeking to identify and respond to local contingent factors in a way which involves managing environmental change in a positive way as well as merely producing plans and other policy statements.

Conservation area designation also includes financial aid for building repair costs, principally from either Secretary of State grants or through a 'town scheme'. A town scheme is an arrangement whereby the DoE and the local authority concerned each allocate money annually for grants for the repair of buildings included in the scheme. Each grant is made at a rate of 25 per cent local authority, 25 per cent DoE. Policy favours the concentration of grant aid in a limited area so as to maximize its impact and encourage other owners to carry out repairs as rising property values make it more economic for them to do so (DoE, 1977b, para. 110).

5.3.6 Operation groundwork

A more recent measure concerned with environmental improvement on the urban fringe is the Countryside Commission's *Groundwork* initiative. Following earlier Countryside Commission innovations aimed at local scale urban fringe problems, e.g. the Bollin Valley study, exercises in Hertfordshire, Barnet and Havering, the first 'Operation Groundwork' project was established in 1981 on an experimental basis in St. Helens and Knowsley, Merseyside. In 1983, five other areas in the North West were incorporated into the Groundwork scheme: Salford and Trafford; Rochdale and Oldham; Wigan; Rossendale; and Macclesfield. Groundwork is essentially an urban fringe land rejuvenation programme centred on a co-ordinated effort by public, private and voluntary agencies to achieve a linked series of environmental improvements. The involvement of a variety of bodies in the Groundwork approach can be understood as a contingent response to a complex organizational situation where a number of agencies are relevant to the achievement of the Groundwork objectives. In seeking actively to involve these bodies, the Groundwork strategy is moving towards the direct management of change in pursuit of environmental improvements. The approach involves a partnership between implementing agencies, and integrated action over several policy areas, e.g. Groundwork links two well-established public sector programmes, countryside management and derelict land reclamation. The six Groundwork projects currently operational are co-ordinated by Groundwork Trusts, charitable organizations composed of a committee of Trustees representing the interests contributing to the project e.g. representatives of the Countryside Commission, participating local authorities, and others from private and voluntary sectors, and supported by a small full-time staff to carry out day-to-day work. The Trusts have three broad functions (Groundwork Trust, 1983):

1 to co-ordinate a programme of large scale land reclamation schemes and countryside projects . . . .
2 to promote a broad range of environmental improvement schemes through co-operation with industry, private landowners and voluntary groups . . . .
3 to encourage a wider appreciation of issues affecting land and the landscape, and opportunities for countryside recreation.

In addition to involving a variety of local agencies within the Groundwork areas, the approach has also attracted backing from central government, e.g. both the Ministry of Agriculture and the Nature Conservancy Council have provided full-time advisory support to Groundwork North West. Central government has also provided special capital allocations, the bulk of which flow to the local authorities engaged in derelict land clearance and other countryside projects.[13] The six Trusts were initially overseen by a small Countryside Commission unit established to develop and promote 'Groundwork North West'. More recently, the idea of extending the Groundwork approach by establishing a national body to encourage the work of local trusts in urban fringe areas has received Ministerial support. To this end, the Groundwork North West unit was replaced in 1985 by the Groundwork Foundation, charged with developing the Groundwork initiative on a national basis.[14]

Within certain urban fringe areas the Groundwork approach strives to offer a focus for environmental improvement, explicitly building on existing local authority programmes. One viewpoint is that the Trusts occupy what the Director of the Groundwork Foundation has called 'the interstices between conventional programmes', complementing the role of established agencies (Davidson, 1983, p. 233), although the Association of Metropolitan Authorities are concerned that the work of the Trusts will simply duplicate that of existing local authority programmes.[15] A further area of uncertainty concerns the relationship between development plans and Groundwork projects. Both structure and local plans are required to include policies and proposals for the improvement of the physical environment, through for instance the reclamation of derelict land (DoE, 1984g, para. 4.34), but the links between the development plans covering the Ground-work areas and the work of the Trusts remain to be investigated. It is, however, clear that local plans can be suitable vehicles for area environment improvement policies and proposals, and that the Groundwork approach (or something like it) offers a new channel for the implementation of such proposals. For example, a series of river valley local plans have been pre-pared in Greater Manchester by the now abolished county council and affected district councils, e.g. the Tame Valley Local Plan, prepared by Greater Manchester MCC and the metropolitan boroughs of Oldham, Stockport and Tameside; the Medlock Valley Subject Plan, prepared by Manchester City Council, Oldham MBC and Tameside MBC (Healey, 1984). These plans typically cross district boundaries, enforcing joint work-ing, and contain policies and proposals to reclaim derelict land, protect existing open land, and provide more opportunities for outdoor recreation. They would, therefore, seem to provide an ideal policy context for implementation initiatives such as Groundwork, although environmental improvement in the Greater Manchester river valleys is already underway via existing programmes. Elsewhere, a similar project to Groundwork in Staffordshire, 'Operation South Cannock', has been directly responsible for implementing some of the proposals in Staffordshire CC's South Cannock Land Renewal Local Plan (on deposit, June 1984).[16] Perhaps the most likely role for Groundwork Trusts on a national basis is as a novel implementing device in conjunction with appropriate development plan policies, e.g. for the protection of open land. Unless the Trusts continue to be supported by special capital allocations from central government, the success of individual Trusts in this role will largely depend on the degree to which they can attract private and voluntary financial and other support.

5.3.7 Waste disposal plans

Waste disposal plans are prepared by waste disposal authorities (county planning authorities in England and district planning authorities in Wales) under the *Control of Pollution Act 1974*. Although primarily dealing with matters such as operational waste management rather than with the development and use of land, some aspects of waste disposal (such as tipping proposals) will clearly have land use implications and therefore relate to land use planning policies. Strategic guidance for waste disposal plans may be provided by the structure plan, i.e. identifying relationships to other planning matters such as mineral extraction or setting out the broad criteria to be applied in identifying disposal sites. More detailed site-specific guidance may be provided by local plans, either in a wide-area subject plan or by the incorporation of waste disposal proposals into more general local plans (DoE, 1984g, para. 5.12).

Waste disposal plans represent a planning/strategy document prepared to facilitate local authority service provision rather than to foster environmental improvement or conservation policies. Nevertheless, they may have important land use implications depending on the waste disposal strategy adopted and the availability of tipping/disposal sites. The waste disposal plan prepared by Rhymney Valley DC (Mid Glamorgan), for example, sets out the Council's strategy to dispose of waste at the least cost to the community with due regard to the safeguarding of the environment and the use of waste as a resource. With these criteria in mind, the plan considers alternative methods of waste disposal and identifies possible major future sites for the most cost-effective of these methods, controlled land fill (Rhymney Valley DC, 1984). Waste disposal plans must hence balance both economic and environmental considerations within the overall requirement to ensure the adequacy of arrangements for waste disposal.

5.3.8 Summary

There are a wide variety of area designations concerned with conservation and the environment. Some, notably conservation areas, are closely linked to statutory local plans where these are being prepared, but others are largely independent initiatives (e.g. Groundwork) or longer standing well-established designations with their own management planning vehicles (e.g. National Parks). These special area designations often carry resources for implementation, and so are important for securing change on the ground. Groundwork, with its special derelict land capital allocation, is a prime example. Where this is the case, it is particularly important for structure and local plans to be centrally involved in these initiatives unless they are not to become increasingly peripheral policy documents, formulating proposals but lacking effective control over the resources necessary for implementation. Where policies of development restraint are involved, however, the ordinary process of development control (in the form of relatively more restrictive policies in structure and local plans) is still likely to be of central importance in achieving policy aims, since few additional powers are available in designated rural areas (Blacksell and Gilg, 1981, p. 123).

In addition to the various formal measures identified above, it is worth noting that structure plan rural strategies have also contributed to the proliferation of environmental designations. This is similar to the various area-wide policies put forward in statutory local plans (see Chapter 4). For example, the Gwent Structure Plan not only has parts of a National Park and AONB within its boundary, but also establishes other special zones, e.g. Landscape Improvement Areas, Special Landscape Areas, Recreational and Amenity Areas, and Agricultural Priority Areas (Gwent CC, 1978). The rationale for these priority areas is, of course, the greater policy sensitivity which results from their establishment.

5.4   Urban development and regeneration

The variety of the special policy areas concerned with the protection, conservation and improvement of the environment is matched by that of a number of measures concerned with the *control* or *promotion* of urban development in specific locations. Two of these measures, new towns and Green Belts, are long-established planning tools which are essentially vehicles for regional or sub-regional policy, but which also have important implications for local plans in their detailed implementation. More recent initiatives, such as Industrial Improvement Areas (IIAs), focus closely on specific aspects of urban development against a background of a declining economic base and attendant social problems, spatially concentrated within the inner urban areas. The development of national policies for the inner city, and in particular the establishment of partnership arrangements between local and central government, has altered the context for local planning in districts such as Liverpool. Finally, initiatives such as Enterprise Zones and Urban Development Areas have continued the trend of defining special policy areas, and this seems set to continue with recent proposals for 'simplified planning zones'.

5.4.1   New Towns

New Towns and Green Belts are complementary planning tools which together have been used to progress a general policy of urban containment. Green Belts have provided a limiting influence on urban growth, which has been accommodated instead through planned town expansion or by the creation of New Towns (see Hall *et al*.,1973a, b). This relationship is perhaps best seen in the plan for the Greater London region prepared by Abercrombie in 1944, which envisaged the establishment of a number of New Towns beyond a Green Belt encircling the capital. In practice, the two policies have been characterized by markedly different statutory provisions and means of implementation. This, in turn, means that their local implications are reflected in different ways in local plans.

The New Towns programme was initiated by the *New Towns Act 1946*[17] following the earlier success of Letchworth and Welwyn and the 1945 Reith Report (Schaffer, 1970). This legislation gave the appropriate Minister the power to designate any area of land in Great Britain as a New Town site and to appoint a development corporation charged with the layout and development of the town. Subject to planning and financial controls being retained by central government, each corporation has broad powers to borrow money and to acquire (if needs be compulsorily), develop and dispose of land. Since 1946 a total of 32 New Towns have been established, eight of which are in a ring around London. The last (Central Lancashire) was designated in 1970. The programme is now being run down, and this is reflected in recent legislation.[18] Government withdrawal is due partly to a shift in planning policy which recognizes the problem of the inner urban areas, where diminishing population and physical and economic decline are key issues. In this context, planned programmes of population dispersal to relieve urban congestion are seen as no longer appropriate. A further factor, however, is the present government's 'privatization' policies, which have led to cuts in finance for further development in New Towns, and to pressures for the large-scale realization of development corporation assets (see, for instance, Orchard-Lisle, 1980).

The planning of New Towns proceeds largely independently of the development plan system. The overall strategy for the development of the town is normally set out in a 'master plan' produced by the development corporation. This document is subject to processes of publication, objection, public inquiry and endorsement by the Secretary of State. The master plan establishes the strategic framework within which the town is to be developed, together with an account of the objectives and constraints which have governed the formulation of proposals and the policies to be pursued in achieving them. The master plan provides a framework for the preparation by the corporation of detailed development proposals, which must be submitted to the Secretary of State for approval.[19] There is provision for local planning authorities to be consulted at this stage. Planning permission for development in accordance with approved proposals is given by a special development order,[20] which covers development by the corporation and development by others on land formerly held by the corporation. In practice, these arrangements cover the bulk of new development. The permission given by the Order may be withdrawn at any time by the Secretary of State. Planning control in New Towns is thus under the direct control of the Secretary of State, although it is worth noting that local planning authorities retain the power to approve or refuse applications for *unapproved* development. These planning provisions are usually supplemented by substantial land ownership by development corporations, commonly extending to up to 80 per cent of the designated area (Grant, 1982).

Within the designated area of a New Town, therefore, it is the master plan rather than the county structure plan which settles the main pattern of development. Similarly, DoE advice stresses that generally local plans will not be needed in New Towns, particularly those built on 'greenfield' sites. However, Circular 22/84 acknowledges that they may be appropriate in New Towns containing substantial existing built-up areas (i.e. which will not be radically affected by the development corporation's proposals), but even here they will only be 'needed' where significant areas are likely to be developed or redeveloped (DoE, 1984g, para. 5.4). The possibility of substantial development within the designated area of a New Town not involving the development corporation seems unlikely, and to the extent that this is so, the scope for statutory local plans in New Towns appears to be rather limited.

5.4.2 Green Belt

In contrast to the planning of New Towns, the other key tool of urban containment policy (Green Belt) is progressed almost wholly via the development plans system, with local plans being particularly important since it is here that Green Belt boundaries are defined. The Green Belt concept has a long history, dating back to the turn of the century and concern with the various problems created by the rapid growth of London (Mandelker (1962); Thomas (1970); Munton (1983); Elson (1986). The first Green Belt had specific statutory backing in the *Green Belt (London and Home Counties) Act 1938*, which empowered the authorities involved to buy land on the fringe of London so as to control its development. Implementation of Green Belt policy by land purchase, however, proved prohibitively expensive, and it was not until the 'nationalizing' of development rights by the *Town and Country Planning Act 1947* that Green Belts could be established on a widespread scale. Around London, eight authorities were involved in the Green Belt proposed by Abercrombie in 1944, and by 1958 all the relevant county development plans had been approved by the Ministry, thereby formalizing his proposals. Green Belts were finally recognized as an element of national planning policy by central government in 1955 and 1958 (after most of the Green Belt around London had been approved). The policy statements involved, Ministry of Housing and Local Government Circulars 42/55 and 50/57, remain extant and have recently been reissued with DoE Circular 14/84. During the late 1950s and 1960s many local authorities submitted Green Belt proposals to central government but these were only gradually incorporated into statutory development plans due partly to the delay caused by the degree of government involvement in the system. Four categories of Green Belt emerged; only one of these related to fully approved Green Belt in a development plan, the others being variously labelled 'interim', 'submitted' and 'sketch'. By 1974 only about one-third of the area subject to Green Belt restraint was fully approved (Gault, 1981, p. 8).

For the new development plan system proposed by PAG and established by the *Town and Country Planning Acts* of 1968 and 1971, Green Belt policy has proved a difficult issue to handle within the context of the county/district split between strategic and detailed plan preparation. The problem is summarized by Gault (1981, pp. 10–11) as follows:

Green belt policy provides one of the few genuine interfaces between the structural, or strategic, and the local levels. On the one hand, the green belt is widely accepted as being a strategic policy tool for controlling the growth and development of large built-up areas; on the other the existence of a green belt has profound implications for detailed planning at the district level. Local planners and politicians may find it difficult to reconcile the requirement for strategic restraint with the needs and aspirations for growth of their local community, with the result that there could be considerable conflict between the county and district authorities.

Such disagreements have indeed arisen, often focusing on the development plan scheme or around the certification stage in the local plan preparation process, where counties rule on the conformity or otherwise of district-prepared local plans to the structure plan, e.g. the disputes between Hereford and Worcester CC and Bromsgrove DC over the latter's Wythall and Bromsgrove Local Plans.[21] In several cases, these disputes have been referred to the DoE, usually on an informal basis (Healey, 1979b). The problems arising from the county/district division of responsibility are particularly acute when it comes to the procedural question of the establishment and definition of Green Belt boundaries. Counties usually regard Green Belt as a key structure planning issue and as an element of strategic planning policy it should properly be dealt with in structure plans. But to be capable of effective implementation the Green Belt boundaries have to be clearly defined, and structure plans cannot do this since they can only show schematic boundaries. Precise boundary definition is a task for local plans, but questions remain over the most suitable type of local plan and which authority should be responsible for its preparation. Two main possibilities have emerged, each with various advantages and drawbacks (see Hebbert and Gault, 1978; Gault, 1981). For these options, boundary definition is by:

1 a Green Belt subject plan prepared and adopted by the county planning authority; or
2 a series of district-wide district plans prepared and adopted by the constituent district planning authorities.

The subject plan approach offers advantages of speed of preparation and county-wide continuity, but risks infringing DoE advice that the subject plan topic must have only limited interactions with other matters and that neither these nor the topic itself will suffer from being planned in isolation (DoE, 1984g, para. 3.16). Green Belt subject plans 'may' be used to settle boundaries, but as consideration of Green Belt boundaries is often closely related to other matters, 'a more general local plan may be more appropriate, particularly in areas where other local plans are proposed and are being prepared or are soon to be placed on deposit' (DoE, 1984g, para. 3.16). Nevertheless, various counties have pursued the subject plan option, e.g. Cheshire CC, Derbyshire CC, Greater Manchester MCC, Merseyside MCC, Staffordshire CC and Warwickshire CC have all adopted Green Belt subject plans. In some cases these plans have attracted objections from the district authorities concerned on the grounds that their scope for expansion is being unduly restricted. Manchester City Council for example has objected to the *principle* of county-adopted local plans as well as to the location of the Green Belt boundary, which threatens to enclose development land owned by the city (Healey, 1984).

The DoE's advice on the preparation of subject plans clearly favours the setting of Green Belt boundaries in district plans, which will usually be prepared by lower-tier district authorities. This approach, however, will possibly be cumbersome and lengthy, leading to uncertainty in the interim, and there is also the danger of variations in Green Belt policy between the districts involved. Districts demonstrate a wide range of attitudes to Green Belt, which must militate against a consistent approach. In any event, Munton points out that in the case of the metropolitan Green Belt around London only a proportion of districts are preparing district-wide local plans that will allow complete definition of the Green Belt boundary. Comprehensive definition of the boundaries of the metropolitan Green Belt around London would necessarily involve 62 districts and London boroughs, but only 33 are producing plans to cover the whole of their areas and not all of these will be seeking boundary definition as a main aim (Munton, 1983, p. 46).

To summarize, New Towns and Green Belt as complementary aspects of containment policy have very different relationships to the development plan system and to local plans in particular. The planning of New Towns is largely independent and self-contained; whilst local plans are of central importance in defining Green Belt boundaries, questions over what sort of plan to prepare and by whom remain unresolved. The inherent nature of Green Belt as a strategic planning policy requiring detailed local specification ensures that it remains a focus for dispute between county and district authorities.

5.4.3 Industrial Improvement Areas

Recent measures concerned with urban development and regeneration have focused more closely on specific issues and problems. *Industrial Improvement Areas* (IIAs) for instance represent an extension of the concept of area improvement first developed in the context of housing rehabilitation. The first IIA was established on a non-statutory experimental basis by Rochdale MBC in 1974. The primary goal of the programme was the removal of industrial obsolescence and the carrying out of environmental improvement, which in turn was expected to help promote economic activity and employment. The strategy of the local authority concentrated on a programme of 'pump-priming' investment in land assembly, infrastructure and environmental improvements, with progress thereafter dependent on private sector investment. A DoE-commissioned study of the Rochdale project confirmed that the IIA programme was likely to achieve a major improvement in the area's economic performance and environmental quality (Roger Tym and Partners *et al.*, 1979). The success of the Rochdale experiment led to general statutory provision for IIAs in the *Inner Urban Areas Act 1978*. This allows authorities designated under the Act to declare industrial or commercial improvement areas, to make loans and grants for the improvement of specified amenities, and to make grants for building conversion or improvement.[22] The amount of grant under this latter provision was originally linked to the job creation potential of the scheme involved, thereby emphasizing the employment objectives of these improvement areas, but the *Local Government, Planning and Land Act 1980* has removed this requirement.[23] Much of the funding for industrial and commercial improvement areas has been derived from the Urban Programme, which was itself recast by the 1978 Act (see below).

Some notable successes have been reported in achieving the linked IIA economic, environmental and amenity objectives, e.g. Gateshead MBC's Derwenthaugh IIA (Allen *et al.*, 1983), but the implementation process has not always been without its problems, particularly where grant take-up by the private sector firms involved has been low, e.g. Wandsworth LB's Queenstown Road IIA (Tayler, 1982). In relation to industrial and commercial improvement areas progressed by Sheffield City Council, Hammersley and Williamson (1985) suggest that such action can be very cost-effective in terms of job maintenance/creation, but that environmental improvement will be a slow process conducted largely at the cost of the local authority. Improvement areas can also be used as a means of managing change in these areas, enabling a smoother transition to take place from, for example, mainly industrial uses to service industries. Hammersley and Williamson (1985, p. 10) thus conclude that improvement areas have been successful as a 'small-scale local palliative' designed in response to economic and environmental problems, but point out that such local action is no substitute for regional and national economic policies aimed at reducing unemployment. As far as local plans are concerned, the land use proposals to be applied within IIAs may be included, but local plans should not designate IIAs (DoE, 1984g, para. 4.2). The DoE's reasoning here is the same as that applied to HAAs or GIAs, i.e. areas where special facilities or grants are available should be designated through the enabling legislation rather than via the local plan preparation process. This seems reasonable enough, particularly since IIA action, with its emphasis on environmental improvement and land assembly, can be expected to be more 'planning led' than GIA or HAA proposals, which will usually be the remit of housing departments. Where statutory backing for IIA proposals is needed, e.g. to aid land assembly through compulsory purchase, these proposals can always be included in a statutory local plan. Nevertheless, IIAs and local plans can proceed independently, in that there is no requirement to prepare a local plan prior to IIA declaration and local plans themselves cannot be used to declare IIAs, and this emphasizes the free-standing nature of these area-improvement policies.

5.4.4 Partnership

The *Inner Urban Areas Act 1978*, which gave legislative backing to the idea of industrial and commercial improvement areas, represented the outcome of a wide review of national inner city policy begun in 1976. The problems of urban deprivation had been formally recognized in 1968 with the introduction of the Urban Programme and the payment of grants to local authorities under the *Local Government Grants (Social Needs) Act 1969*. The Urban Programme was not directed expressly at inner areas or at economic problems, being mainly concerned with specifically social projects such as the establishment of community facilities. However, a series of research projects, culminating in the three DoE-sponsored Inner Area Studies, had shown that inner area problems were of such a magnitude that a more drastic and specific response was required. Inner area problems were found to be both complex and highly inter-related, and demanded an organized and co-ordinated policy response. The Liverpool study for example called for a 'total approach' to the inner area problems of physical decay, economic decline and social disadvantage (DoE, 1977a). The government response to the Inner Area Studies was contained in the White Paper 'Policy for the Inner Cities', which set out a number of proposals designed to strengthen inner area economies, improve the physical fabric, alleviate social problems, and secure a new balance between the inner areas and the rest of the city region in terms of population and jobs (DoE, 1977d, para. 26). This latter aim reflected a view that policies of large-scale decentralization such as the New Towns programme, which had been pursued since 1945 and were designed to reduce urban congestion, were at least partly responsible for the newly discovered problems of the inner cities (*ibid.*, para. 24). The key strategy in the new inner city policy was the concept of 'Partnership' arrangements between central government and selected local authorities, which were designed to bring about a new unified and co-ordinated approach to the complexity of inner area problems. This innovation reflected the assumption that existing responsibility for urban policy was spread across too wide a range of institutions who had few opportunities to develop a co-ordinated response; Partnership was thus to provide a permanent forum for collaboration between such agencies and organizations as the relevant district councils, county councils, area health authorities, the Manpower Services Commission, and various central government departments such as Environment, Education and Science, Health and Social Security, Industry and the Home Office (Parkinson and Wilks, 1983).

The legislative framework for this initiative was provided by the 1978 Act, which allows the Secretary of State to specify any local authority as a 'Designated District' if satisfied that a special social need exists in any inner urban area which could be alleviated by the exercise of the Act's powers.[24] These included new powers to make grants and loans available, to declare IIAs, and to adopt local plans in advance of structure plan approval.[25] The

Secretary of State can also specify 'special areas' requiring a concerted effort to alleviate the conditions therein; these form the Partnership areas.[26] In addition to the two statutory categories of 'Designated' and 'Partnership' authorities established by the Act, a third non-statutory designation has emerged – 'Programme Authorities'. This represents an informal category falling between the two statutory designations in terms of priority for resources and action. There are currently (1985) nine Partnership Authorities, 23 Programme Authorities, and a further 16 Designated Districts.[27] Funding for the policy derives from two sources: an expanded Urban Programme, which has also been extended in scope to cover industrial, environmental and recreational provision as well as social projects; and a redirection of main spending programmes (especially housing, education, transport and the Rate Support Grant) towards the inner city. Priority in Urban Programme allocations has been given to Partnership Authorities. As a vehicle for the general strategy of co-ordination, Partnership and Programme Authorities prepare Inner Area Programmes (IAPs) which are intended to cover both policy and expenditure issues. These programmes represent an extension of the concept of topic investment programmes, introduced earlier for housing and transport. The IAPs, however, broke new ground in that for the first time central government requested local authorities to tackle a specific complex problem, by producing what amounted to a corporate plan covering a number of different spending departments (Nabarro, 1980).

Statutory local plans appear to have played little part in practice in the development of IAP policy and expenditure commitments, or indeed in the inner area initiative in general. This reflects a widespread professional feeling that while local plans have a clear role in urban fringe locations, where the main planning issues are restraint or the detailing of structure plan allocations, they are less appropriate in inner areas where existing land uses and development promotion provide the planning context. Adcock (1979) identifies four key difficulties limiting the effectiveness of local plans within the inner areas:

1 The usual role of local plans in defining/allocating land use is largely irrelevant given the diversity of agencies involved and the uncertainties over their medium/long-term spending programmes.
2 It is difficult, if not impossible, for the planner to gain the willing and meaningful participation of other municipal spending departments such as housing, education or social services, which are the main agencies for action in the inner areas.
3 The timescale of the statutory local plan production process is too long and is in stark contrast to the timescales involved in the formulation of other public spending programmes. Adcock (1979, p. 41) comments that this is the main reason why local plans have had so little influence on

the inner area initiative, 'the timing of (local plan) preparation and approval must be brought into line with the expenditure programmes to which the plans will relate'.

4 The problem of evaluating the resource implications of local plans is particularly serious in the inner areas, where the technique of phasing to achieve flexibility merely creates the danger of blight; hence, short-term expenditure-linked planning seeking to explicitly manage change by combining proposals and resources is the only way forward.

Many local authorities have developed non-statutory approaches to local planning in inner areas as a contingent response to these problems, and these have been readily adapted to service Partnership arrangements where these have been established. The best documented example of this process is Liverpool, where prior to the introduction of the Partnership the goal of establishing corporate priority for particular local plan areas tended to be overriden by the need to adhere to city-wide policies and by the volatile political context. A statutory local plan was commenced but abandoned (Duerden, 1978). The subsequent establishment of the Partnership has emphasized the development of immediate programmes rather than an overall strategy in statutory or non-statutory form (Chape, 1978), so that the main planning concern has been with the basic land resource and its place within the Partnership's topic-based programmes (Hayes, 1981). In a similar context, Birmingham City Council has evolved informal 'Inner Area Studies', short- and medium-term packages of policies and proposals rolled forward annually, in an attempt to identify worthwhile schemes in advance of resource availability (Urwin and Wenban-Smith, 1983). Wenban-Smith contrasts the situation in Birmingham's inner areas with the relatively orderly process of preparing a statutory local plan for a peripheral development area (Sutton Coldfield), and comments that paradoxically 'it seems to be the areas where change is largely the result of public sector intervention which lend themselves least to statutory plans' (Wenban-Smith, 1983, p. 204). Adcock highlights Wirral MBC's use of conventional local plans to provide a framework for the preparation of an IAP. However, these plans are non-statutory, and were at an advanced stage when the authority received programme status. The availability of these plans was fortuitous, since the Partnership and Programme Authorities were given less than one year to submit the initial IAPs to the DoE. In a Partnership context, however, existing statutory frameworks may not be so readily integrated; Allan (1978, p. 116) comments that the Lambeth Partnership 'appears to be of the opinion that the Development Plan should accommodate all that the Partnership wishes to pursue rather than viewing the Plan as a very useful framework in which it can develop its Programme . . . the two processes must be brought together in a balanced and coherent fashion and so far this has not happened'. Lambeth's first IAP was prepared without reference to the emerging Borough Plan, whose policies had already been politically agreed as an interim guide to development in the Borough. This emphasizes the point that although IAPs are essentially spatially-oriented investment programmes, they are not formally linked to statutory plans. Unless statutory local plans can meet Partnership requirements (especially the budget cycle), opportunist or entrepreneurial informal local planning approaches are likely to be more suitable in providing a planning input to IAPs. These could also be used to provide a local community perspective on IAPs, i.e. as vehicles for consultation and participation. Non-statutory vehicles will tend to be more appropriate anyway given the various problems associated with the use of statutory plans in inner areas. In practice the 1977–8 inner city policy initiatives, and the Partnership arrangements in particular, have been criticized for failing to achieve changes in the traditional working relationships between local authorities, central government departments and other public agencies (Parkinson and Wilks, 1983). Experience of the Birmingham Partnership indicates that although Partnership has helped to strengthen existing links between agencies and provided a valuable opportunity for central government to work alongside local organizations, many interagency links are *ad hoc* or absent and the opportunity for collaboration between Partners has not yet been fully exploited (University of Aston Management Centre, 1985). There have also been charges that the IAPs have not benefited from sufficient public participation (Green, 1978), and that there was widespread ignorance of the policy initiative on the part of industry in the designated areas (Purton and Douglas, 1982).

5.4.5 Enterprise Zones

Partly in response to these problems and in order to emphasize the role of the private sector, the Conservative administration elected in 1979 introduced two new measures in the *Local Government, Planning and Land Act 1980: Enterprise Zones* (EZs) and *Urban Development Corporations* (UDCs). The idea of Enterprise Zones was first suggested in 1977 by Hall, who outlined what amounted to a market solution to inner city problems; as an experiment he advocated that 'small, selected areas simply be thrown open to all kinds of initiative, with minimal control . . . [to] create the Hong Kong of the 1950s and the 1960s inside inner Liverpool or inner Glasgow' (Hall, 1977 quoted by Taylor, 1981, p. 424). There has since been substantial debate on the likely efficacy of this proposal,[28] but the idea was readily taken up by the Conservatives while in opposition and established as a policy measure by the Chancellor's 1980 Budget Speech. In the process of implementation, the original link with inner city policy soon evaporated, with EZs being viewed not as an element of existing urban or regional policies, but as an experiment to see how a reduction in state intervention could benefit the economy. The DoE (1981b, p. 3) defines the purpose of EZs as follows:

The idea is to see how far industrial and commercial activity can be encouraged by the removal of certain tax burdens, and by relaxing or speeding up the application of certain statutory or administrative controls . . . EZs are not part of regional policy, nor are they directly connected with other existing policies such as those for inner cities or derelict land. The sites chosen will continue to benefit from whatever aid is available under these policies.

In any case, the initial criteria established for the siting of EZs (particularly the requirement that the sites be around 500 acres in size) severely restricted the available inner city options, e.g. in the Midlands there were no inner city sites of this size available. There was also a certain amount of 'policy drift' regarding the concessions that would be available in zones. Initially, the EZ package had included exemption from some non-basic health, safety and pollution controls and from employment protection legislation. However, the Department of Employment refused to countenance the introduction of exemptions from the *Employment Protection Act*; the Health and Safety Executive objected to the lowering of those statutory requirements which it was responsible for enforcing (Taylor, 1981). In the transition from concept to implementation, the emphasis of the EZ package has shifted from deregulation to the provision of enhanced financial assistance, so that the measure can be seen as 'one of the many small area initiatives currently seeking to stimulate local economies' (McDonald and Howick, 1981, p. 33). The following benefits are now available, for a period of ten years from the date that each EZ is established, to both new and existing industrial and commercial enterprises in the zones:

1 The simplification of land use planning procedures; developments that conform with the planning scheme for each zone do not require individual planning permission.
2 Exemption from rates on industrial and commercial property.
3 A 100 per cent allowance from corporation or income tax on capital expenditure on industrial and commercial buildings.
4 Exemption from industrial training levies.
5 Administrative benefits, such as fewer requests by government for statistical information and the priority processing of applications for special customs facilities.

Apart from the promised administrative changes, these concessions were all formalized in legislation in 1980 and 1981, with the *Local Government, Planning and Land Act 1980* establishing the procedures for the designation of EZs, the adoption of the planning scheme and the exemption from general rates.[29] This includes provision for central government to make up in full the loss of rate revenue suffered by EZ authorities. As Taylor (1981) comments, this has led to the Treasury becoming involved in detailed negotiations over EZ boundaries in an attempt to limit the cost of the scheme to central government. In these discussions the Treasury has sought to exclude existing industry, and this has had the effect of breaking up EZs into a number of smaller sub-zones principally comprising vacant or derelict land. A further factor encouraging the fragmentation of EZs is the shortage of suitably-sized single sites.

A total of 25 EZs, each with a lifespan of ten years, have now been designated throughout England, Wales, Scotland and Northern Ireland. A first round of designations in 1981 created 11 zones, with a further 14 being established by a second round in 1983–84.[30] Zones commonly consist of more than one site, the extreme cases all belonging to the second round of designations: Workington (six sites), Tayside and N.E. Lancs (seven sites) and Milford Haven (13). The EZs established by the two rounds differ significantly in terms of both location and physical characteristics. First-round zones are typically located in urban areas, although not necessarily within the inner city; the Swansea and Clydebank EZs are on the urban/conurbation fringe. In the second round, however, zones have also been established in more rural areas, e.g. the Workington EZ in Cumbria, Invergordon EZ in the Highlands of Scotland. Physically, first-round zones, although varying widely in their characteristics, 'all consist in large part of vacant and under-used land. In most of the zones the present position is the result of the decline of earlier industrial or commercial activity which has often left behind seriously damaged land' (Roger Tym and Partners, 1982, p. ii). In contrast, second-round zones include a relatively larger number of industrial estates and greenfield sites than derelict areas or vacant buildings, implying that they can be developed relatively quickly and cheaply. This reflects a central government criterion that second-round zones should be capable of early development, and suggests that a shift in emphasis has taken place in terms of the role of EZs as an instrument aimed at physical and economic decay (Lloyd, 1984b).

Although the EZs are directed explicitly at the private sector as the main agency for development, the consultants commissioned to monitor the progress of the EZ experiment have reported that the public sector is taking a prominent role in progressing the zones. Indeed, referring to the early zones designated in 1981, 'the first move in bringing land forward for development will generally be made by public sector landowners, who do not manage their land solely according to market criteria' (Roger Tym and Partners, 1982, p. iv). This relates partly to the extensive existing public sector landholdings within the first-round zones, and partly to the way in which local authorities and development agencies have used the EZ initiative as an opportunity to establish concerted economic development programmes. Botham and Lloyd (1983) comment that, in those zones in which most of the land is privately owned, the absence of a publicly-sponsored land assembly mechanism may act to retard development. One of the criteria for second-round designations is that the local authority or development corporation involved should be willing to co-operate closely with the private sector to ensure the continuous release of land for development and promotion and marketing of the zone (cited by Lloyd, 1984b). Here central government is recognizing the role that the public sector has played in progressing the first-round designations. In several of the first-round zones, substantial public expenditure had already been incurred prior to EZ designation in order to render the sites fit for development. The Swansea zone, for instance, is in the Lower Swansea Valley, for many years the focus of a major land reclamation effort by the local authority. Designation of the EZ has simply fitted in to the wider plans of the City Council for the Valley, which envisage a series of five separate but linked 'parks' extending from the EZ (dubbed the 'enterprise park') via the 'leisure', 'riverside' and 'city' parks to the 'maritime' park at the river mouth (Swansea City Council, 1981). Paradoxically, despite the free-market ideology of the EZ measure, active public sector involvement appears to be a key factor in the successful development of the zones, although this activity may bear little relation to the EZ concessions and will probably use resources diverted from elsewhere (McDonald and Howick, 1981). It is worth noting that other local authorities without EZs designated under the 1980 Act have nevertheless mounted analogous initiatives. Stockport MBC, for instance, established an 'economic enterprise area' comprising the bulk of the town centre in 1980, thereby predating EZs. This initiative utilizes no special planning powers or financial incentives, building instead on co-operation between local authority, the local business community, statutory undertakers and government departments in order to achieve its objectives. These are: stimulating public and private investment; enabling private development through land reclamation and infrastructure provision; the creation of new employment; and promoting environmental improvement (Hargreave and Kirkpatrick, 1982). This approach is grounded not so much on the removal or relaxation of controls but on the concentration of local authority staff and finance on the defined area and the involvement of local industrialists and business concerns,[31] representing a contingent response to limitations on public sector resources in relation to policy objectives.

As far as the statutory development plan system is concerned, the principal innovation of the EZ initiative is the simplified planning scheme for each zone. Schemes are prepared for prospective EZs by local authorities (district councils or London boroughs), new town corporations or urban development corporations on the invitation of the Secretary of State. These authorities then decide what concessions to offer to prospective developers, but the scheme must be in accordance with the terms of the invitation. Draft schemes must be given adequate publicity; those who may be expected to want to make representations on the scheme must be made aware of their entitlement to do so; such persons should be given adequate opportunity to make representations; and these must be considered by the authority provided they are 'made on the ground that all or part of the development specified in the scheme should not be granted planning permission in accordance with the terms of the scheme'.[32] Following this process the authority may adopt the scheme. There is a six-week period after publication of the notice of adoption during which the validity of the scheme can be questioned, after which the Secretary of State may designate the area to which the scheme relates as an EZ. These procedures are obviously modelled on those laid down for the preparation and adoption of statutory local plans, but there is no provision for a public local inquiry and hence no independent assessment of the scheme's proposals or of the strength and validity of objections made by occupiers of land excluded from the scheme (Grant, 1982). Nor is there any requirement for conformity with operative statutory plans. Indeed, the EZ scheme takes precedence over both structure and local plans; authorities must review any structure or local plans for the area and prepare any necessary alterations.[33] In the case of local plans, this process of review may include repeal and replacement. Circular 22/84 advises that new local plans will only be required for EZ areas where it is important to make proposals for (or to establish a development control framework for) development falling outside the terms of the EZ scheme (DoE, 1984g, para. 5.10). After designation of the zone there is no further provision for appeal, so that any proposal conforming to the scheme will bypass the protective mechanisms covering third-party interests (Purton and Douglas, 1982). The EZ scheme is thus a unique planning instrument which has been variously compared to a hybrid of a local plan and a planning permission (McAuslan, 1981) and to a form of special development order (Grant, 1982).

After designation of the EZ the scheme has the effect of granting planning permission for such development as is specified in the scheme. This has been done in two ways. The Swansea EZ scheme, which is based in part on the North American 'zoning' schemes, specifies the types of development that can take place automatically, while other schemes give a general grant of permission and set out exclusions to this. Whichever approach is taken, planning permission granted in EZ schemes may be subject to conditions, limitations, and reserved matters; schemes can also specify sub-zones where the full range of planning permission granted by the scheme is inappropriate, e.g. to protect the amenity of an adjoining residential area. Schemes often exclude development from the general grant of planning permission where specific control is needed, e.g. to protect health and safety or for pollution control; and other classes of development may also be excluded to reflect local circumstances, e.g. mineral development, buildings over a certain height, scrapyards, retail development over a certain size.[34] The retailing issue has proved particularly controversial, with local authorities seeking to restrict large-scale retail development (particularly hypermarkets) in their zones in the face of opposition from the DoE.[35]

The consultants who have monitored the early progress of the EZ experiment report that in practice the simplified planning regime set out in the schemes has had little deleterious effect on either the quality of new development in the zones or their land use structure (Roger Tym and Partners, 1984). The standard of new development reflects the substantial involvement of public sector agencies as developers and landowners; the continuing need for building regulation approval and the use of standardized building materials and construction techniques; the requirements of investors for attractive development; continuing informal contact between developers and planning authority, and the fact that much new development had been applied for prior to EZ designation and so was subject to the usual planning procedures. Moreover, as regards the overall development of the zones the types of use that have occurred have not as yet varied significantly from what established planning policy would have allowed without the EZ scheme. This is not wholly surprising given that local planning authorities can be expected to try to ensure that EZ schemes only grant permission for development that would have been permitted anyway (Purton and Douglas, 1982). A further factor relates to the physical and economic characteristics of the areas chosen as the first round EZs and the relative weakness of planning control therein as a restraining influence on economic growth and development. As two members of the firm of consultants monitoring the zones (McDonald and Howick, 1981, p. 33) have pointed out, this is:

because local authorities there are at their most anxious to stimulate employment; and the environmental impact of industrial and commercial development is if anything favourable, since congestion is hardly a problem, and environmental conditions are poor. There has been no attempt to apply the Enterprise Zone experiment to areas where strong development pressures conflict with the fulfilment of planning objectives.

This must limit the extent to which useful conclusions can be drawn about the effect of a simplified planning regime in other planning contexts, e.g. of restraint. This is of relevance since central government has recently proposed to extend the EZ scheme idea to other locations which would be designated as 'simplified planning zones' (see below). The monitoring consultants recognize the possibility of extending EZ schemes to other areas, but caution that this would not be without problems, e.g. as a result of a more intricate and mixed pattern of land uses or a greater need for environmental controls, both of which would lead to more problems safeguarding third-party interests (Roger Tym and Partners, 1984). It is in any case worth emphasizing along with Lloyd (1984a) the extent to which EZ schemes bypass traditional planning procedures, particularly through the absence of independent assessment during scheme preparation and the lack of any 'conformity' link to established planning policy. In turn, this latter point is an aspect of the DoE's wider assertion (1981) that EZs are not part of existing regional policies or of those for inner cities or derelict land. EZs are an autonomous measure and are *intended* to be so. This very free-standing nature of the EZ measure has allowed them to be portrayed as part of central government's economic and employment policy as politically appropriate.[36]

Though Hall (1984, p. 297) concluded that the EZ experiment 'may prove to have relatively little impact on the way we conduct our planning system on the remaining 24 million hectares of the United Kingdom', the recent proposals for 'simplified planning zones', based on the EZ simplified planning regime, suggest otherwise. The idea was first mooted in its present form by the Secretary of State for the Environment in September 1983; speculating that there could be a good case for extending the simplified form of planning arrangements applying in EZs to other defined areas, he suggested that:[37]

this could be done by having a system by which local authorities could give general planning permission for specified categories of development in particular zones . . . in such areas the need for an individual planning application would be obviated.

Subsequently a consultation paper on simplified planning zones (SPZs) was issued by the DoE (1984a; see Lloyd, 1985, for a review of the background to the proposal). Referring to the value of the EZ planning schemes, the consultation paper proposed a system whereby local planning authorities would prepare schemes for SPZs within their areas that once adopted would have the effect of granting planning permission for development specified in the scheme. SPZs would have a limited lifespan (five to ten years); revocation within this period would be a possibility. Schemes could include conditions, limitations and reserved matters, although these should be minimized, and hazardous processes and substances would be excluded from the permission granted. SPZs would be inappropriate in designated National Parks, AONBs, Green Belt or conservation areas. The DoE noted that schemes could be framed in one of two ways: where the development for which permission is automatically granted is set out in the scheme (based on the Swansea EZ scheme); or where all development receives an automatic grant of planning permission except that specifically excluded by the scheme. The Swansea approach would be most appropriate where a specific type of development was required, while the latter would be better suited to circumstances where the primary aim is to secure investment in an area and a variety of types of development would be acceptable. Procedurally, schemes would incorporate consultations and participation, with provision for a public local inquiry, or written representations to consider objections; the Secretary of State could have reserve call-in powers, e.g. to be used where a proposed SPZ was in conflict with regional or national policies. Other possibilities suggested by the proposals paper are for each SPZ scheme to be submitted to the Secretary of State for approval, or for schemes to be incorporated into statutory local plans, being prepared either as part of the plan or as a proposed alteration. The usual local plan procedures would then apply, with the Secretary of State having call-in powers.

Responses by the local authority associations and professional bodies to the consultation paper have focused critically on the underlying and unproven premise that planning control is responsible for creating delay and uncertainty in the development process, and have pointed out that to base the SPZ proposals on the claimed success of the EZ 'experiment' is at best premature.[38] The RTPI, for example, has suggested that the SPZ idea is likely to be of little use for areas of mixed land uses such as the inner cities, whose complexity would militate against the devising of a practical scheme; there would still be difficulties in co-ordinating development in areas of multiple ownership. Other bodies have emphasized that removal of planning control by itself will do little to stimulate development; SPZ schemes should be seen as part of a wider local authority programme of positive development promotion (National Housing and Town Planning Council (NHTPC), Town and Country Planning Association (TCPA)). Worries have also been expressed at the threat to local autonomy in the presently highly discretionary area of development control represented by the procedural proposals for scheme approval or call-in by central government (Association of Metropolitan Authorities (AMA), Royal Town Planning Institute (RTPI), NHTPC). Responses to the consultation paper, however, have not been wholly negative, with both the AMA and the RTPI supporting the suggestion that local authorities should have some means of translating local plan development proposals into planning permission. For the RTPI the way forward is not via the SPZ scheme but through the simpler mechanism of allowing local planning authorities to apply for, and obtain permission for, development within their areas which accords with the adopted local plan. At present, regulations only allow the local planning authority to grant itself planning permission where it is either the landowner or the prospective developer.[39] According to the RTPI, removal of this restriction would formalize and strengthen local plan designations, thereby providing a degree of certainty lacking in the development plan system. It would also provide a convenient means of development promotion. At the same time, the need for primary legislation, the loss of public opportunity to comment on development proposals, and further central government intervention in local development issues, all of which are entailed in the SPZ proposals, would be avoided. Both this suggestion and the DoE's proposals to link SPZ schemes to local plans would have the effect of giving greater weight to statutory local plan proposals. In terms of the strategic planning hierarchy, both suggestions would strengthen the presently weak link between local plan statements (Level 3) and implementation (Level 5) and are thus to be welcomed. However, statutory local plans that are up-to-date are not universally available, and in these situations the benefits of either approach would have to wait on local plan preparation or modification.

Nevertheless linking SPZs with local plan proposals in some way is surely preferable to the alternative of the free-standing designation of SPZs without reference to the development plan system, since this would do little to foster closer links between statutory local plans and development activity. Following the 1984 consultation paper, the government announced in July 1985 that new legislation is to be introduced to permit the setting up of SPZs. Under this legislation local planning authorities will have powers to introduce SPZs, but will also be required to consider proposals for the establishment of SPZs initiated by private developers, while it is intended to give the Secretaries of State reserve powers to direct the preparation of proposals for a SPZ. These latest proposals however make no reference to the way in which SPZs are intended or expected to relate to approved and adopted development plans.[40]

5.4.6   Urban Development Corporations

In addition to EZs, the *Local Government, Planning and Land Act 1980* also makes provision for the designation of *Urban Development Areas* with a view to securing their economic, social and physical regeneration, and for the establishment of *Urban Development Corporations* (UDCs) to oversee that regeneration. The introduction of UDCs with their specific urban development remit reflected a shift from the earlier policy position that 'local authorities are the natural agencies to tackle inner city problems' (DoE, 1977d, p. 8) towards a recognition of the need for a single function agency unencumbered by the broader and diverse local authority responsibilities, and also embodied a greater emphasis on private sector funding and involvement than had been the case in earlier phases of inner city policy.

Within their designated areas UDCs are largely autonomous agencies similar in many respects to the New Town development corporations. Interestingly, the possibility of using such an agency to tackle inner area problems had been considered by the 1977 inner city policy White Paper but rejected on the grounds that in the inner-city development context 'it is important to preserve accountability to the local electorate' (DoE, 1977d, p. 8). The concept was revived nonetheless by the incoming administration in 1979, who were faced with the general assumption that the Partnership approach was characterized by the exclusion of local action groups and entrepreneurs from the bureaucratic bargaining process, resulting in a lack of drive, purpose and ability to generate new investment in the areas concerned (Cox, 1980). This indicates something of a change in attitude in central government, emphasizing the advantages of the UDC approach (single-minded management, industrial promotion expertise, development experience) at the expense of the disadvantages associated with the weakening of local control. Subsequently, the UDCs designated under the 1980 Act have been criticized for their largely independent style of operation and lack of political accountability to a local electorate.

An urban development area may be designated by the Secretary of State if he considers that 'it is expedient in the national interest to do so', and for the purposes of regenerating such an area a UDC may be established.[41] In England the areas designated as urban development areas must be within a metropolitan district or an inner London borough. There is no requirement for prior consultation with affected local authorities (although UDCs are required to prepare a code of practice governing consultation with the relevant local authorities)[42] or for a public local inquiry. However, designation is by statutory instrument, which must be approved by both Houses of Parliament, and these procedures offer the chance for objections to the proposed designation to be heard. The wide-ranging remit of UDCs is 'to secure the regeneration of its area' by 'bringing land and buildings into effective use, encouraging the development of existing and new industry and commerce, creating an attractive environment and ensuring that housing and social facilities are available to encourage people to live and work in the area.'[43] To these ends UDCs have extensive powers to acquire, manage, reclaim and dispose of land and property; to carry out building and other operations, to seek to provide basic infrastructure services such as water, electricity, gas and sewerage; to carry on any business or undertaking; and to 'generally do anything necessary or expedient' for the purpose of achieving their main objective.[44] There are also wide powers of land acquisition (if needs be compulsorily), together with a provision allowing the Secretary of State to order that specified land in the development area held by local authorities, statutory undertakers and other public bodies be vested in the UDC.[45] Finally, there are extensive planning powers. The UDC may submit proposals for the development of land within the area to the Secretary of State; once approved by him following consultations with relevant local planning authorities, a special development order may be made granting automatic planning permission for any development of land in accordance with the approved proposals.[46] These arrangements are similar to those governing development by the New Town corporations. Grant (1982, p. 543) comments that the development proposals may be submitted in a loose- and broad-brush form, and that the usual forward planning requirements of survey, consultation and participation are here largely absent.

The UDC may also be made responsible for the exercise of development control within the area,[47] although development plan functions are retained by the local planning authorities involved. However, there would be little point in a local planning authority preparing a local plan for part or all of an urban development area given the absence of effective control over the implementation of its provisions via the development control process. Circular 22/84 (DoE, 1984g, para. 5.7) confirms and enlarges upon the subservient role envisaged for structure and local plans in relation to UDC proposals:

In producing their strategic proposals an urban development corporation should have regard to existing structure and local plans which remain the responsibility of county and county or district planning authorities respectively. In the event of a conflict between an important urban development corporation proposal and the relevant structure or local plan, the urban development corporation may ask the Secretary of State to direct that an alternative be prepared to that structure or local plan.

Following the 1980 Act two UDCs were established in 1981, one for London Docklands (the LDDC) and one for Merseyside (the MDC, whose area comprises three non-contiguous sites in Bootle, Liverpool and the Wirral). The LDDC area includes a first round EZ on the Isle of Dogs, for which the Corporation rather than the local authority (Tower Hamlets LB) has responsibility. Special development orders granting planning permission for approved proposals were made in 1981, as were the orders transferring the development control function from the local planning authorities involved to the Corporations; a number of orders vesting publicly held land in the two UDCs have also been made. The two Corporations together were allocated £66 million for their first year of operation 1981–82 out of a total provision for expenditure on inner city regeneration of £224 million (the remaining £158 million went to the Urban Programme).[48] The fact that 29 per cent of the 1981–82 inner city policy provision went to the two Corporations is a measure of the extent to which resources are now being concentrated on relatively small areas within two inner city locations. Some £52 million (78 per cent) of the total UDC allocation of £66 million went to the LDDC, no doubt partly reflecting the fact that the Docklands area at 5000 acres is much larger than that designated on Merseyside (800 acres). Other key differences between the two areas are as follows (Nabarro, 1981):

1   The LDDC area has a residential population of more than 50,000 compared to less than 500 in the MDC area.
2   The MDC area is 80–90 per cent publicly owned by the local authorities, British Rail, and the Mersey Docks and Harbour Company; there are also major public holdings in the London Docklands but also substantial areas in private ownership.
3   Much of the MDC area is redundant/vacant land; the LDDC area has similar features but also contains a large number of established industrial and commercial concerns.
4   The MDC has announced its intention to consider all of the land within its area for acquisition and development, whilst the LDDC's strategy is to selectively consider sites for vesting/acquisition and development.

Despite these differences, Nabarro points out that the two UDC areas share common 'supply-side' problems of development, such as poor ground conditions, legal and ownership constraints, and a perceived public sector inability to provide development opportunities to attract the private sector. The UDC approach may well overcome difficulties such as these on the supply side but is unlikely to offer much hope of an improvement in terms of the private sector demand for land in the inner areas. This conclusion is likely to be more applicable to the MDC area than to the LDDC area, given the latter's regional location and proximity to the City of London.

The differences between the characteristics of the two UDC areas also help to account for differences in terms of the degree of local opposition voiced to the establishment of the two Corporations. The MDC passed through the Parliamentary designation procedure unopposed (one objection was made by the union NALGO but subsequently withdrawn), but the LDDC attracted objections from the GLC, the Dockland boroughs, and various community groups. The common theme to these petitions was the loss of local autonomy that would result from designation of the LDDC, a largely independent body not politically accountable to a local electorate. The LDDC was established none the less, but opposition to and criticism of their proposals has continued (Page, 1985). Local hostility is directed at the LDDC's autocratic style of operation as well as focusing on specific issues such as the LDDC-promoted residential development by private sector housebuilders,[49] or the proposal to develop a STOLport (a short take-off and landing airport trading on proximity to the City). Local opposition to the latter has led to the production by community groups of an alternative set of 'local plan' proposals in the form of a 'people's plan' for the Royal Docks (Newham Docklands Forum, 1983). This proposes that the Docks be kept open for a variety of port-related activities. The plan's immediate purpose was as a submission to a public inquiry into the STOLport proposal,[50] but there are longer-term ambitions to integrate the plan into a statutory local plan in preparation by Newham LB. The plan recognizes, however, that this step would not guarantee implementation given the LDDC's wide-ranging development, planning and land-owning powers, and this fact is emphasized by the DoE's position in Circular 22/84 quoted above.

In spatial terms, the two designated development areas do not significantly reduce or fragment the role or scope of the development plan system on a national basis. However, the principle of the UDC approach clearly marks a threat to the statutory land use planning system, the long-established use of similar agencies for the more discrete task of New Town development notwithstanding. If further and more extensive urban development areas were to be designated, with UDCs enjoying the wide powers of the 1980 Act, then the effectiveness of the formal development plan system will be further reduced. In particular, the UDC approach involves the preparation and implementation of investment and development strategies for inner-city or other areas in a way which potentially ignores the wider implications of those developments. There are as yet no firm plans to designate further urban development areas, but the principle is established and further designations are clearly under consideration.[51] In this way the introduction of UDCs has a parallel with the initially limited EZ 'experiment' which, rapidly declared a success, has readily generated the SPZ proposal to extend the loosening of planning control involved in the EZ scheme. However, SPZs *could* be integrated into the development plan system; UDCs are free-standing agencies with no statutory obligations to conform to existing or future development plan provisions. Local planning authorities retain plan-making powers, but any such plan production risks being a hollow exercise given UDC development control powers and the special development order provisions. Much, however, depends on the operational 'style' adopted by individual UDCs and their interpretation of their role under the 1980 Act. The theoretical potential for autonomy from the statutory development plan system does not rule out by any means UDCs working closely with other agencies to achieve common objectives, and there is some evidence to suggest that this is what is happening on Merseyside (Nabarro, 1981; Boaden 1982; Adcock, 1984).

The objectives and style of operation of the urban development corporations are similar in some ways to the Glasgow Eastern Area Renewal (GEAR) project. There is, however, one important difference: GEAR is a partnership operation which involves amongst others central government, local authorities, the Scottish Development Agency, the Housing Corporation and the Manpower Services Commission (Leclerc and Draffan, 1984). Political accountability is thus an integral part of the structure, and ensures that the wider implications of the problems and policies for the area will be addressed by the bodies responsible for the wider area.

5.4.7 Task Forces and Urban Development Grant

As the UDCs were being established the urban riots in Liverpool and other United Kingdom cities in the summer of 1981 stimulated further *ad hoc* policy measures. Funding for the Urban Programme was increased, while the Secretary of State for the Environment was made 'Minister for Merseyside' since the worst rioting had occurred in Liverpool. As such, he was supported by an inter-departmental 'task force', established on a purely informal basis and intended to '. . . bring together and concentrate the activities of central government departments and to work with local government and the private sector to find ways of strengthening the economy and improving the environment in Merseyside.'[52] This would seem to be no more than a duplication of the co-ordinating activity supposedly being undertaken by the Liverpool Partnership. The potential value of the task force initiative remains obscure; although it has not yet been declared a 'success', the introduction to the Partnership areas of similar 'city action teams', partly modelled on the Merseyside task force, was announced in 1985.[53]

At the national level, financial advisors from the private sector were seconded to the DoE for a year to form the Financial Institutions Group

(FIG), charged with developing new approaches and ideas for urban regeneration. The proposals from FIG which have received the most enthusiastic support from government are those which involve co-operation between the public and private sectors, the best example of which is the Urban Development Grant (UDG). Based closely on the American Urban Development Action Grant, UDGs were introduced in 1982 as a low-cost but high-profile measure designed to encourage private sector investment in inner city areas. Grants are paid to support specific development proposals put forward jointly by local authorities and private developers acting in partnership, and are designed to 'lever' private investment by providing just sufficient public funding to turn a negative project appraisal into one which can be expected to yield a reasonable return. The aim is thus to ensure that developers involved with projects on which UDG is to be paid have an opportunity for profit but also face the normal risks of development (Jacobs, 1985). Payments of UDG largely derive from the Urban Programme but some comprise derelict land grant. Experience in Yorkshire and Humberside shows that while participating local authorities have related UDG applications to existing policy frameworks, e.g. UDG has been used as a means of securing a particular type of project or the development of a particular site, little emphasis has been given to the specific socio-economic benefits expected from grant submissions (Mason and Whitney, 1985a, b). Nationally, there is evidence to indicate that some regions such as London and the West Midlands are achieving greater success in attracting UDG funding than others; 'the incentive of UDG appears to be insufficient to break this traditional investment pattern' (Goodhall, 1985, p. 42). Goodhall suggests that this trend is likely to be consolidated unless greater consideration is given to the problems of different areas and to the criteria used to direct the grant to areas of need. In fact, the success of the policy is commonly measured in terms of the 'leverage ratio' between private and public money invested. By 1984, UDGs totalling £65 million had been awarded, which is expected to bring forward some £285 million of investment from the private sector; however, this measure of 'success' ignores non-financial benefits, project externalities, and other relevant public investment, e.g. on infrastructure (Alderton, 1984). A further problem in evaluating the UDG policy is assessing the extent of substitution, whereby public grant merely replaces private investment that would have been forthcoming anyway.

5.4.8 Summary
The various measures concerned with urban development and regeneration reviewed here once again demonstrate the tendency apparent in housing and conservation-oriented initiatives of limiting the availability of special funding, or the application of special powers or policies, to more or less restricted areas. Also notable is the degree of independence from the

statutory development plan system in general, and from local plans in particular, that most of these measures display. Green Belt is the only policy that actually requires a local plan to be prepared to enable it to be effectively implemented. The other policy tools concerned with urban development are effectively free-standing, with their own empowering legislation; there are links to statutory local plans, but these have been established in a variety of ways. Sometimes the relationship is specified by statute (e.g. the requirement that local plans be reviewed, and if necessary altered, repealed or replaced to take account of EZ scheme provisions), by circulars (which usually emphasize the discrete character of such initiatives in terms of the legislation involved, e.g. 22/84 stresses that local plans cannot designate IIAs). Sometimes this relationship has evolved through practice (as seems to be the case with local plans and IAPs), whilst the actual nature of the relationship varies between initiatives, even where these are apparently similar. New town development corporations are wholly independent of the local planning authority as far as their own proposals are concerned by virtue of the new town master plan and associated special development orders. In urban development areas the UDC also benefits from special development orders granting planning permission for approved proposals but, in addition, exercises the general development control function; local planning authorities here retain all forward planning powers, but local plan proposals are expected to give way to those of the UDC. Similarly, EZ schemes take statutory precedence over local plan proposals but there is at least a possibility that SPZs will be linked to local plans so as to foster implementation. In any case, the actual policy content of EZ schemes is partly determined locally and to this extent will tend to reflect prevailing planning policies. Industrial Improvement Areas are formally established outside of the development plan system but local plans can include relevant land use policies and proposals. Like other policy measures IIAs neither *need* a local plan nor *can* they be designated in a local plan, despite the obvious common focus of interest – industrial land use policy.

Together, this plethora of centrally-inspired policy initiatives (which are also sometimes centrally designated) hints at a somewhat limited and jaundiced view of the local plan on the part of central government, since the search for new policy forms is occurring outside the development plan framework rather than within it. This view of statutory local plans is not limited to the DoE, however, as the use of informal local planning vehicles particularly by inner city local authorities seems to suggest. The concern with resources that these approaches commonly show raises a further issue relating to the integration of local plans and other formal initiatives. This is the relationship between local plan proposals and various public sector topic-oriented investment programmes.

5.5 Investment programmes

The gradual introduction and development of the relatively sophisticated structure and local plan system in the 1970s was matched not only by the growth in various free-standing policy initiatives and experiments reviewed above, but also by the establishment of planning and programming devices in other areas of public policy-making. These policy plans and programmes were introduced on a topic or sectoral basis by central government in order to guide and control policy development and subsequent expenditure at the local authority level. The two most significant examples are Transport Policies and Programmes (TPPs), introduced in 1975, and Housing Investment Programmes (HIPs), first used in 1978. The preparation of IAPs by Partnership and Programme Authorities under the *Inner Urban Areas Act 1978* reflects the same trend. All of these devices have a strong expenditure component; each consists typically of a plan or strategy statement explicitly linked to a budget, representing an attempt in each particular policy area to relate plan proposals to their cost and to the availability of resources.

5.5.1 Housing Investment Programmes

Housing Investment Programmes, for example, comprise three elements: a narrative strategy statement setting out the overall local housing situation and the general approach adopted; a more detailed numerical statement of housing 'need' giving information about levels and trends in population, households and housing stock; and a financial statement relating to past expenditure and also making spending proposals. The forward financial planning contained in this last element had a four-year timescale when the HIP system was first introduced, but this has since been reduced to two years. The HIP system was, according to the DoE, designed 'to enable local authorities to present co-ordinated analyses of housing conditions in their area and to formulate coherent policies and programmes of capital spending on public housing'; the new system allowed local authorities to produce strategy/spending solutions to local housing needs, albeit 'within the framework of national policies and resources available'. HIP submissions it was promised 'will strongly influence the pattern of allocations and national priorities for housing expenditure; they will enable the Government to allocate resources according to comprehensive assessments of need' (DoE, 1977c, paras. 3 and 4). HIPs are submitted by the local housing authorities (in England the district councils, London boroughs and, prior to its abolition in 1986, the GLC) to the DoE, who following scrutiny of the bids make spending allocations for the coming year. These allocations do not represent a grant, but constitute a notification of the amount which local authorities may spend in that year; this is financed by borrowing at interest. There are similar systems for the allocation of housing capital expenditure in Wales and Scotland. The HIP system originally sought to integrate the planning of *local* housing strategies and programmes with the *central* allocation and control of investment, but the subsequent evolution of HIPs has stressed the latter aim at the expense of the former (Paris, 1979). HIPs are now widely viewed as an effective means of implementing public expenditure cutbacks, both through cuts in allocation and underspending by the local authorities. As Leather (1983, p. 215) comments, 'HIPs have evolved into a narrower mechanism for short-term financial control and the imposition of national housing policy objectives at local level'. The effectiveness of HIPs in controlling housing investment has led to the introduction in 1981 of a similar but wider framework for the control of local authority capital expenditure (the block grant system, established by the *Local Government, Planning and Land Act 1980*).

5.5.2 Transport Policies and Programmes

However, HIPs themselves were based on a similar programming device in the transport field, the Transport Policy and Programme (TPP). These programmes were originally required to be submitted annually by the highway authority as a bid for Transport Supplementary Grant (TSG). The submission took the form of a statement of the highway authorities' objectives and policies for transport for the following 15 years; a statement of its expenditure on transport in the preceding year, and a proposed budget for transport expenditure on physical infrastructure and subsidy for the next year and an estimated budget for the succeeding four years. The short-term policies and proposals set out in the structure plan were intended to form the basis of the TPP which, in effect, was a five-year rolling programme for investment into transport infrastructure, transport rolling stock and subsidy which is rolled forward annually.

Since 1984 in England (1983 in Wales) the situation relating to TSG has changed. TPPs are no longer required to be submitted, and TSG is available only for capital highways expenditure of more than local significance in England and on schemes in excess of £5 million in Wales. However, bids for TSG for such highway schemes must be justified by supporting evidence and most authorities continue to provide this in the form of a TPP.[54] Bids for grant support for local highway schemes and other transport proposals are now made under the normal bid for capital allocation for the county councils as a whole.[55] In Scotland, grant aid for highway and transport proposals is provided through the Rate Support Grant. The bid for the highways and transport allocation is made by the regional councils through the TPP which is now required on a four-yearly cycle, although most regions update their TPPs annually.

In being linked to a specific grant in England and Wales, therefore, TPPs differ from HIPs where the allocation simply constitutes permission to spend.

5.5.3 Inner Area Programmes

As noted, the introduction in 1979–80 of Inner Area Programmes (IAPs), prepared by designated Partnership and Programme Authorities, marked a new development in local planning since they required a corporate approach to a single (but complex) problem. Stewart (1983) notes that although the policy process varies widely from authority to authority, IAPs usually comprise an overall strategy statement, incorporating an assessment of issues and problems and a statement of policies, together with a specific costed programme of projects to be funded out of the Urban Programme and to be carried out over a period of from one to three years. The idea of the 1977 inner cities policy White Paper, of 'bending' main programmes favourably towards the inner areas, has been difficult to achieve, partly because of the initial emphasis in the 1979–80 IAPs on Urban Programme administration, and subsequently because of a lack of central government commitment to main programme review. Thus, the DoE in a 'lead' role has more often had to resort to negotiations to sort out inter-departmental issues rather than fronting a collaborative and central government wide policy review process; 'Partnership has not been the forum for the review of central government policies as they impinge on the inner areas . . . [this feature] has been one of the main causes of local government disillusionment with the inner-cities initiative' (Stewart, 1983, p. 207). The IAPs have thus been largely concerned with Urban Programme allocations, which have increasingly favoured economic/infrastructure development, with alternative 'social' projects becoming less important. Central government has been anxious to involve the private sector in IAPs, but this has only reinforced the emphasis on infrastructural support for economic development. At the same time, private sector involvement in inner areas policy is taking place, but generally outside of Partnership, e.g. the work of the Financial Institutions Group and the subsequent introduction of Urban Development Grant. Stewart thus characterizes the inner cities policy of which IAPs are a part as a channel by which central government offers a specific grant to approved programmes with a strong economic development orientation (although there is also social/community interest). In this respect, the 1977–78 inner city policy initiative has come to resemble the other local authority investment programmes, demonstrating in IAPs the usual relationships of bid/allocation and expenditure control rather than a genuine joint attempt to foster policy development. The inter-related nature of complex 'wicked' public policy problems is ignored. In Christensen's terms (1985), a 'known technology' (the investment programme) is being applied in a situation where there is only limited agreement or commitment to the policy goals involved within government.

5.5.4 Development plans and investment programmes

What implications does the establishment and use of these programmes have for the statutory development plan system in general and local plans in particular? The fact that programmes have an explicit resource component means that they are of potentially great value in the implementation of structure and local plan proposals, but the realization of this potential depends on the existence of close links between development plan proposals and those put forward in the programmes. The need for a policy overview to be established by development plans, and their suitability for this task, can be summarized in terms of the differences between statutory plans and investment programmes:

1 *Timescale.* Development plans provide a longer policy perspective than the investment programmes, particularly where the short-term budgetary component is concerned, thus enabling long-term goals and objectives to be set. The programmes in contrast can be more specific on the timing of expenditure.

2 *Scope.* Structure and local plans together provide an important spatial perspective, relating overall land supply/location to the policies and programmes contained in HIPs and TPPs. Development plans are thus the obvious vehicles to ensure the integration of land use and transport developments. Structure plans in particular derive from a more comprehensive policy approach than that taken in the topic programmes and so will be essential, e.g. in providing data for the strategy statements of HIPs and TPPs, and in ensuring that the resultant programmes are compatible. All three investment programmes reviewed above are essentially short-term (annual review), resource-orientated planning systems, emphasizing financial decisions, cost effectiveness, management and implementation (Murie and Leather, 1977), and as such are hardly suitable vehicles for the development of strategic policy.

3 *Preparation processes.* Development plan preparation involves statutory provision for formal consultation and participation, together with a public local inquiry in the case of local plans. Investment programmes on the other hand are typically drawn up with little or no public consultation in a process which does not encourage open policy debate, e.g. as in the case of some IAPs. This makes such programmes unsuitable vehicles for the development of long-term or strategic policy as opposed to the short-term incremental roll forward of established projects and commitments. Indeed, Bayliss (1975) suggested that there was a possibility that TPPs will replace structure plans as the main policy instrument for local transport due to their greater currency and budgetary significance, although given the recent (1984) change in emphasis of the TPP this is now less likely.

Certainly the strategic planning model developed in Chapter 2 emphasizes the dominant role of the development plan. The model requires that the sectoral rolling programmes developed at Level 4 (resource allocation) are based on the general policies and resource allocation proposals set by higher levels in the hierarchy. In other words, such programmes may formulate phasing proposals and a relative distribution between development projects, but would be constrained by the absolute allocation of resources and development given in the levels above. The statutory local plan is not wholly adequate to this task, particularly where resources are concerned, but this reflects the organizational context within which such plans are framed as much as being a problem with the local plan itself. Nevertheless, the DoE considers that structure and local plans provide 'a useful framework' for those aspects of investment programmes which have a bearing on the development and use of land; 'the preparation of development plans and other plans and programmes should stem from a common planning process, so that assumptions, for example in HIPs, are consistent and the policies and proposals are compatible'. Integrated transport and land use planning is particularly important, and development plans 'enable transport policies to be set in a wider context and within a longer timescale than TPPs and PTPs [Public Transport Plans] and their preparation should be fully integrated' (1984g, para. 5.1). The relationship envisaged here between development plans and TPPs in particular reflects the subordinate role for such programmes suggested by theory, where they are based on and constrained by higher level policies, and confirms once again the hierarchical model of policy control informing the DoE's view of the land use planning system.

In practice, however, this model is but imperfectly realized as far as the various topic investment programmes are concerned. There are two key problems: first, problematic district–county relationships have in some cases hindered integration between district-prepared HIPs and housing policy in the county structure plan; second, the strategy statements which are a shared feature of HIPs, TPPs and IAPs have all been down-graded in importance since the programmes concerned were introduced, thereby emphasizing financial matters at the expense of policy content. To take the district–county issue first, this has been raised in relation to HIPs rather than TPPs since the counties have responsibility for the preparation of both TPPs (highways) and structure plans (strategic land use and transport planning). It is the districts, however, who as the local housing authorities, prepare HIPs potentially isolating the counties whose main influence on housing policy is through the structure plan. Yet structure plans are not adequate vehicles for setting strategic housing policy; as with local plans, DoE advice has emphasized the need to restrict structure plan policies and proposals to those concerned with development and land use, and this will tend to exclude issues of housing finance, access or tenure except as supporting justification. Such matters are central to HIPs, and in this sense HIPs have been viewed as potentially a much more adequate vehicle for effective housing planning (Murie and Leather, 1977), although the experience of HIPs in practice rather belies this early optimism (Paris, 1979). The danger, however, is that the preparation of HIPs by different district housing authorities will lead to policy incompatibilities within a single county. Experience with the HIP system shows that in the early years districts took differing attitudes towards the role of the county in housing policy, with some hostile to county involvement and opposed to structure plan policies affecting their own housing strategies, the most common county role being the provision of data for the HIP needs statement. This provides counties with a means of obtaining an involvement in district housing policy formulation, but obviously falls short of meaningful consultation over policy issues (Bramley, Leather and Murie, 1980). In many areas, however, the relationship was judged to be working well, with the county providing the constituent districts with information, ensuring that neighbouring districts' policies and programmes were related, and that strategic planning considerations were being taken into account. Where there were problems, these tended to reflect pre-existing hostility to county involvement in general rather than specifically in relation to housing. This point can be seen in Oxfordshire, where the county structure plan proposed an overall policy of restraint in the Oxford area, a strategy which Oxford City Council has implicitly rejected in its HIP submissions (Bowie, 1979) and tested in local plans; certification problems have resulted.[56] Where fundamental disagreements do exist, close collaboration over HIPs is unlikely and in any case can hardly be expected to resolve wider strategy conflicts.

The second difficulty affecting the integration of programme and development plan is the tendency of resource considerations to dominate the programme planning process at the expense of other considerations such as needs or service development priorities (Hambleton, 1983). In TPPs central government favours cost effectiveness and schemes which will show a relatively high rate of return, and this has reduced both the demand and the need for a comprehensive strategy statement to preface specific proposals. The significance of IAPs as a vehicle for innovative *policy* has been reduced by the failure to incorporate main programmes and the subsequent emphasis on bids for (and spending of) Urban Programme funding. Stewart concludes that Partnership 'has become a policy cul-de-sac in which it is either unnecessary or inconvenient to discuss key policy issues', partly because of widespread agreement among participants that the emphasis on economic development is right. New developments in inner city policy are thus taking place outside of Partnership (EZs, UDCs, UDG, the Merseyside task force and city action teams), but these innovations and the resources they command all serve to down-grade the established IAP procedures (Stewart, 1983, pp. 212–213). HIP strategy statements were reduced in scale for the 1980–81 submissions, but in any case it soon became clear that evidence of serious housing need supported by strategies to ameliorate this would not lead to additional allocations nor even safeguard existing programmes, the HIP system overall being revealed as resource-led rather than need-led

(Leather, 1983). The idea of HIPs as a planning document for local housing policy has thus almost completely disappeared, the emphasis now being on tightly controlled expenditure allocations in a framework provided by national housing policy. Similarly, TPPs now deal only with highway schemes of more than local importance. The general tendency to reduce the strategy or policy content of these programmes clearly risks further weakening the links to the statutory development plan system. This will be particularly true for HIPs, already subject to the tensions that may be involved in district–county relationships and where the local bid seems to play only a limited role in the resource allocation process. What seems to be happening in terms of the strategic planning hierarchy is that central government (Level 1) is 'short-circuiting' the planning process by making district resource allocation decisions at Level 4. This is done through HIPs, TPPs and IAPs, whose local needs content seems to be of relatively little importance. To the extent that this is so, the development plan system is left out in the cold, the main implication of this being an explicit separation of structure and local plan proposals from the public sector resources partly necessary for their implementation.

5.6 Conclusions

The various plans, programmes and policy initiatives reviewed here address a range of inter-related social, economic and environmental problems. The complexity and 'wicked' nature of these problems and issues is rarely reflected in policy form, which is commonly limited to a particular topic, spatial area or solution. Partnership reflects a perception of the complex and highly inter-related set of issues comprising the 'inner city problem', but it is doubtful if this is an adequate response in terms of co-ordination across traditional policy areas or of levels of funding. At the same time, it is clear that many of these initiatives are concerned with the management of change in their areas with respect to their particular policy objectives. This is reflected for instance in attempts to involve a variety of bodies and local interests through consultation and joint working, e.g. National Park Plans, Operation Groundwork. Local contingent factors will also be taken into account in these consultative and collaborative processes of policy formulation and implementation.

For local plans and local planning, the fragmentation of policy initiatives arguably represents a further (itself complex) contingent factor which – given that such measures often have resources associated with them – may nevertheless be significant and valuable at the local level. To 'tap' such initiatives however may entail a shift away from statutory local plan production towards a more entrepreneurial style of local planning. Again the local planner is concerned to manage change in the environment.

All the various policy measures and initiatives reviewed here have significant land use and development implications, and yet (with the exception of Green Belt) they are to varying degrees independent of the statutory development plan system and the local plan, being typically established by policy/administrative initiative and/or legislation outside of the *Town and Country Planning Acts*. Otherwise these measures share few common features. Table 5.1 summarizes this variety, emphasizing the number of initiatives involved and their differing formal bases and aims. The table also shows that each separate measure usually has an associated planning, policy or strategy document, raising the question of how decisions contained therein are reflected in (and themselves reflect) local plan policies and proposals. For many of the policy measures concerned, this relationship is often unsatisfactory. Generalizations are hampered by the wide variety of measures involved and differences in both the ways such links are specified (e.g. in statute, by circular, administrative evolution) and in the nature of the relationships themselves. However, in terms of the strategic planning hierarchy, the various initiatives reviewed above can be seen as freestanding policy elements incorporating aspects of the third, fourth and fifth levels (i.e. local planning, resource allocation, implementation). In addition to statements of policy, many come complete with resources for the implementation of their proposals and so can be expected to be important in securing change on the ground. It is clear that as a result these policy vehicles potentially have a useful role to play in complementing and implementing land use policies and proposals formulated by the development plan system. Their relationship to this system, however, often remains obscure. The danger here is that their relatively independent nature (i.e. the lack of *formal* links to local plans or to higher level strategic proposals) will lead to the development of policies and proposals in the geographical and policy areas concerned which are either not to be found in approved statutory plans or which actively conflict with such plans.

The relatively independent character of such measures as UDCs or Partnership designation, together with the tendency to focus on discrete and specific 'problem' issues, weakens the ability of local plans to secure effective co-ordination. The effects of this *policy* fragmentation are aggravated by the practice of limiting the application of the resultant proposed 'solutions' to defined and often restricted geographical areas; by the short-term and action-orientated timescale commonly involved; and by the organizational context within which the planning system as a whole operates. The trend of setting spatial limitations on policies can be seen in virtually all of the measures reviewed above with the exception of investment programmes. Sometimes justified in terms of the economies of scale resulting from the concentration of effort that the area-based policy offers (e.g. GIAs), the approach also has the benefit for an expenditure-conscious central government of restricting costs without necessarily damaging political significance, e.g. EZs. For the local plan though, the area policy approach has the disadvantage of introducing what amounts to formal single-topic 'plans' *with an explicit spatial component* which somehow the local plan must seek to integrate or ignore. To the extent that these areas have been established within the plan area this represents an effective limitation on local plan content, both in terms of *policies* and in terms of the definition of *zones* to which they will apply. The fact that area policies such as those for housing, industrial or commercial rehabilitation/improvement areas cannot be established via local plans further weakens the plan's scope. The inclusion of such land use proposals in local plans, however, will paradoxically act to strengthen the area policy itself.

Further, the recent rapid introduction of such measures as Partnership, EZs, UDCs, and UDG and the focus on implementation that they represent, also acts to circumvent the local plan, with its emphasis on a ten-year timescale and realistic proposals, i.e. those carrying firm financial commitments. In any case, such commitments will be difficult enough to obtain in a period of public sector expenditure restraint. These points apply in particular to the inner areas, often the focus for special area policy measures, and where the main agents of change are usually drawn from the public sector. The problem is that the new initiatives typically relate to the short- rather than medium- or long-term, and require an equivalent local response, e.g. local authorities were given less than one year to submit the initial IAPs to the DoE. Where such short-term funding becomes available, the danger is that this will be spent on non-priority but short-notice projects. This phenomenon (termed 'burning-off') has been variously identified in the context of short-term HIP allocations (Leather, 1983) and in relation to inner city local planning (Urwin and Wenban-Smith, 1983). Only exceptionally can conventional local plans be expected to be available to provide an appropriate policy framework e.g. Wirral MBC's use of (non-statutory) local plans in IAP preparation. In response, as noted above, Birmingham City Council for example has abandoned statutory local plans in the inner areas and developed a rolling programme of informal Inner Area Studies to co-ordinate initiatives and to identify worthwhile schemes in advance. The result is a shift in local planning emphasis, 'away from the development control and longer-term land use orientation of statutory plans towards annual rolling investment programmes', which is claimed to avoid the dangers of short-term expediency (Urwin and Wenban-Smith, 1983, p. 52). This emphasizes in turn the disadvantages involved with the statutory local plan preparation and adoption processes in terms of timescale and speed of response, and the fact that partly for this reason the formal local plan will not always be the most suitable vehicle for the development of local planning policies.

Finally, the organizational context within which the British planning system operates is one of great complexity, characterized by the administrative separatism of the agencies involved and the absence of 'horizontal' links

Table 5.1   *Formal plans and programmes*

Policy area	Initiative	Formal basis	Aim	Documents
Housing	General Improvement Areas	*Housing Act 1969*	Housing rehabilitation and environmental improvement through provision of grants	Designation documents, policy documents
	Housing Action Areas	*Housing Act 1974*	Housing rehabilitation in areas of housing stress	Designation documents, policy documents
	Priority Neighbourhoods	*Housing Act 1974* (provision now repealed)	Housing rehabilitation as a holding operation pending GIA or HAA action	Designation documents, policy documents
Conservation	National Parks	*National Parks and Access to the Countryside Act 1949*	Conservation and recreation	National Park Plans
	Lee Valley Regional Park	*Lee Valley Regional Park Act 1966*	Recreation and leisure	Lee Valley Park Plan
	Areas of Outstanding Natural Beauty	*National Parks and Access to the Countryside Act 1949*	Conservation and recreation	Statements of Intent, management plans
	Heritage Coasts	Non-statutory Countryside Commission designation	Conservation and recreation	Managements plans

Policy area	Initiative	Formal basis	Aim	Documents
Conservation (cont.)	National and Local Nature Reserves	*National Parks and Access to the Countryside Act 1949; Wildlife and Countryside Act 1981*	Research and preservation	Agreements, non-statutory plans
	Sites of Special Scientific Interest	*National Parks and Access to the Countryside Act 1949; Wildlife and Countryside Act 1981*	Research and preservation	Voluntary agreements
	Conservation Areas	*Civic Amenities Act 1967; Town and Country Planning Act 1971*	Preservation/enhancement of special architectural and historic interest	Informal proposals, but preferably included in a statutory local plan
	Groundwork	Non-statutory Countryside Commission designation	Land reclamation, environmental improvement, education and recreation	None
	Waste Disposal Plans	*Control of Pollution Act 1974*	Waste disposal strategy	Waste disposal plans
Urban development and regeneration	New Towns	*New Towns Act 1946, 1981*	Town development in pursuit of dispersal policy	Master plan, detailed development proposals subject to Secretary of State approval
	Green Belt	See DoE Circulars 42/55, 50/57, 14/84	Development restraint	Structure and local plans

cont.

Table 5.1 (cont.) Formal plans and programmes

Policy area	Initiative	Formal basis	Aim	Documents
Urban development and regeneration (cont.)	Industrial and commercial improvement areas	*Inner Urban Areas Act 1978*	Local economic development, building rehabilitation and environmental improvement	Designation and policy documents
	1977–78 inner city policy; Partnership, Programme and Designated districts	*Inner Urban Areas Act 1978*	Inner city regeneration	Inner Area Programmes
	Enterprise Zones	*Local Government, Planning and Land Act 1980*	Small area economic development	Enterprise Zone schemes
	Urban development corporations	*Local Government, Planning and Land Act 1980*	Urban development	Development proposals submitted to the Secretary of State
	Task force, city action teams	Central government inter-departmental initiative	Co-ordination of central and local government and the private sector: inner-city economic development and environmental improvement	None

Policy area	Initiative	Formal basis	Aim	Documents
Urban development and regeneration (cont.)	Financial Institutions Group	Central government inner-city policy initiative	Provide new approaches to urban regeneration	None
	Urban Development Grant	Policy initiative: grants paid under Urban Programme and Derelict Land funding	Stimulate private sector investment in development/construction	None
Investment programmes	Housing Investment Programmes	DoE procedure undertaken in context of central government controls over local authority capital expenditure	Planning and distribution of housing capital expenditure approvals	HIP submissions
	Transport Policies and Programmes	DoT procedure undertaken in context of central government controls over local authority capital expenditure	Planning and distribution of capital expenditure allocations for highway schemes	TPP submissions
	Inner Area Programmes	DoE procedure	Planning and distribution of inner city funding (largely Urban Programme)	IAP submissions

between them. At the local level, physical land use planning is bedevilled by the failure of central and local government to produce, respectively, statements of national or corporate policy, and by the difficulties of securing co-operation and compatibility between the numerous sectors of central and local government. This latter point is hardly surprising given the complex inter-relationships which exist between central and local government, between county and district authorities, between the nationalized industries and the *ad hoc* agencies of the state and local government, and between the public and private sectors (Bruton, 1984). In terms of the co-ordination of local plans with the other area-based policy initiatives, these are usually the responsibility of an agency other than the local authority planning department. This may be another municipal department (e.g. responsible for designation and progression of HAAs or GIAs) but a large number of other formal agencies are also involved here (National Park Authorities, the Countryside Commission, the Nature Conservancy Council, Groundwork Trusts, New Town Development Corporations, UDCs). Central government departments may also be directly concerned, e.g. on Partnership committees (which have themselves reflected differences in the degree of emphasis different departments are prepared to give to inner city policy), or in debates about the policy content of EZs (DoE, Department of Employment, Treasury, Health and Safety Executive). This context of fragmentation and complexity clearly militates against co-operation or 'horizontal' policy integration at the local level, as the difficulty in securing a 'unified approach' to inner area problems through Partnership has shown. Statutory local plans, with their land use orientation, cannot reasonably be regarded as an appropriate vehicle for this task. In local plan areas where several different public sector agencies are involved, perhaps the best that can be hoped for is that local plans will accurately reflect the policies of these agencies, perhaps set out in other strategy or planning documents, which have a bearing on the development and use of land; and that these will be integrated with the development control policies of the local planning authority itself.

Notes

1   *National Parks and Access to the Countryside Act 1949*, s.5.
2   As recommended by the Report of the National Park Policies Review Committee (the Sandford Report) and accepted by the Secretaries of State in DoE (1976) *Report of the National Park Policies Review Committee*, Circular 4/76.
3   *Local Government Act 1972*, Schedule 17, paras. 18 and 19.
4   'Locals only policy to go despite panel support', *Planning*, No. 433, 28 August 1981, p. 7.
5   *National Parks and Access to the Countryside Act 1949*, s.9.
6   *Ibid.*, s.6(4).

7   *Ibid.*, Part III. The term 'National Nature Reserve' is given statutory recognition by the *Wildlife and Countryside Act 1981*, s.35.
8   *National Parks and Access to the Countryside Act 1949*, s.21.
9   *Ibid.*, s.23; *Wildlife and Countryside Act 1981*, s.28.
10 *Town and Country Planning Act 1971*, s.277. The conservation area designation was introduced by the *Civic Amenities Act 1967*.
11 *Town and Country Planning Act 1971*, s.277(8).
12 *Town and Country Planning Act 1971*, s.277B.
13 See 'Operation Groundwork boosts derelict land grant campaign', *Planning*, No. 434, 4 September 1981, p. 12.
14 See 'Foundation mooted for fringe link', *Planning*, No. 578, 20 July 1984, p. 1; 'Foundations of optimism for fringes', *Planning*, No. 596, 23 November 1984, pp. 8–9; 'Groundwork set to go nationwide as foundation looks for funding', *Planning*, No. 612, 5 April 1985, p. 16; 'Groundwork moves from original base', *Planning*, No. 640, 18 October 1985, p. 8.
15 See 'Groundwork suspected by city authorities', *Planning*, No. 595, 16 November 1984, p. 6; 'Groundwork suspicion is not justified', Letter to Editor, *Planning*, No. 598, 7 December 1984, p. 1.
16 See 'Shifting the groundwork to other areas', *Planning*, No. 577, 13 July 1984, pp. 8–9.
17 Now superseded by the consolidating *New Towns Act 1981*.
18 For example, the *Local Government, Planning and Land Act 1980* confers powers on the Secretary of State to direct that development corporations or the Commission for New Towns pay money to him, and power to reduce the designated area of any New Town.
19 *New Towns Act 1981, s.7.*
20 *Town and Country Planning (New Towns) Special Development Order 1977* (SI 1977, No. 665).
21 See 'Green belt threat: is it strategic?', *Planning*, No. 480, 6 August 1982, pp. 12–13.
22 *Inner Urban Areas Act 1978*, ss.4, 5 and 6; Schedule to s.4(1).
23 *Local Government, Planning and Land Act 1980*, s.191.
24 *Inner Urban Areas Act 1978*, s.1.
25 *Ibid.*, ss.2, 3, 4 and 12.
26 *Ibid.*, ss.7 and 8.
27 *Partnership Authorities*: Birmingham, Gateshead, Hackney, Islington, Lambeth, Liverpool, Manchester, Newcastle, and Salford. *Programme Authorities*: Blackburn, Bolton, Bradford, Brent, Coventry, Hammersmith, Hull, Knowsley, Leeds, Leicester, Middlesbrough, Nottingham, Oldham, Rochdale, Sandwell, Sheffield, Sunderland, Tower Hamlets, S. Tyneside, N. Tyneside, Wirral, Wandsworth and Wolverhampton. *Designated Districts*: Barnsley, Burnley, Doncaster, Ealing, Greenwich, Haringey, Hartlepool, Langbaurgh, Lewisham, Newham, Rotherham, St. Helens, Sefton, Southwark, Walsall and Wigan. Source: DoE.
28 See, for instance, the papers by Hall, Harrison, Massey and Goldsmith in the *International Journal of Urban and Regional Research*, Vol. 6, No. 3, 1982.
29 *Local Government, Planning and Land Act 1980*, s.179, Schedule 32. The other concessions were contained in the *Finance Act 1980* (100 per cent corporation

and income tax allowances) and the *Employment and Training Act 1981* (exemption from industrial training levies).

30   *First-round designations*: Salford/Trafford, Wakefield, Dudley, Hartlepool, Corby, Tyneside, Speke, Isle of Dogs, Swansea, Clydebank and Belfast. *Second-round designations*: Middlesbrough, N.E. Lancs, N.W. Kent, Rotherham, Scunthorpe, Telford, Wellingborough, Workington, Glanford, Delyn, Milford Haven, Invergordon, Tayside and Londonderry.

31   Richard Hargreave was involved in the development of the Industrial Improvement Area concept while at Rochdale, and Stockport's Economic Enterprise Area clearly represents an extension of the IIA approach. See 'Enterprise', *Planning*, No. 356, 22 February 1980, pp. 8–9.

32   *Local Government, Planning and Land Act 1980*, Schedule 32, para. 2(3).

33   *Ibid.*, para. 23.

34   See 'Enterprise Zones', *Journal of Planning and Environment Law*, April 1984, pp. 218–219.

35   See 'No shopping threat in Enterprise Zones', *Planning*, No. 396, 28 November 1980, p. 8. Taylor (1980) reviews the negotiations between local authorities and the DoE over the retail floorspace limits to be set in the schemes for the first round EZs, above which normal planning procedures would operate.

36   The projection of the EZ measure as an aspect of economic and employment policy dates from July 1980 and an opposition motion to censure the Government for its unemployment record (see Taylor, 1981; Warren Evans, 1981).

37   Secretary of State's Address to the Town and Country Planning Summer School, 1983.

38   See 'Fatal flaws in zonal strategy', *Planning*, No. 590, 12 October 1984, pp. 8–9, and 'Papering over the cracks in zone solution', *Planning*, No. 591, 19 October 1984, pp. 8–9, for a review of these responses.

39   *Town and Country Planning General Regulations 1976* (SI 1976, No. 1419), Regulations 4 and 5.

40   White Paper *Lifting the Burden*, Cmnd. 9571, presented to Parliament by the Minister without Portfolio, July 1985, para. 3.6(i).

41   *Local Government, Planning and Land Act 1980*, ss.134 and 135.

42   *Ibid.*, s.140.

43   *Ibid.*, s.136.

44   *Ibid.*

45   *Ibid.*, ss.141 and 142.

46   *Ibid.*, s.148.

47   *Ibid.*, s.149.

48   *Hansard, House of Common Debates*, Vol. **998**, cols. 603–610, February 9 1981.

49   See 'Dockland housing row breaks out', *Planning*, No. 590, 12 October 1984, p. 1.

50   See 'Noise snag provokes call for new inquiry', *Planning*, No. 583, 24 August 1984, p. 1; 'Councils combine in call for new inquiry', *Planning*, No. 584, 31 August 1984, p. 1; and 'Fight lost on airport', *Planning*, No. 612, 5 April 1985, p. 1.

51   'Dr Ridley prescribes a UDC dose', *Planning*, No. 676, 11 July 1986, p. 1.

52 See 'No Heseltine sparkle in Merseyside show', *Planning*, No. 440, 16 October 1981, pp. 6–7.
53 See 'More task forces to tackle inner cities', *Planning*, No. 600, 11 January 1985, p. 1; 'Partners sorted out by Jenkin's cat-like tread', *Planning*, No. 608, 8 March 1985, p. 1.
54 Department of Transport (1984), *Transport Supplementary Grant*, Circular 3/84.
55 See Department of the Environment Circulars 9/83 and 6/84, *Capital Programmes*.
56 See 'Certification wait over on Oxford plans', *Planning*, No. 530, 5 August 1983, p. 1.

6 Non-statutory approaches to local planning

6.1 Introduction

The discussion so far has pointed to the fact that the statutory local plan is only one of a number of policy vehicles which, taken together, enable the process of local planning in Britain to be carried through. The particular remit of the statutory local plan is with issues of land use and development. Other formal tools based in legislation have also been developed, which although focusing primarily on topic or policy issues such as housing or the environment also have land use implications (Chapter 5). One important group of local planning instruments, however, remains to be discussed. This is the non-statutory material prepared by local planning authorities either instead of, or as well as, their statutory local plans. These instruments take a wide variety of forms, but share the common characteristic of lacking an explicit legislative foundation.

Although the indications are that the local authorities find such documents to be useful, central government has taken a less sanguine view of their continuing role within the local planning process, and the Department of the Environment is firmly set against their use. Current advice in Circular 22/84 (para. 1.13) is that 'where there is a need to devote resources to the preparation of proposals for the use of land, these should be settled in the statutory plan' rather than some other type of planning document. A role is recognized for non-statutory supplementary planning guidance, in the form of development control practice notes, development briefs and detailed or sketch layouts for housing or open space (*ibid.*, para. 1.14). This advice is a re-statement of the line taken by the Department in the earlier development plan Circulars 55/77 and 4/79; the latter, for instance, judged that the preparation of informal land use plans by local authorities was 'inappropriate' (para. 4.3). This attitude towards non-statutory plans appears to be founded on a concern with the public credibility of planning and with the protection of individual property rights and land values. The statutory local plan-making process enshrines a formal right of objection, whereas in preparing informal plans authorities are free to use whatever procedures they think suitable. However, the statutory procedures for local plan preparation leave the formal decision on whether or not to modify the plan as a result of objections with the local planning authority (Chapter 3), attracting the charge that local authorities are effectively judge and jury in their own case.

Official advice on the use of non-statutory local planning material needs to be seen in context. Prior to the gradual emergence of structure plans from the mid-1970s onwards, DoE advice was rather different, for informal plans were actively recommended to local authorities in Circulars 102/72 and 122/73 as devices to meet the demand for housing land.[1] The present DoE attitude dates from Local Plans Note 1/76, which recognized that 'in recent years, many planning authorities have had to work with informal plans'. However, 'now that all counties either have an approved structure plan or are working towards one, statutory local plans should be prepared' (DoE, 1976, para. 10). Current advice, therefore, is aimed firstly at authorities who are still using such 'interim' documents. In addition, there are also indications that many of the practitioners involved in formulating and implementing local planning policies are disenchanted with statutory local plans, and in consequence are falling back upon a wide range of non-statutory instruments. A recent discussion paper prepared by a working party of the District Planning Officers' Society (DPOS), for instance, notes that 'there has . . . been a dramatic swing away from the preparation of statutory plans towards the use of other means of making development decisions' (1982, p. 3). The advice of Circulars is aimed as much at authorities who have recently produced statutory local plans and found them wanting in some respect as at those who have found over many years that informal planning documents are well-suited to their needs.

Pointers to practice in individual local authorities are readily available. Both Sheffield and Birmingham City Councils, for instance, have developed non-statutory instruments linked closely to resources for implementation, while also preparing statutory plans as appropriate. Local planning practice in Sheffield is explicitly concerned to make the optimum use of all available resources and thus seeks through both statutory plans and informal mechanisms to influence the capital and revenue programmes of other departments and to guide the allocation of a variety of other resources, including the Urban Programme, manpower services programmes, and EEC grants via the Regional Development Fund (Bajaria, 1982; Fudge, 1984b). Birmingham City Council has developed a three-tier system of local planning:

1 Statutory local plans for areas of major change led by the private sector (e.g. the city centre).
2 Inner area studies, informal local plans developed from Partnership documents to provide a non-statutory means of co-ordinating major change led by the public sector.
3 Outer area statements for the more stable outer suburbs, identifying existing development control policies and major proposals.

This approach recognizes the quasi-legal strength of statutory plans in the face of development pressures, particularly at appeal; the need to match annual Partnership budget cycles and the needs of urban renewal with a suitably responsive planning framework; and that simple policy statements for the stable surburban areas where relatively little change is anticipated will usually be sufficient (Middleton, 1982). In Liverpool, it is claimed that the statutory system is unable to meet the dual requirements of effective implementation and management; local plan preparation has come to be seen as (Hayes, 1981, p. 77).

a sterile, abstract process with no political credibility, no ability to command resources for implementation, no corporate commitment within the authority, a slow and wasteful activity and an inflexible instrument frequently overtaken by changes in local and (increasingly) central government policy.

Local planning in the city is wholly non-statutory and is primarily concerned to service the spending programmes of the local-central government Partnership. East Lindsey DC, faced with a situation where pressures for change are in total substantial, though widespread and individually small scale, has devised a policy framework comprising seven statutory local plans for the main towns; informal village plans for the 13 major villages; and development guidelines for minor villages (33 approved to date). The scale and intensity of local planning problems has dictated the choice of policy vehicle, a robust and contingent approach which has led to considerable coverage of settlements within the district over a short period (Hewitt, 1985). In Norfolk, local plans are seen as inappropriate vehicles for the formulation of rural settlement policies; these are currently being progressed, within the statutory framework provided by the county structure plan, through informal guidelines for individual villages prepared in consultation with parish councils in the way envisaged by PAG. Advantages of relevance and responsiveness are claimed to accrue (Shaw, 1982). Not all authorities, however, are following either wholly or predominantly non-statutory approaches to local planning. In contrast to Norfolk, where only one statutory local plan is in progress,[2] Hertfordshire is developing a comprehensive statutory coverage composed of the county structure plan and district-wide local plans progressed by the ten constituent district councils. This strategy, it is argued, 'is justified within the operational context of planning for restraint in the county' (Griffiths, 1978, p. 18), though a role is also recognized for informal plans and supplementary guidelines where no strategic interest is involved. At the national level, on the basis of a review of stated local authority intentions in current development plan schemes, Bruton (1983a) has shown that informal instruments feature significantly in the planning system in England. Many of the non-statutory plans and policies recorded in these schemes have been in existence for some years; it is also clear that the majority of authorities in the country will continue to prepare and update them. A similar situation has been reported in Scotland, not-withstanding the statutory requirement that Scottish local authorities achieve total local plan coverage (Angus, 1985). This picture, of the widespread use of informal documents, is confirmed by other recent work. Fudge *et al.*, (1983, p. 3), for instance, note that in addition to the statutory plans being progressed by their sample of 58 authorities 'a large amount of non-statutory activity was also being pursued. Most authorities have some commitment to preparing informal plans, policy frameworks or develop-ment briefs in circumstances where they felt statutory plans to be inapprop-riate'. Similarly, Pountney and Kingsbury (1983a, p. 148) point out that the need for specifity in plan policies is often met through supplementary guidance: 'all six [planning] departments involved with the study used guidance notes and/or development briefs of one sort or another'. Lastly, the studies of Greater London by Field (1983), and of Birmingham, Bristol, Leeds and Leicester by Farnell (1983), both record much non-statutory material being used by the local planning authorities concerned. This evi-dence from both practice and research indicates that, contrary to consistent DoE advice from 1976 on, local planning authorities do not consider them-selves restricted to the statutory local plan for either the formulation or implementation of their planning policies and proposals. On the contrary, informal planning documents appear to be widely used, a fact which sup-ports the view that the majority of local planning authorities are acting contingently.

As a first step in providing basic information on the roles played by informal instruments, a questionnaire survey of local planning authorities in England was undertaken in March 1983 (Bruton and Nicholson, 1984a). The survey covered the metropolitan and non-metropolitan district and county councils, together with the London boroughs, and achieved an overall response rate of 70.6 per cent (Table 6.1).

This chapter draws on the information provided by this survey in order to outline both the *extent* and *nature* of local authority non-statutory local planning activity. It begins by exploring, in greater depth, the debate bet-ween the advocates of statutory and informal approaches to local planning, focusing on the relative advantages and disadvantages that are claimed for each. This is followed by a review of the various *forms* of non-statutory material and of the *scale* on which such policy documents are being used by local planning authorities in England. The links between statutory local plans and non-statutory documents are then discussed, considering both the extent to which statutory plans have made use of policies previously set out in non-statutory instruments ('upgrading'), and the scale on which local authorities, for a variety of reasons, have lapsed their intentions to take plans in preparation through to statutory status ('abandonment'). These two processes emphasize that specific policies can find expression in both statu-tory and non-statutory local planning documents at different stages of their

Table 6.1   *Survey response rates by type of authority, England*

Type of authority	*Number of authorities*	*Number of respondents*	*Response rate (%)*
Metropolitan counties*	7	4	(57.1)
Shire counties	39	34	(87.2)
Metropolitan districts	36	18	(50.0)
Shire districts	296	210	(70.9)
London boroughs	33	24	(72.7)
Total	411	290	(70.6)**

Source: Bruton and Nicholson, 1984a.
* Includes the Greater London Council.
** 290/411 = 70.6 per cent.

effective life. Finally, the procedures that authorities have used during the production of their informal documents (public participation, consultation, adoption by committee or council) are outlined, and the various functions or roles that non-statutory instruments play in developing and progressing local planning policies are discussed.

6.2   Statutory versus non-statutory local planning

The evidence cited above indicates that non-statutory local planning instruments are in widespread use. However, this does not necessarily imply a wholesale drift away from the statutory local plan. Indeed, now that the structure plan framework is virtually complete, the process of statutory local plan deposit and adoption is proceeding apace (see Figure 4.1, p. 130). Nor is the statutory plan without its supporters in the professional literature. Thompson (1977a, p. 148), for instance, discussing the approach taken by the London Borough of Camden to the production of its District Plan, notes that:

some [local planning authorities] are succumbing to the allure of the non-statutory plan. Tempting though the freedom from statutory entanglements may be, this approach will not resolve critical problems such as blight, and its champions may find the going hard on appeals. Camden played the game and opted for a statutory plan.

Other advantages claimed for the statutory plan by the DPOS (1978 para. 3.42) are that

(a) the statutory system creates the planning service;
(b) it provides a means of public scrutiny of local authorities' policies;
(c) it provides formal safeguards for those whose interests are affected;

(d) it imposes a beneficial discipline;

(e) local plans provide a basis for corporate decisions and an aid to co-ordination;

(f) statutory status provides a firm basis for investment and other decisions by public authorities;

(g) it helps where matters go before the Secretary of State for decision.

Some of these advantages may in practice be less significant than they seem; (e) and (f), for instance, imply that local plans are central and necessary to the spending decisions of both other local authority departments and external public agencies respectively. Yet this is far from being the case (see Chapter 4). Also, the strength of the 'formal safeguards' (c) is restricted, partly by the fact that local planning authorities are free to reject if they so choose any of the Inspector's recommendations. In any case, some of these advantages are not *necessarily* unique to the statutory local plan; a programme of public participation during the preparation of an informal plan, for example, could also provide a means of public policy scrutiny (b). Moreover, *informal* plans must also be taken into account by the Secretary of State at appeal as an 'other material consideration' (g), the issue *vis-à-vis statutory* plans being the relative weight given to their policies. Circular 22/84 (para. 1.13) suggests that informal plans will be given 'little weight', although greater weight will be given to supplementary planning guidance 'when it has been prepared in consultation with the public and has been made the subject of a council resolution' (*ibid.*, para. 1.15). There is evidence to suggest that individual Inspectors do not always follow the Departmental line in disregarding the provisions of informal plans, particularly where procedures such as participation or adoption have been used in the preparation of these documents (Bruton and Nicholson, 1984c).

In any case, the status of a *statutory* local plan at appeal, as part of the development plan, will be limited to that of one material consideration among many, so it is not surprising that in a later discussion paper a working party of the DPOS (1982, p. 4) pointed to 'a growing catalogue of cases where decisions have been made on planning appeals which have . . . gone against the policies and proposals of an adopted plan where there are no obviously compelling reasons to do so'. Finally, as far as 'creating the planning service' (a) is concerned, Healey (1983, p. 99) concurs with the DPOS that the statutory plan is simply to be used as part of a package of instruments to solve particular problems, and from this argues 'that the role of planning departments is an issue of a much more general nature than the question of the selection of appropriate tools for a particular task'. It would seem to be very difficult to identify a set of advantages associated with the use of the statutory local plan which are both incontrovertibly and uniquely linked with the statutory processes *per se*.

The bulk of professional comment on the statutory local plan, however, emphasizes the drawbacks of the system rather than its advantages. This criticism of the local plan focuses variously on questions relating to preparation, adoption and amendment procedures, plan content, and the treatment of resources. The 1978 DPOS report (para. 3.2) provides both a good example of such comment and a useful summary of the issues involved. The main criticisms are that

(a) the system is slow and expensive;
(b) it is unduly constrained; and
(c) it is inflexible.

An early indication of the different timescales involved in statutory/non-statutory work is given by Couch (1978, p. 9) whose 1978 survey of local planning practice found that informal plans took 22 months on average to reach adoption, with formal plans taking 31 months. Couch however argues that this apparent advantage of informal plans may not be as substantial as it first appears, since a statutory local plan 'on deposit' has a similar status to an adopted informal plan. Moreover, in some circumstances statutory plans can be produced quickly; Salford City Council for example has placed statutory local plans on deposit within 14 months of beginning work with only limited staff resources (Gilbert *et al.*, 1982; Shields, 1983). This clearly weakens arguments for non-statutory local planning documents which refer to the speed with which they can be prepared relative to statutory plans. However, on average the time taken for a statutory local plan to reach the deposit stage is 36 months (Bruton, Crispin and Fidler, 1982). In any event, the DPOS (1978, para. 3.7) accept that statutory plans take longer to produce than informal documents, contending that 'the time taken in preparing local plans is a main reason why many local planning authorities are pursuing the use of alternative instruments such as a local policy statement, statements of corporate policy, briefs, and informal studies'. Indeed, the pace of local plan preparation will often be much slower than the timescale of development decisions, so that authorities will perforce have to make difficult choices as to whether decisions should be made on the basis of draft proposals or to delay until the plan is adopted (*ibid.*, para. 3.4). A working party of DPOS members has recently advocated the abolition of the public local inquiry in a bid to meet professional dissatisfaction with the statutory system stemming from this source (District Planning Officers' Society, 1982). In contrast, informal instruments may be produced as and when needed with no predetermined programme of participation. However, while the absence of procedural requirements can be an advantage, in terms of speed of preparation, amendment and low cost, it may also subsequently prove to be a disadvantage, especially if the plan or policy is being used at appeal.

The constraints on the statutory system identified by the DPOS under (b) above are primarily those set by the Department of the Environment in circulars and other advice rather than the legislative limit imposed by the Act or regulations, e.g. that subject plans can only be prepared for topics which can be treated in isolation. the result of this and other DoE advice is that 'the system does not give as much freedom to select the types and contents of local plans as it appears to offer' (DPOS, 1978, para. 3.14). Other limitations include restrictions placed on the local plan preparation process, e.g. as to presentation, where the requirement that the proposals map must define the site of each proposal creates difficulties where the details of a sizeable development are not finalized. Problems of inflexibility cited under (c) include the difficulties inherent in the requirement that a local plan must conform generally to the structure plan, and that the procedure for reviewing a local plan is exactly the same as that for preparing it in the first place. Summarizing the effects of these factors, the DPOS comment that the lengthy statutory procedures (DPOS, 1978, para. 3.33):

make it extremely difficult to react readily to changed circumstances, because of the problems associated with conformity with, and review of, structure plans. Add to those the limitations on freedom of type and content of local plans, and we find it no wonder that even before most authorities have completed their first local plan, many are switching the emphasis to alternatives to statutory plans.

The DPOS are here emphasizing that it is the various disadvantages associated with the use of the statutory plan which have encouraged authorities to make use of non-statutory 'alternatives'. In order to explore how local planning authorities actually using informal instruments viewed these issues, authorities were asked in the questionnaire survey referred to earlier to identify advantages and disadvantages in their non-statutory approaches to local planning. Of the 290 responding authorities, 278 (96 per cent) used informal material of some kind;[3] the relative attractions and drawbacks of non-statutory instruments as perceived by these authorities are set out in Tables 6.2 and 6.3, respectively.

It is clear from Table 6.2 that many authorities regard the relative procedural ease with which informal instruments can be prepared and (to a lesser extent) amended, as a significant advantage over the statutory plan. Here, the bulk of authorities emphasize the *shorter timescale* within which informal material can be brought into use. As a result, they can be produced quickly so as to meet a particularly urgent local need, e.g. to reach and support a decision on an important planning application (City of Lincoln), to meet rapidly changing circumstances such as the establishment of amusement arcades in prime shopping locations (Scunthorpe BC), or to identify land for housing or employment (Kettering BC). In sum, 'they allow for speedy preparation and amendment to reflect changing standards or ideas without prejudicing the established broader context, nor entailing cumbersome and costly procedures for adoption/amendment/rejection' (Hereford and Worcester CC).[4] Many authorities point to the savings in staff time which result from speed of preparation, as well as or rather than responsiveness, e.g. Oswestry BC, whose local plans section comprised one officer in

Table 6.2 *Advantages identified by English local authorities in the use of non-statutory local planning instruments*

Advantage	County		District		Total	
	No.	(%)*	No.	(%)*	No.	(%)*
Form	11	(31)	87	(36)	98	(35)
Content	9	(25)	61	(25)	70	(25)
Preparation procedures	14	(39)	164	(68)	178	(64)
Amendment procedures	6	(17)	47	(19)	53	(19)
Other	4	(11)	55	(23)	59	(21)
Not applicable** no response	11	(31)	53	(22)	64	(23)
Number of responding authorities	36		242		278	

Source: Bruton and Nicholson, 1984a.

* Expressed as a percentage of the respective number of responding authorities.

** 'Not applicable' includes: statutory and non-statutory plans not perceived as alternatives; no particular advantages; not tested at appeal; no statutory local plans to judge against.

Table 6.3 *Disadvantages identified by English local authorities in the use of non-statutory local planning instruments*

Disadvantage	County		District		Total	
	No.	(%)*	No.	(%)*	No.	(%)*
Use at appeal	3	(8)	59	(24)	62	(22)
General status	6	(17)	55	(23)	61	(22)
Procedures	4	(11)	30	(12)	34	(12)
Other	7	(19)	26	(11)	33	(12)
Not applicable** no response	23	(64)	134	(55)	157	(56)
Number of responding authorities	36		242		278	

Source: Bruton and Nicholson, 1984a.

* Expressed as a percentage of the respective number of responding authorities.

** 'Not applicable' includes: statutory and non-statutory plans not perceived as alternatives; no particular advantages; not tested at appeal; no statutory local plans to judge against.

1983. Overall, the emphasis on the timescale of informal plan production suggests that the statutory procedures *per se* are not seen as problematic. Instead, the stress on speed implies that it is the form in which these procedures are specified by regulations, together with the logistical complexities which are imposed on the statutory plan preparation process as a result, which causes many local planning authorities to opt for the informal alternative.

One-third of authorities point to the *form* of non-statutory material as a significant advantage over statutory plans, stressing issues of flexibility and suitability to local circumstances and needs. Thus, non-statutory instruments can be tailored to meet particular situations – rather than, as with statutory local plans, attempting to make the situation fit the legal framework set by statute and regulations (Warwickshire CC). Specific instances include: the use of topic-based rather than area-related plans to identify land for housing and employment (Kettering BC); a similar concern with articulating authority-wide policy statements as opposed to those relevant only to the local plan area (City of Salford); the desire to deal with topics such as recreation or countryside planning, which encompass many points of planning interest but involve few policies or proposals suitable for inclusion in a statutory plan (Nottinghamshire CC); and the need to deal with a mineral industry (china clay) requiring a longer term planning strategy than possible through the statutory vehicle (Cornwall CC). In Middlesbrough, low development pressure and substantial municipal land ownership mean that the local authority is not concerned merely to provide a framework for the control of development, as envisaged in 1965 by the Planning Advisory Group, but is now seeking financially to assist development or to act as developer itself. The Borough Council has evolved a decision-making process which considers policy issues and each year formulates a programme linked to available resources. Land and land use are considered as key elements in policy formulation, with implementation of individual proposals by way of development briefs (Southerton and Noble, 1982). The effectiveness of this process has resulted in the abandonment of three intended statutory plans (one action area, two district); the issues and problems identified by these documents are now being tackled by other means. For instance, the North Middlesbrough District Plan was formerly intended to deal with the redevelopment of industrial land and the rejuvenation of a local housing area, but these problems are now being resolved via the policy process and individual briefs including Industrial Improvement Area designations, joint development schemes between the Council and private industry, and recent designation as an Enterprise Zone. In short, because of particular circumstances within the Borough and the depressed condition of the local economy, Middlesbrough's annual policy-making process offers a more relevant, quicker and effective approach to the area's problems than that provided by the statutory local plan. The Borough is acting contingently.

Linked to the question of form is that of the *content* of informal local planning material; one-quarter of authorities identify this factor as an advantage of non-statutory documents over the formal local plan. Two particular aspects emerge from the questionnaire returns. The first of these concerns the ability of non-statutory instruments to deal with a specific topic or area at a level of detail inappropriate to a statutory local plan, and emphasises the supplementary role of informal material. Thus, non-statutory instruments are seen as interpreting statutory local plan policies at a localized level (Vale Royal DC), or as having a hierarchical relationship to the statutory vehicle, rather than being an alternative (Brighton BC). These points are wholly compatible with DoE advice on the role of supplementary planning guidance. In contrast, the second aspect of the 'content' advantage is more radical, for it relates to the ability of non-statutory documents to include policies not specifically related to land use (Humberside CC); subjects can be included which the DoE would not find acceptable in a statutory plan (East Northamptonshire DC). Informal plans may consequently be more comprehensive, as both land use and non land use policies can be included in the one document (South Derbyshire DC). Specific examples include the preparation of a rural settlement policy by Cherwell DC in order to deal with the dominant issue of housing, DoE advice having ruled out a subject plan, and the use of informal countryside management plans by Suffolk CC, as these 'are not considered to be compatible with the regulations governing statutory local plans' (though, where appropriate, land use policies and proposals will be incorporated into district council local plans).

A number of themes are subsumed within the 'other' category of Table 6.2. Several authorities point out that informal plans are a means of avoiding political conflicts with other planning agencies, mostly in the context of county–district relations. Humberside CC, for instances, notes that non-statutory documents are more acceptable to the districts, while for Essex CC the use of such material 'is more expedient politically since it meets less district council opposition, and is therefore a more effective means of getting strategic planning policies implemented through local decisions'. A second theme emphasizes the usefulness of informal material as interim documents pending structure plan completion or the production of more detailed local plans (Cherwell DC, Mid Suffolk DC, and Spelthorne BC). As a result, councils have been able to make decisions on matters which probably would have otherwise had to be deferred pending the adoption of a local plan. As noted above, this interim role has been recognized by the DoE in the past, in the context of housing land availability, but this advice has changed substantially as the structure plan framework has developed. However, it is worth pointing out that, in some areas, uncertainties with regard to the statutory strategic framework have only recently been resolved. For instance, amendments made by the Secretary of State to the Central Berkshire structure plan have created doubts surrounding the location of major housing allocations; as a result, Wokingham DC has continued to use informal plans on an interim basis. The structure plan allocations were clarified in February 1983, and the Council, as a consequence of this and of continuing DoE advice and warning against the use of non-statutory local plans, have now decided to pursue a statutory approach to the detailed local allocation of housing development.

Finally, a third theme within the 'other' category concerns the use of informal material as a means of communicating developing policy. The audiences identified are various: parish councils (Lewes DC), the general public (Hambleton DC), or elected members (Lincoln City). Fudge *et al.* (1983) identify similar roles for statutory plans; informal documents are presumably more suited to this function as a consequence of their flexibility, ease of amendment, and lack of procedural constraints.

In sum, the various advantages identified by local authorities in their use of informal planning documents are characterized by a variety of *pragmatic* or contingent responses to local planning issues and problems. Similar considerations also inform the disadvantages which authorities associate with their use of such instruments (Table 6.3). These drawbacks are seen primarily to revolve around the possibly inferior status of informal documents when compared to statutory local plans, and to the relatively low weight that may be given to such material by the Inspectorate when used at appeal. General fears over 'status' include the possibility that both the public and elected members will have less confidence in non-statutory material (Chelmsford BC, Rushcliffe BC), and that informal documents 'do not provide as comprehensive, integrated and well tested policies as found in statutory plans' (Crewe and Nantwich BC). Potential implementation difficulties are also recognized, particularly with regard to land assembly and the use of compulsory purchase orders (Mansfield DC). Several authorities also note that other development agencies – including the county council, statutory undertakers, government departments and private sector developers and applicants – may be less co-operative or responsive towards informal work than to statutory plans (Charnwood BC, Melton BC, Bromsgrove DC).

Many authorities give a more specialized expression to these general doubts over the 'status' of informal documents by referring to the dangers of poor performance at appeal. This concern is a reflection of the importance of the development control process, and particularly of appeal decisions made on refusals of planning permission, in giving local plan policies practical effect. These fears are based on the unfavourable references in DoE Circulars concerning non-statutory plans, 'which one day the Inspectorate may be obliged to enforce' (East Yorkshire BC). Experience to date may be inconclusive, but some have little doubt as to the likely future standing of non-statutory material at appeal: 'although there is no evidence of the DoE Inspectorate placing less weight on non-statutory plans at present, it is felt that this is increasingly likely in the future' (East Devon DC). Paradoxically, it is clear from the survey returns that these fears of poor standing at appeal are often seen to relate to the truncated procedures used in the preparation of non-statutory instruments, the same factor which is widely identified as an advantage. For instance, 'non-statutory plans are likely to be given less weight at appeal and by other authorities because they have not been through the statutory objection–inquiry–adoption procedures' (Swale BC); statutory plans provide a more stable basis for decision-making and carry more weight in appeal cases having gone through the recognized preparation processes, including public inquiry (Amber Valley DC). Clearly, the fact that there are no prescribed procedures for preparing planning documents outside the statutory development plan system can be something of a two-edged sword. If non-statutory plans are prepared, advantages of flexibility, responsiveness and relevance accrue, but at the risk that such documents will not be supported at appeal.

A number of authorities indicate that the procedures involved in statutory local plan preparation and adoption are felt to be valuable in their own right; ensuring strength at appeal is here only a secondary consideration. In these cases, it is the possible absence of certain procedures from the preparation of informal material which causes concern. A few authorities point to public participation in this respect (Liverpool City Council, Durham City Council), but the bulk of responses here emphasize the lack of any right to object, which in the statutory plan preparation process is enshrined in the public local inquiry (PLI). On this point authorities are at one with the DoE, whose concern to foster the statutory system is partly based on the degree of protection that it affords to individual property rights and land values by the formal procedures in general, and the PLI in particular. Authorities variously point to the danger of alienating the public by removing the right of objection (Bury MBC); to the fact that non-statutory documents do not necessarily provide an opportunity for debate, criticism or testing of plan policies by the public (West Sussex CC, Arun DC, Wansbeck DC); and to the absence of any satisfactory way of resolving serious disputes should these arise (Humberside CC). The lack of an independent test at a PLI may also affect plan 'credibility', particularly where major land use issues are to be decided (Warwickshire CC, Charnwood BC), while there may also be a temptation for the local planning authority involved to ignore relevant objections or concerns in the absence of such a check. At least one authority seems to have experienced difficulties in objecting to non-statutory plans prepared by adjoining authorities (South Derbyshire DC).

This emphasis on the PLI as a valuable part of the statutory local plan preparation process echoes the responses to the recent discussion paper produced by a DPOS working party (1982). One of the key propositions of this paper called for the abolition of the PLI, but this recommendation has since met with little professional support.[5] Many planning practitioners appear to view the PLI as necessary, and see its absence in the case of informal plans as a significant drawback. In fact, recent research indicates that the time and resource implications of the PLI may well be enough of a disincentive to cause local planning authorities to resort to the statutory procedures only where formal support is considered necessary, e.g. in the face of complex conflicting interests where clearly established statutory policies will hopefully be supported on appeal. In other less complex situations, where the conflicts of interest are less pronounced, non-statutory approaches to local planning will be favoured since they are more flexible and less demanding of time and financial resources (Bruton, Crispin and Fidler, 1982). Clearly, the relative balance between statutory and non-statutory elements in any local planning strategy must be considered within the context of local problems and issues; an important factor in such deliberations is likely to be the required strength of policies at appeal.

As with Table 6.2, the 'other' category of Table 6.3 includes a number of themes. Several authorities, for instance, identify a range of essentially practical problems with their use of informal instruments, such as lack of comprehensiveness (Scarborough BC) or inconsistencies in the treatment of different areas of the same district (Leicester City Council, Bromley LB). Others – primarily county councils – complain that the lack of legal requirements to consult or to obtain a certificate of conformity with the structure plan means that there is a greater risk, with non-statutory documents, of non-conformity with structure plan policies and of policy conflicts emerging between one district and another (Devon CC). Put another way, non-statutory documents provide district councils with a means of expressing policies where these diverge from those advocated by the county and which therefore could not be progressed on a statutory basis without running into problems of conformity. Since informal material does not usually reach the stage of certification or public inquiry, such contradictions cannot be challenged, either by the county council or members of the public (Cleveland CC); however, 'it is hoped that Inspectors will carefully seek out the status of such plans and their compliance, or otherwise, with structure plan policy' (Wiltshire CC). These county fears cannot be regarded as surprising. After all, as one county authority points out, statutory local plans in the development plan scheme allow the county greater control and involvement in the timing and content of local plans (Lincolnshire CC). Research on county-district relations underlines the role of such formal documents as 'bargaining counters' (Leach and Moore, 1979, p. 168). Bruton (1980) for instance has shown how the preparation of a development plan scheme and the production of local plans was used as part of a process of 'distributional bargaining' to establish the relative levels of responsibility between district and county. If districts opt to pursue a non-statutory approach, then the county's formal 'weapons' for ensuring district compliance with structure plan policies are effectively debased or nullified.

A third 'other' theme emphasizes that uncertainty can be a concomitant of flexibility given the susceptibility of informal material to local political changes (Southampton City Council). Such changes may lead to a rapid turnover of non-statutory policies (Christchurch BC), but, in any case, such material is vulnerable to political pressure for modification or amendment, and is equally open to politically expedient misinterpretation (Rotherham MBC, Lewes DC).In Liverpool, the non-statutory nature of these instruments has created the danger of a lack of medium- or long-term commitment by the City Council, which has been able to renege on informally agreed land use proposals. Hayes (1981, p. 80) points out that the volatile nature of political priorities has to be regarded as part of the context for local planning in Liverpool; 'Land use planning issues can, and do, become the subject of intense political debate. Allocation in a plan is not necessarily a guarantee of future political intent to achieve implementation'.

This review of perceived advantages and disadvantages in the use of informal local planning documents has necessarily been somewhat impressionistic. Many of the questionnaire returns make use of such terms as 'credibility' or 'effectiveness' without specifying the context that is involved, and such responses have simply been reported. Nevertheless, some general points may be made.

Central government advice as to the unsuitability of informal material is seldom referred to and judging by the widespread extent of non-statutory activity is largely ineffectual. Yet many authorities are aware of the Secretary of State's views, especially with regard to the likely attitude of the Inspectorate when faced with informal material at appeal. In contrast, the main concern of many authorities appears to be a pragmatic interest in evolving or choosing a contingent local planning strategy to deal 'effectively' with local situations, problems and issues. These are various: staffing levels, county–district relations, the need (urgent or otherwise) for policy frameworks, and required strength at appeal are all factors which may influence the 'choice' between statutory or non-statutory elements. Thompson's (1977a, p. 145) premise that 'to be effective a local plan must be formulated firmly within the confines of its operational context' also applies to local planning strategies as a whole.

Finally, the various attractions and drawbacks of informal material are not separate but closely linked. The absence of statutory preparation procedures, for instance, may well lead to a shorter production time and a saving in cost, but will also tend to raise doubts over 'status'; in any case, the formal procedures (especially the right to object) may be felt to be valuable in their own right. Similarly, the very flexibility and responsiveness of informal material may also create uncertainty over the authorities' commitment to the policies involved; local political changes may lead to these fears being realized. Clearly, whether informal material is seen as valuable or unsuitable primarily depends upon the *local* operational context in which it is to be used. Individual local authorities can be expected to use whatever form of planning instrument – statutory or non-statutory – appears to them to be most appropriate.

6.3 Forms of non-statutory local planning

Authorities choosing the non-statutory approach to local planning are free from the various limitations on form and content affecting the statutory plan, and may design their own approach to the planning issues they face. As a consequence, non-statutory documents come in great variety. In some ways, this is analogous to the variations in statutory local plan content discussed earlier, but is here greater since the guiding framework of legislation is absent. Despite this variation, several classes of informal local planning instruments can be identified on the basis of their distinctive function and content (Bruton and Nicholson, 1983a). These are:

1 Development control practice/policy notes

Function: provide statement of local authority policy and guidance on detailed aspects of development control.

Content: various, including car-parking standards, residential layout and density, housing extensions, advertisements, shop fronts, together with general development control policies.

Notes: approved by DoE as part of supplementary planning guidance provided kept publicly available, published separately from the policies and proposals of the statutory development plan for the area, and consistent with the structure plan and any local plan for the area. See para. 1.14 of 1984 Memorandum on Structure and Local Plans (22/84).

2 Development briefs

Function: provide detailed guidance on the form a development should take on a particular site or sites. May also be used to establish framework for development, including desired land uses and disposal terms where the authority is landowner. Other designations possible depending on content, e.g. 'planning brief', wherein planning considerations for site development are set out. May also be prepared in response to individual planning applications as 'appraisal documents' etc., setting out in full the local planning authority's stance to the proposed development.

Content: site specification (location, area, ownership), constraints and services, often together with an indication of preferred development form, mass, and design.

Notes: as above, approved by DoE in order to assist preparation of planning applications by developers. Such briefs may take various viewpoints, e.g. planning, architectural, developer; content varies accordingly. See para. 1.14 of 1984 Memorandum on Structure and Local Plans (22/84).

3 Informal local plans

Function: similar to statutory district or action area local plans; spatially based, concerned with land use. May be prepared to give more detail for selected areas in context of existing local plans; as interim documents in advance of formal adoption, either as plans or 'insets'; or as alternatives to statutory local plans.

Content: relate to a wide variety of settlements, including villages, areas of new development or redevelopment, and inner city areas. Spatial coverage may be entire district or specific area.

4 Single topic-based frameworks

Function: equivalent to statutory subject plans. May be used to relate structure plan policy to district issues, to establish policy on matters requiring attention in advance of formal plan preparation, and to decide policy on topics whose significance or complexity does not merit a formal local plan.

Content: various, topics dealt with include tourism, shopping, conservation, land, housing, and industry.

Notes: replace subject plans where DoE have advised that the topic cannot adequately be dealt with in isolation.

5 Other policy frameworks

Function: policy formulation on issues which do not require or merit a statutory local plan, either throughout the local authority or in a particular area; or to 'fill in' the background to individual settlement policies.

Notes: unlike (4), deal with a number of policy issues.

6 Local plan briefs

Function: to amplify structure plan by relating its objectives and policies to the more detailed treatment appropriate to particular local plans. May be prepared in advance of structure plan approval to aid in the drawing up of local plans concurrently with the structure plan. May also have other designations, e.g. 'District Planning Statement'.

Content: elaboration of the planning framework established by relevant structure plan information, policies and proposals.

Notes: see para. 3.43 of 1984 Memorandum on Structure and Local Plans (22/84)

7 Informal joint studies

Function: bridge 'knowledge' gap between strategic and local levels, e.g. county problems in formulating structure plan housing policies when it is not the housing authority. Aim is mutual benefit.

Content: topic-based research into local issues such as housing, shopping or rural settlements.

Notes: all involve joint working and resource sharing between tiers.

8 Special area work

Function: establish design criteria and grant availability for areas such as GIAs, HAAs, and CAs. Also provision of information.

Content: area description and scope for future renovation/rehabilitation work.

9 Design guides

Function: statements of authority policy on aspects of architectural/ engineering design for use by developers.

Content: analysis of preferred design styles for specific elements (e.g. estate roads) or general ('townscape').

Notes: distinguished from (1) by content (design rather than technical), together with advisory nature.

As indicated, some of these classes of instruments are officially recognized as having a valid role to play within the statutory development plan system. Classes 1 and 2, for instance, together constitute supplementary planning guidance, while the value of local plan briefs in aiding the preparation of statutory local plans within the structure plan context has also been recognized by the DoE. Similarly, the various (non-statutory) documents prepared for GIAs, HAAs, etc., are linked to the area-based approach to local problems increasingly common in housing and other legislation (see Chapter 5). The remaining classes, however, have no official recognition, either explicit or implicit. Instead, the majority are actively proscribed, for the strictures of the Department on the use of non-statutory instruments relate specifically to those documents in Classes 3 (informal local plans), 4 (single topic-based frameworks), and 5 (other policy frameworks). Accordingly, if official advice were being heeded, relatively little non-statutory activity will fall within these groups compared to those which are recognized as legitimate.

Table 6.4 sets out the use made of the various classes of informal instruments identified above by type of authority. Development control practice/ policy notes is a significant area of activity (15 per cent of the total number of documents) and are especially important for the London boroughs and the shire districts. A limited number are compendiums of policies, such as Welwyn Hatfield DC's 'Development Control Standards and Criteria', or the 'Statement of Planning Policy' prepared and successively refined by Nottingham City Council. The majority, however, deal with single issues, common areas being residential extensions, car-parking standards, open-space provision, radio masts, and advertisement policies. In total, an

Table 6.4 *Types of non-statutory local planning instruments in use by type of authority, England, March 1983*

Type of Authority	Development control practice/policy note	Development brief	Informal local plan	Single topic-based framework	Other policy framework	Local plan brief	Informal joint study	Special area work	Design guide	Unspecified	Total
Metropolitan counties (%)*	1** (1)	8 (11)	4 (6)	5 (7)	2 (3)	29 (41)	5 (7)	8 (11)	4 (6)	5 (7)	71 (2)
Shire counties (%)	43 (8)	84 (17)	130 (26)	44 (9)	31 (6)	78 (15)	14 (3)	38 (7)	12 (2)	34 (7)	508 (11)
Metropolitan districts (%)	52 (9)	193 (33)	145 (24.8)	64 (11)	9 (1)	1 (0.2)	8 (1)	81 (14)	21 (4)	10 (2)	584 (13)
Shire districts (%)	476 (17)	648 (23)	854 (30)	220 (8)	141 (5)	60 (2)	35 (1)	283 (10)	98 (3)	19 (1)	2834 (63)
London boroughs (%)	88 (18)	214 (44)	57 (12)	22 (4)	8 (2)	14 (2.8)	0 –	26 (5)	61 (12)	1 (0.2)	491 (11)
Total (%)	660 (15)	1147 (26)	1190 (26)	355 (8)	191 (4)	182 (4)	62 (1)	436 (10)	196 (4)	69 (2)	4488 (100)

Source: Bruton and Nicholson, 1984a.

*Row percentages based on the number of instruments in use by each type of authority.

** Figures refer to number of instruments.

extremely wide range of topics is covered, with local policies being variously established on such diverse matters as guest houses, redundant churches, sunblinds, pig and chicken sheds, beach huts, multiple occupation, stables and thatched roofs. More general policies have also been developed to guide control decisions on office development, non-retail uses in shopping areas, mineral developments, and retail warehousing. Local planning authorities are clearly capable of responding contingently and flexibly to particular local planning issues, though as a result policy notes may verge on the exotic, e.g. a study on 'chicken slaughtering' in preparation by Birmingham City Council.

Development briefs (Class 2) are widely used to establish desired uses and standards on individual sites, accounting for over a quarter of the total number of instruments. They are particularly important for the London boroughs and metropolitan districts (44 and 33 per cent of documents, respectively). Liverpool City Council, for instance, has produced some 60 site development briefs over the last five years, while Barnet LB has 36 'planning briefs' relating to particular sites. Development control notes and development briefs together (as supplementary planning guidance) account for over 40 per cent of all informal instruments, a figure which rises to 62 per cent for the London boroughs.

Other non-statutory instruments, however, are also extensively used. Informal local plans (Class 3), for example, despite the antagonism of the DoE, are even more numerous than development briefs. However, the pattern of their use is different. While relatively few are used by the London boroughs and the metropolitan counties, they are especially significant for the metropolitan and shire districts and the shire counties, which together are responsible for 95 per cent of these instruments. In addition, use is also made of documents which primarily address specific policy issues (Classes 4 and 5) rather than problems within particular local areas, with the bulk again being used by the (especially shire) districts. The topics covered by these policy frameworks include: camping and caravans (a 'subject plan' (*sic*) prepared by South Lakeland DC); local economic development ('The Economy of Durham', Durham City Council; 'Industrial Development Policy', Hyndburn BC); tourism and related issues ('Tourism Study', South-ampton City Council; 'Tourism in Cambridge', Cambridge City Council); land availability (for industry, Melton BC; for housing, Bracknell DC); and Green Belts (primarily boundary definition, Southend-on-Sea BC, Mole Valley DC, and Essex CC). Localized planning issues may also be dealt with by these instruments, as the following examples illustrate; 'Specialist Schools and Colleges in Cambridge', awaiting approval by Cambridge City Council; 'Residential Homes for the Elderly', Penwith DC; and two com-plementary documents prepared by Cornwall CC to provide a 'Short Term Development Plan' and 'Long Term Strategy' for the china clay industry. These local policy statements, whether based on a single issue or on a number of topics, are clearly of use as vehicles for general expressions of policy on such frequently occurring planning issues as land availability and retailing, but can also be employed to deal with local planning problems of a very specific character.

The remaining classes of informal instrument are less significant. Local plan briefs, for example, represent 4 per cent of all non-statutory documents in use; as would be expected, they are mainly produced by the counties, with the shire and metropolitan counties together accounting for just under 60 per cent of these instruments. Special area work in contrast is largely limited to the districts, for of the 436 documents recorded here, 390 (89 per cent) are used by the districts and London boroughs. This material deals primarily with local proposals for the carrying through of such national initiatives as conservation areas, general improvement areas, and housing action areas, as well as dealing with the declaration of such areas; it usually has a strong spatial component. Much of the design work available is also to be found at district level, dealing with such aspects of the environment as shop fronts, house extensions, footpaths, tree and shrub planting, and fencing and hedgerows.

Overall, it is the shire districts which make most use of informal material, with 63 per cent of instruments; the other groups of authorities, with the exclusion of the metropolitan counties, each account for some 10 to 13 per cent of the total. These crude figures, however, are misleading, for they fail to take into account variations in the number of authorities of each type. The reason for the numerical dominance of the shire districts lies largely in the fact that 210 such authorities responded to the survey as against 24 London boroughs, 18 metropolitan districts, and 34 shire counties. A more accurate assessment of the varying extent to which authorities of different types use non-statutory instruments can be drawn from Table 6.5, which gives *average* rates of use for each type of authority. The figures for adopted and deposited local plans are calculated on a different base (*all* authorities of each type) from those relating to the use of informal instruments (*responding* authorities only). The table gives a rather different picture from Table 6.4, for the shire districts are now revealed as the most parsimonious users of informal material. Instead, it is the metropolitan districts which on average are the most intensive users of non-statutory instruments; this feature recurs in the figures for supplementary guidance, informal local plans, and policy frameworks. The counties, though this applies less to the shires than the metropolitan authorities, are only limited users of supplementary guidance or informal plans; their relatively high overall average results from their use of local plan briefs on a larger scale than the districts. The London boroughs match the metropolitan districts in terms of the use of supplementary planning guidance; both groups of authorities also have relatively well-developed statutory plan frameworks. Finally, the table graphically illustrates the overwhelming numerical dominance of informal documents over statutory local plans, although this cannot be taken to imply that formal

Table 6.5   *Average number of local plans and informal local planning instruments by type of authority, England*

Type of authority	Adopted and deposited local plans March 1985*	Supplementary planning guidance (Classes 1 and 2)**	Informal local plans (Class 3)	Policy frameworks (Classes 4 and 5)	All non-statutory instruments (Classes 1–9)
Metropolitan counties	2.0	2.2	1.0	1.7	17.7
Shire counties	1.4	3.7	3.8	2.2	14.9
Metropolitan districts	2.9	13.6	8.0	4.0	32.4
Shire districts	1.2	5.3	4.0	1.7	13.5
London boroughs	1.5	12.6	2.4	1.2	20.4
Total	1.4	6.2	4.1	1.9	15.5

Source: Bruton and Nicholson, 1984a; DoE schedules of local plan deposit and adoption to March 1985.

* Gives average number of adopted and deposited local plans for *all* authorities of each type, March 1985.

** Average number of non-statutory instruments for *responding* authorities for each type only, March 1983.

plans are a relatively insignificant element in local planning strategies. A statutory local plan is likely to be a more significant and robust instrument for local planning and development control purposes than a comparable informal instrument, precisely because of its statutory form.

The dates of completion of informal instruments in use at the time of survey are given in Table 6.6. Over 10 per cent date from before local government reorganization in 1974. Particularly important here are informal plans, accounting for 60 per cent of pre-reorganization instruments; over a quarter of the total number of these documents date from before April 1974. This may reflect the survival of informal 'interim' plans prepared following the advice of Circulars 102/72 and 122/73 on housing land availability, but in any case is an indication that successive DoE advice to either lapse informal plans or proceed them to statutory status has not been completely followed. Another factor which may account for the dominance of informal plans in the pre-reorganization category is that they usually contain more comprehensive statements of authority policy than other instruments, and should, as a result, have a longer lifespan. The fact that few development briefs were prepared prior to 1976, for instance, does not necessarily mean that they were seldom produced or used in this period, but

Table 6.6   Date of completion of non-statutory local planning instruments in use by type of instrument, England, March 1983

Type of instrument	Date of completion								No response unknown	Total
	Pre-April 1974	Ap. '74– Dec. '75	Jan. '76– Dec. '77	Jan. '78– Dec. '79	1980	1981	1982	1983†		
Development control note	38**	93	103	111	68	70	101	19	57	660
(%)*	(5.8)	(14.1)	(15.6)	(16.8)	(10.3)	(10.6)	(15.3)	(2.9)	(8.6)	(14.7)
Development brief	28	44	117	210	143	151	161	40	253	1147
(%)	(2.4)	(3.8)	(10.2)	(18.3)	(12.5)	(13.2)	(14.0)	(3.5)	(22.0)	(25.6)
Informal local plan	311	105	171	180	83	140	88	23	89	1190
(%)	(26.1)	(8.8)	(14.4)	(15.1)	(7.0)	(11.8)	(7.4)	(1.9)	(7.5)	(26.5)
Single topic-based framewk	34	28	65	66	36	43	34	9	40	355
(%)	(9.6)	(7.9)	(18.3)	(18.6)	(10.1)	(12.1)	(9.6)	(2.5)	(11.3)	(7.9)
Other policy framework	18	12	27	31	21	35	33	10	4	191
(%)	(9.4)	(6.3)	(14.1)	(16.2)	(11.0)	(18.3)	(17.3)	(5.2)	(2.1)	(4.3)
Local plan brief	0	19	53	33	15	26	20	4	12	182
(%)	(–)	(10.4)	(29.1)	(18.1)	(8.2)	(14.3)	(11.0)	(2.2)	(6.6)	(4.1)
Informal joint study	5	3	2	12	10	10	17	1	2	62
(%)	(8.1)	(4.8)	(3.2)	(19.4)	(16.1)	(16.1)	(27.4)	(1.6)	(3.2)	(1.4)
Special area work	58	42	56	64	35	43	48	15	78	436
(%)	(13.3)	(9.6)	(12.8)	(14.7)	(8.0)	(9.9)	(10.3)	(3.4)	(17.9)	(9.7)
Design guide	10	24	26	33	12	14	24	7	40	196
(%)	(5.1)	(12.2)	(13.2)	(16.8)	(6.1)	(7.1)	(12.2)	(3.6)	(23.5)	(4.4)
Unspecified	12	6	9	15	7	2	0	0	18	69
(%)	(7.4)	(8.6)	(13.0)	(21.7)	(10.1)	(2.9)	(–)	(–)	(26.1)	(1.5)
Total	514	376	629	755	430	534	523	128	599	4488
(%)	(11.5)	(8.4)	(14.0)	(16.8)	(9.6)	(11.9)	(11.7)	(2.9)	(13.3)	(100.0)

Source: Bruton and Nicholson, 1984a.

*Row percentages based on number of instruments of each type in use.

**Figures refer to number of instruments.

†Part year only.

more likely reflects a poor 'survival rate'. Development briefs have a relatively short 'shelf life' as a result of their comparatively specific content, and tend to date quickly either through a change in circumstances (for example, a reversal of policy) or through completion of the development that they were formulated to promote. More recently, half of the documents produced in 1982 are components of supplementary planning guidance, with informal plans of much less significance. This could be an indication that authorities are taking note of DoE advice, but probably also reflects a growing need for supplementary guidance now that statutory local plans are coming into more general use.

This overview of the extent of local authority non-statutory planning activity identifies both the *scale* on which such material is used and points to particularly significant *types* of instruments. Clearly, local authorities make extensive use of informal documents, particularly of informal local plans, development briefs, development control notes, various policy frameworks, and special area work. Indeed, these figures are likely to be underestimates of the scale of use, given the number of authorities who did not respond to the questionnaire survey. Applying the average rates given in Table 6.5 to the total number of authorities in each group suggests that the total number of instruments is actually in the region of 6500 (assuming that non-responding authorities make comparable use of informal instruments). However, a caveat must be added here. These aggregate figures are interesting in their own right, but the actual significance of this material, in terms of the links to statutory plans and the manner in which it is actually used within local planning strategies, remains to be assessed.

6.4 Upgrading and abandonment

The distinction between statutory and informal material must not be allowed to hide the possibility that individual documents and the policies contained therein can belong to both groups of instruments at different stages of their effective life. Circular 22/84 (para. 3.12) for instance, suggests that where an existing informal plan is needed for development control purposes, priority should be given in the development plan scheme to its adoption on a statutory basis. This may be either as a new local plan or as part of a plan already in the scheme. The policies within an informal plan may thus be incorporated within a subsequent local plan and thereby assume statutory force. On the other hand, documents once intended as statutory plans may be used informally if the statutory preparation and adoption procedures were never completed. Fudge *et al.* (1983), for example, comment that a number of documents once intended as subject plans are now in use on a non-statutory basis, following DoE advice that the topics involved could not properly be treated in isolation and were, therefore, inappropriate for the statutory subject plan treatment.

To explore these relationships, authorities were asked in the question-naire survey to list any of their statutory local plans which made use of previously informal material, and to comment on the manner of inclusion. A number of possibilities were suggested here, e.g. included as 'insets' in a wider plan; several non-statutory documents aggregated to form one statut-ory plan; or one informal plan upgraded to statutory status. The results, summarized in Table 6.7, shows that at the time of survey in March 1983 some 58 local plans were based on or included previously informal material. This represents 34 per cent of local plans adopted by March 1983. Thirteen plans use non-statutory material as 'insets', e.g. the Folkestone and Hythe District Plan (Shepway DC) which incorporates the Hythe Town Centre Plan and the Sandgate Study by this means. Less significant methods are aggregation or single plan upgrading; in the case of the latter, several councils appear to have followed the lines suggested by the DoE in taking their informal material to statutory status. The Church District Plan (Hynd-burn DC), for instance, was initially prepared as a non-statutory document but was subsequently revised and adopted following approval of the North East Lancashire structure plan. Many of the documents upgraded in this way are now in use as action area plans, such as the Prescot Town Centre plan (Knowsley MBC), the Bexleyheath Town Centre plan (LB of Bexley), and the Broughton Astley Central Area plan (Harborough DC). A larger number of local plans incorporate informal material by 'other' means, which primarily refers to the selective inclusion of policies in local plans, e.g. the East Hertfordshire district plan and the Tamworth district plan both cover the whole local authority area involved and make use of those policies and proposals in earlier informal plans which were still considered to be relevant.

Authorities were also asked if they *intended* to incorporate any of their currently informal documents in future statutory plans. Overall, authorities intend to use some 20 per cent of their informal instruments in statutory plans in one way or another, with the various methods of insets (5 per cent of all instruments), aggregation (6 per cent) or 'other' means (6 per cent)

Table 6.7   *Use of non-statutory material in statutory local plans, England, March 1983*

	County		District		Total	
Method of upgrading	No.	(%)	No.	(%)	No.	(%)
'Inset'	1	(33)	12	(22)	13	(22)
Aggregation	0	(–)	8	(15)	8	(14)
Single-plan upgrading	0	(–)	10	(18)	10	(17)
Other means	2	(67)	25	(45)	27	(47)
Total	3		55		58	

Source: Bruton and Nicholson, 1984a.

relatively important. Single-plan upgrading is relatively insignificant in numerical terms (2 per cent, 95 plans), but each of the documents included here will, if authority intentions are realized, eventually be in use as a statutory plan. In contrast, the other identified upgrading methods essentially represent only partial contributions to such plans. In terms of the types of informal instruments involved in upgrading intentions, the three groups expressly proscribed by the DoE, informal plans and single/multi-topic policy frameworks, make up 58 per cent of this material. This could simply mean that authorities are beginning to follow Departmental advice in progressing their informal plans to statutory status. However, single-plan upgrading accounts for only one-tenth of these documents, and this indicates that more elaborate strategies, involving the use of insets, aggregation or the selective incorporation of policies, are being followed in a bid to raise existing informal material to an acceptable statutory level.

The use of informal material in statutory local plans is only one side of the relationship between these two types of local planning activity. The corollary is the informal use of documents which it was once intended to progress to statutory status, but which – for a number of reasons – were never so adopted. Table 6.8 shows the extent to which such 'abandoned' statutory plans are now in use informally. A total of 126 documents are involved here, the bulk of which were once intended to be district plans. The number of abandoned subject plans is relatively small, while only a handful of instruments once intended as action area plans are now in use informally. These abandoned plans are mainly in use as informal local plans (75 per cent of the total number of abandoned documents), single-topic frameworks (10 per cent), and other policy frameworks (5 per cent). The bulk of intended district plans are not surprisingly used as informal local plans (93 per cent), while the majority of intended subject plans are now being progressed as single and other policy frameworks (85 per cent). While one-half of action area plans are in use as informal plans, a further 37 per cent (6 documents) have been carried on as development briefs.

Table 6.8 *'Abandoned' local plans in use as non-statutory instruments, England, March 1983*

Type of plan	County No.	(%)	District No.	(%)	Total No.	(%)
District	12	(52)	78	(76)	90	(71)
Action area	2	(9)	14	(13)	16	(13)
Subject	9	(39)	11	(11)	20	(16)
Total	23		103		126	

Source: Bruton and Nicholson, 1984a

Authorities were asked to give some brief indication of their reasons for abandoning the intention to proceed these documents as statutory local plans. A wide range of relevant factors were identified. For 'intended district plans', for instance, a number of authorities pointed to various resource constraints, in terms of time, staff or cost, which militated against taking the document concerned through the statutory stages to adoption (Tunbridge Wells BC, South Shropshire DC). For other authorities, changes in the overall local planning strategy altered or removed the rationale for taking certain documents through to adoption. For Broxtowe BC, for example, a recent recognition of the need for a borough-wide plan has meant that two draft district plans which had progressed to deposit will not now be adopted as they stand but will instead be incorporated into the Broxtowe Local Plan. Elsewhere, local plan preparation programmes have been revised downwards in the light of informal DoE advice and that contained in the now cancelled Circular 23/81 *Development Plans*, which recommended that local plans should only be prepared where clearly needed. The resultant 'abandoned' documents have, in some cases, remained in use as non-statutory instruments (Rotherham MBC). Furthermore, various changes in the strategic (structure plan) context have necessitated complementary changes in local plan policies and proposals. For example, an 'interim local plan' prepared by Newcastle upon Tyne City Council in 1978 has been replaced by a new local plan which seeks to identify additional housing sites in line with the requirements of the Tyne and Wear structure plan, approved in 1981. The requirement under the 1971 Act that local plans could not be deposited or adopted before the approval of the structure plan concerned (now amended by provisions in the *Inner Urban Areas Act 1978* and the *Local Government, Planning and Land Act 1980*) has also created difficulties. Authorities have found that delays in structure plan approval have meant that development decisions have unavoidably been taken on the basis of *informal* documents. The cumulative effect of these decisions has, in some cases, been to remove the need for formal adoption, since by the time of structure plan approval the policies involved had been largely implemented (Tamworth BC). Other reasons cited for the 'abandonment' of statutory district plan intentions include: the absence of contentious issues requiring resolution at a public local inquiry (Waveney DC, Middlesbrough BC); the approval of land use and layout proposals via detailed planning applications (Great Grimsby BC); the need for a rapidly available planning framework capable of looking beyond the normal (10 year) local plan period (Nuneaton and Bedworth BC); and the effects of the recession on key proposals for retail development (Tendring DC).

To turn to 'abandoned' subject plans, a number of authorities point to the influence of DoE advice that the proposed subject could not properly be dealt with in isolation and, therefore, was inappropriate for statutory subject plan treatment (industry, Glanford BC; housing, Wycombe DC). North

Yorkshire CC found this advice borne out when attempting to consider Green Belt issues independently from those concerning housing and industrial land; as a result, a policy statement on Green Belts in the county is an informal rather than statutory document. Other authorities indicated that the non land use policy content of the documents involved meant that the statutory subject plan treatment was inappropriate (Staffordshire CC), particularly where management issues were involved and flexibility was required (Babergh DC). For example, a statutory subject plan determining land use issues was not required by Tyne and Wear MCC in setting recreation and countryside management principles for the Derwent Valley, given on the one hand a primarily non land use management role and, on the other, the fact that most of the land involved was held by the local authority. Some counties cited district council opposition as a reason for progressing 'subject' policies on an informal basis (North Yorkshire CC, Essex CC).

Finally, intended action area plans were most often lapsed as a result of being overtaken by events, either through the granting of planning permissions (Staffordshire CC) or the implementation of the main proposals involved (Blyth Valley BC), perhaps before structure plan approval (North East Derbyshire DC). Blyth Valley BC also pointed to the need for a more flexible working arrangement with the one developer involved than that allowed by the statutory format; other authorities cited financial and time factors as reasons for abandonment (Wellingborough BC, East Staffordshire DC, Arun DC).

While these various reasons for the lapsing of statutory intentions are partly peculiar to individual local authorities and planning contexts (e.g. the absence of contentious issues, implementation pre-empting statutory adoption), they can also be related to the national trend towards a reduction in the number of *proposed* local plans (Bruton, 1983a). This trend partly reflects the influence of Departmental advice, either through such devices as circulars or Local Plans Notes, or more informally via direct consultations with individual authorities. In addition, it is safe to assume that this central government advice has been reinforced at local level by the realization that the early local plan programmes were over-ambitious in terms of staff and other financial resources, given the authorities' working experience of the plan adoption system on the one hand and on the other increasing pressures to reduce municipal expenditure. Together, these factors have led to a down-grading of the scale of statutory *local plan* intentions, but not necessarily to a reduction in *local planning* activity; lapsed 'local plans' may still be adopted and used. In Newcastle upon Tyne, for instance, early experience with statutory plans pointed to the lengthy timescale involved and the consequent staff costs. The Little Benton Plan was begun in 1977, but not adopted until 1982; much of the period up to deposit was spent awaiting the results of formal consultations with various council committees, departments and other interests. It was subsequently agreed that statutory plans would henceforth only be used when considered necessary and where no less costly or time-consuming alternative existed, and that a streamlined procedure would be used for any future statutory plans. In a subsequent review of the development plan scheme, a number of statutory plans which had been proposed were deleted in favour of informal action to handle outstanding issues.[6] Such action does not, of course, prevent the future incorporation of informal policies and proposals into statutory plans (possibly wider area) via upgrading.

6.5   Preparation procedures and roles

Many local authorities identify the relative procedural ease with which informal instruments can be prepared as a major advantage of non-statutory documents over formal local plans. Advantages of speed of preparation, economy, relevance and ease of amendment are argued to stem from the freedom from the various procedural stages laid down in the statutory local plan preparation and adoption process. However, this does not mean that as a result non-statutory instruments will necessarily be prepared without the benefit of such basic procedures as public participation, consultation or council adoption. These procedures have a more fundamental part to play in plan preparation than simply the satisfaction or achievement of statutory requirements (Healey, 1983, p. 139):

For both central government and local planning authorities, the most significant role for public consultation in plan preparation is to give support to the planning authority's policies where significant conflicts over land use and development exist. The fact of engaging in an apparently public debate on policy alternatives helps to legitimate the policies arrived at.

Consultations with government departments and other public agencies also lends credence to plan policies, although the difficulties of securing the genuine commitment of such bodies to statutory or non-statutory plan proposals are unlikely to be surmounted by a process which also allows these agencies to protect their own interests. Council adoption will ratify political support but not prevent subsequent changes of policy. For these reasons, and despite the difficulties involved, local authorities will look towards such preparation procedures in an attempt to establish public support for and political commitment to their local planning policies and proposals, irrespective of whether these are being progressed through a statutory local plan or an informal instrument.

Table 6.9 gives a breakdown of the procedures used by local planning authorities in preparing their informal instruments. The information in this table refers to the use of the specified procedures *in any form*; no data were requested on aspects such as the extent of public participation or the number of consultations involved. Overall, the most common procedure is adoption

Table 6.9 *Procedures used in the preparation of non-statutory local planning instruments by type of instrument, England, March 1983*

Type of instrument	Procedures							
	Public participation	Consultations	Adoption by committee/council	Participation, consultations, adoption	Other	None	No response	Total
Development control note	104**	151	573	58	38	6	57	660
(%)*	(15.7)	(22.9)	(86.8)	(8.8)	(5.7)	(0.9)	(8.6)	(14.7)
Development brief	305	607	870	247	121	76	103	1147
(%)	(26.6)	(52.9)	(75.8)	(21.5)	(10.5)	(6.6)	(9.0)	(25.6)
Informal local plan	873	909	1006	810	49	1	163	1190
(%)	(73.4)	(76.4)	(84.5)	(68.0)	(4.1)	(0.08)	(13.7)	(26.5)
Single topic-based framework	146	167	303	107	29	2	38	355
(%)	(41.1)	(47.0)	(85.3)	(30.1)	(8.2)	(0.6)	(10.7)	(7.9)
Other policy framework	70	93	180	46	47	0	3	191
(%)	(36.6)	(48.7)	(94.2)	(24.0)	(24.6)	(–)	(1.6)	(4.3)
Local plan brief	66	90	137	62	35	1	11	182
(%)	(36.2)	(49.4)	(75.3)	(34.0)	(19.2)	(0.5)	(6.0)	(4.1)
Informal joint study	15	34	44	11	8	3	6	62
(%)	(24.2)	(54.8)	(70.9)	(17.7)	(12.9)	(4.8)	(9.7)	(1.4)
Special area work	239	159	289	134	14	66	51	436
(%)	(54.8)	(36.5)	(66.3)	(30.7)	(3.2)	(15.1)	(11.7)	(9.7)
Design guide	42	48	149	11	16	2	30	196
(%)	(21.4)	(24.5)	(76.0)	(5.6)	(8.2)	(1.0)	(15.3)	(4.4)
Unspecified	3	45	20	2	0	0	20	69
(%)	(4.3)	(65.2)	(29.0)	(2.9)	(–)	(–)	(29.0)	(1.5)
Total	1863	2303	3571	1488	357	157	482	4488
(%)	(41.5)	(51.3)	(79.6)	(33.1)	(7.9)	(3.5)	(10.7)	(100.0)

Source: Bruton and Nicholson, 1984a.

*Row percentages based on number of instruments in use.
**Figures refer to number of instruments.

267

by local authority committee or council (79 per cent of documents), followed by the carrying out of consultations with statutory undertakers, interested parties and resident groups (51 per cent) and by public participation (41 per cent). These different procedures, of course, are commonly used together in preparing individual documents. Thus, the figures for each procedure hide the fact that a third of all informal material has been subject to participation, consultations and adoption in combination.

Among the individual classes of instrument, a number of distinct procedural 'profiles' may be identified. Development control notes, for example, are only infrequently subject to participation and consultations (15 and 22 per cent of these instruments, respectively); committee/council adoption is common, but the use of all three procedures together is not. Similarly, development briefs are seldom progressed using all three procedures in combination; although the rates for both adoption and consultations approximate to the average, relatively little use is made of participation. The converse of these points is that a large number of development control notes and development briefs do not appear to have benefited from the use of either consultations (77 and 47 per cent, respectively, of these two groups of instruments lack this procedure) or public participation (84 and 73 per cent) during their preparation. The significance of this point lies in the DoE's stated principle that when supplementary guidance is being taken into account as a material consideration for development control purposes 'the weight to be accorded to it will increase when it has been prepared in consultation with the public and has been made the subject of a council resolution' (DoE, 1984g, para. 1.15). The finding that many 'supplementary' instruments have not been subject to public participation or consultations during their preparation thus casts considerable doubt on the likely status of these documents at appeal. The other procedural criteria affecting weighting – council adoption – does appear to be widely met by development control notes and development briefs alike.

Also interesting is that in contrast to these two accepted components of supplementary guidance, informal local plans are regularly subject to public participation, consultations, and adoption, with these procedures respectively being carried out on 73, 76 and 84 per cent of such documents. Moreover, these three procedures have been used in combination in the preparation of over two-thirds of informal local plans, a rate which is more than double that for all non-statutory instruments taken together. The particularly extensive use of these procedures here, together with the relatively low levels of participation or consultations on control notes and development briefs, raises the possibility that the procedural weightings attached to supplementary guidance by the DoE could also be used to determine the status of other informal material when used in development control. Circular 22/84 (para. 1.13) rules this out, stipulating that any plan containing land use proposals and not subject to the *statutory* procedures of public participation, deposit, inquiry and adoption 'can have little weight for development control purposes'. However, this is no more than a statement of the DoE's views on the acceptability or otherwise of informal local planning documents in development control and at appeal, and ignores the fact that many informal local plans have been subject to some form of participation and adoption. These procedures could easily be regarded as in some way validating the policies and proposals concerned, just as they are accepted as giving greater weight to supplementary guidance.

This point also applies to the single-topic and other policy frameworks, for both groups of instruments maintain high levels of adoption; whilst the rates for participation and consultation are below the levels recorded for informal local plans, they are still in the main higher than those achieved by development control notes or development briefs. The remaining groups of instruments largely reflect the overall average figures, with some variations: 'special area work' has relatively high levels of participation, presumably involving residents affected by GIA or HAA proposals, but a low level of consultations; the figures for design guides (low levels of participation and consultations, but usually adopted by committee/council) probably reflects their essentially advisory status.

Although local planning authorities in England place great emphasis on the procedural advantages of non-statutory instruments, a variety of procedures is commonly used in producing informal material. This tends to confirm the suggestion made earlier that local authority dissatisfaction with the statutory requirements relates not so much to the procedures *per se*, for these are still being widely used in an informal context, but to the specific form in which these procedures are laid down in regulations. Given the emphasis placed on *speed* of preparation as an advantage of the informal local plan, the *timescale* of statutory plan production seems to be a principal factor in determining approaches to local planning. Also relevant are questions of the flexibility and suitability of statutory procedural requirements to local needs. The DPOS for instance comment that the regime for public participation set out in the regulations will not always be appropriate to the circumstances of individual plans, e.g. where the authority's approach is community-based (DPOS, 1978, para. 3.15). Lastly, the widespread use of a number of procedures in informal plan preparation bodes well for the strength and status of these documents at appeal, provided that individual members of the Inspectorate are:

1 made aware of the type and extent of procedures involved in their preparation, and
2 prepared to take a relatively independent line to DoE guidelines on the use of non-statutory local planning material.

Table 6.10 summarizes the various roles played by non-statutory instruments in developing and progressing local planning policies. These roles

Table 6.10   Functions of non-statutory local planning instruments by type of instrument, England, March 1983

Type of instrument	Function										Total
	Development promotion	Determination of planning Applications	Determination of policy on a site	Determination of policy within a specific area	Determination of policy on an issue	Policy co-ordination	Conservation, management, implementation programme	Elaboration of structure plan	Other	No response	
Development control note	155**	482	7	19	66	2	6	14	21	65	660
(%)*	(23.5)	(73.0)	(1.0)	(2.9)	(10.0)	(0.3)	(0.9)	(2.1)	(3.2)	(9.8)	(14.7)
Development brief	533	284	387	78	15	15	23	6	86	160	1147
(%)	(46.5)	(24.8)	(33.7)	(6.8)	(1.3)	(1.3)	(2.0)	(0.5)	(7.5)	(13.9)	(25.6)
Informal local plan	237	598	35	501	53	30	82	69	195	158	1190
(%)	(19.9)	(50.2)	(2.9)	(42.1)	(4.4)	(2.5)	(6.9)	(5.8)	(16.4)	(13.3)	(26.5)
Single-topic framework	63	126	8	33	123	6	33	25	37	57	355
(%)	(17.7)	(35.5)	(2.2)	(9.3)	(34.6)	(1.7)	(9.3)	(7.0)	(10.4)	(16.1)	(7.9)
Other policy framework	27	85	9	43	41	0	17	40	21	24	191
(%)	(14.1)	(44.5)	(4.7)	(22.5)	(21.5)	(−)	(8.9)	(20.9)	(11.0)	(12.6)	(4.3)
Local plan brief	6	17	0	8	9	0	2	147	13	9	182
(%)	(3.3)	(9.3)	(−)	(4.4)	(4.9)	(−)	(1.1)	(80.8)	(7.1)	(4.9)	(4.1)
Informal joint study	18	16	1	7	18	3	0	2	15	4	62
(%)	(29.0)	(25.8)	(1.6)	(11.3)	(29.0)	(4.8)	(−)	(3.2)	(24.2)	(6.5)	(1.4)
Special area work	84	107	6	75	5	1	97	1	138	91	436
(%)	(19.3)	(24.5)	(1.4)	(17.2)	(1.1)	(0.2)	(22.2)	(0.2)	(31.6)	(20.9)	(9.7)
Design guide	89	87	3	1	8	1	2	3	5	52	196
(%)	(45.4)	(44.4)	(1.5)	(0.5)	(4.1)	(0.5)	(1.0)	(1.5)	(2.5)	(26.5)	(4.4)
Unspecified	12	45	5	1	29	0	1	0	1	10	69
(%)	(17.4)	(65.2)	(7.2)	(1.4)	(42.0)	(−)	(1.4)	(−)	(1.4)	(14.5)	(1.5)
Total	1224	1847	461	766	367	58	263	307	532	630	4488
(%)	(27.3)	(41.1)	(10.3)	(17.0)	(8.2)	(1.3)	(5.9)	(6.8)	(11.8)	(14.0)	(100.0)

Source: Bruton and Nicholson, 1984a.   *Row percentages based on number of instruments of each type in use.   **Figures refer to number of instruments.

cover a wide range of local planning activity, from elaboration and interpretation of the strategic structure plan, through policy determination on a variety of issues, specific geographical areas and sites, to the implementation of such policies through the active promotion of development or the operation of the development control system. Overall, this latter function is the most significant, with over 40 per cent of documents being used in the determination of planning applications, and 27 per cent for development promotion. Other important roles are the determination of policy within a specified plan area (17 per cent), on individual sites (10 per cent) and on an issue or issues (8 per cent).

These functions, or specific combinations of these functions, tend to be closely associated with individual types of instruments. Development control notes, for example, are largely used in implementation via either development promotion (23 per cent of notes) or more frequently the processing of planning applications (73 per cent). Special area work is also used most frequently in implementation, through either development promotion or control or the establishment of conservation and management programmes. Policy determination within specific locations is also notable, as a consequence of the fact that these documents are linked to defined areas. A number of roles can be carried out by a single instrument, e.g. the Middleport Policy Report prepared and adopted by Stoke-on-Trent City Council in 1980, which declared a number of HAAs, made proposals for environmental works and traffic management measures, and identified land reclamation and housing redevelopment sites. In contrast to this focus on implementation, local plan briefs are most commonly used to elaborate the structure plan context (80 per cent of documents), although an interesting minor role is their use in development control. Development briefs have a number of significant functions, the most important being development promotion (46 per cent of briefs seek actively to promote desired development), followed by site policy determination (33 per cent) and, to a lesser extent, the processing of planning applications (24 per cent). This variety of roles reflects the fact that briefs may be provided for a number of different purposes depending on site attributes, development interest, and authority policy. As Ratcliffe (1982) notes, development briefs may be either *promotional*, aiming to attract prospective developers; *explanatory*, adopting a neutral position towards sites in which interest has already been expressed; or *regulatory*, stipulating uses, standards and densities for what are seen as sensitive sites. It is clear that briefs have been most often produced for the first of these reasons, development promotion, with the others of less significance; however, a combination of roles cannot be ruled out, for as Ratcliffe notes, different aspects of the same brief may be promotional, explanatory or regulatory in character as circumstances require.

Informal local plans, like development briefs, carry out a number of roles, with the significant functions being use in development control (50 per cent of plans), the formulation of policy within a specific plan area (42 per cent), and development promotion (20 per cent). This set of roles points to the usefulness of many of the documents in this class in the *setting* of policy as well as in its application or implementation, and suggests that informal plans are being widely used as direct substitutes for the statutory vehicle. In a similar manner, single topic frameworks are used in development control (35 per cent of documents), promotion (17 per cent), and to set policy on an issue (34 per cent). The range of issues referred to by these frameworks is considerable; policies are set on such broad topics as residential, industrial and commercial development, together with statements on more local issues such as specialist colleges in Cambridge, or the china clay industry in Cornwall. A specific example is provided by Bedfordshire CC's 'Landscape and Wildlife Strategy' (1980), a county-wide document seeking to promote landscape and wildlife conservation and incorporating a statement of principles, the identification of areas at risk, broad suggestions for treatment, and guidance on the availability of grant-aid and advice. The policies included in this document were later developed and applied to a particular locale in the 'River Valleys Sector Study' (1982), which sets out proposals for landscape improvements and conservation measures to be negotiated with landowners, as well as incorporating a statement of the position on land drainage works. For a particular policy area, therefore, these two documents both *establish* and seek to *progress* policies by putting forward specific proposals for consideration.

This review of the various roles played by non-statutory local planning instruments shows that such documents relate to the entire range of local planning activity, linking the strategic level structure plan with development on the ground. Individual classes of instrument have characteristic functions or combinations of functions which they carry out in linking strategy to implementation. Perhaps most significant is the use of informal documents to fill such roles as structure plan interpretation and the provision of a detailed basis for development control, for these functions are precisely those which the statutory local plan was originally intended to fulfil within the context of local flexibility.

6.6 Conclusions

It is clear that local planning authorities in England are currently preparing and using an extensive array of non-statutory instruments in response to specific local needs. Practice tends to explain or account for this activity by referring to deficiencies with the statutory local plan, which allegedly have driven authorities to make use of the informal 'alternative'. Two common strands of criticism, for example, relate to logistical shortcomings, in terms of the time-consuming and costly procedures involved in statutory plan preparation and adoption, and to more fundamental restrictions on the legitimate form and content of local plans. These are no doubt valid problems facing authorities seeking to prepare and adopt statutory plans, but by themselves these difficulties do little to explain the form of local planning strategies which authorities have actually evolved. In fact, the response of planning practice to these shortcomings has been anything but uniform. The various problems associated with the statutory plan, for instance, have not led to its wholesale rejection; on the contrary, a substantial number of local plans have now been deposited and adopted (see Chapter 4). Moreover, many authorities using informal material appear to be developing markedly different and innovative local planning strategies. Examples include Middlesbrough BC's policy process, whereby land use decisions are taken within a series of annually updated policy and resource programmes, or Birmingham City Council's three-tier system of informal inner area studies, outer area statements and statutory plans where appropriate (Southerton and Noble, 1982; Middleton, 1982). No doubt in some cases dissatisfaction with the statutory plan is a key factor in shaping local planning strategies (e.g Newcastle upon Tyne City Council), but in others there are clearly additional factors at work.

It is here that those theoretical approaches to local planning which emphasize the need to consider the context within which the local planning function is worked out are of most value. In terms of local planning strategies, these approaches suggest that practice has resorted to informal as well as statutory means of progressing policies because it has had to meet a wide range of contingent factors in its efforts to translate more general policies for change into development on the ground. The wide variety of informal documents that have been produced is testimony to the wide variety of contingent factors impinging on the local planning process. These will include the nature of inherited strategic and local policies, required strength at appeal, the perceived need for policy frameworks in terms of timescale, topic and area coverage, the nature of county–district relations, the local political context, the planning issues involved, central government exhortation and advice, resource constraints and opportunities, and the influence of other agencies, public and private, within the plan/authority area. The relative influence of these factors will vary between different local planning contexts, so that the prime determinant of any local planning strategy is likely to be the particular local context or environment which contains the very issues the strategy must deal with.

Central advice will thus be only one element in a range of factors which authorities will take into account in evolving or choosing a local planning strategy to deal 'effectively' with local issues or problems. Moreover, since different authorities will be attempting to deal with different planning contexts, *detailed* central advice (as opposed to more general explanations of the legislation) must necessarily be limited in its relevance. This much was recognized by the Planning Advisory Group, who argued in 1965 (p. 35) that 'the precise form and content of the plans can be left largely to local option'. It can be argued that authorities are doing no more than pursuing the PAG philosophy of local flexibility in using statutory and non-statutory instruments as appropriate to respond to the contingent factors they face.

It is not surprising, that individual authorities have developed different approaches to problems, since a range of situations calls for a range of different responses. Authorities will use whatever form of planning instrument appears to them to be most appropriate. If, for example, the restrictions on the content of local plans are such as to make the statutory plan an unsuitable policy vehicle, then some form of non-statutory instrument will be used. On the other hand, if greater strength at appeal is required, then the balance of considerations may well favour a statutory plan. Thus, statutory and non-statutory documents both have valid roles to play in progressing higher level (structure plan) policies towards implementation, but the validity or appropriateness of these roles will be contingent upon the local context. It is of course possible for a strategy to incorporate a mix of statutory and informal documents as required, and indeed this is exactly what is happening in practice (e.g. Birmingham City Council, East Lindsey DC). Finally, as the local planning context evolves and changes over time, some adjustment in the authorities' response to this context is likely to be required, through for instance the amalgamation of statutory plans in preparation to form a larger area plan, the upgrading of an informal document, or the abandonment of an intention to proceed a plan to statutory form.

Notes

1   Department of the Environment (1972), Land Availability for Housing, Circular 102/72, para. 7; (1973), Land Availability for Housing, Circular 122/73, para. 7.
2   The Horsford/Taverham/Drayton Local Plan, prepared by Broadland DC.
3   A further seven authorities (2.4 per cent of the total number of respondents) had non-statutory instruments in preparation; five authorities (1.7 per cent) reported that they made no use of informal documents.
4   Quotes are taken from questionnaire survey returns unless otherwise cited.
5   Paper given by K. L. Mayoh (Planning and Development Officer, Newcastle under Lyme BC, and Convenor of the District Planning Officers' Society Local Plans Topic Group) to a one-day seminar on 'Local Plans – the Need for Radical Change' organized by the Association of District Councils and the District Planning Officers' Society, 22 March 1983.
6   Questionnaire response by Newcastle upon Tyne City, citing report to Newcastle upon Tyne City Council Development, Planning and Highways Committee, 'The Work of the City Planning Department, Statutory Local Plans', 12 November 1981.

7 Local planning and implementation
I: The development control process

7.1 Introduction

The basis of the British system of land use planning is that, in general, no development of land should take place without the prior consent of the local planning authority, granted in the form of a planning permission. When making a decision on a planning application the local planning authority is given broad statutory discretion by the *Town and Country Planning Act 1971* (s.29(1)) which requires that local planning authorities should '. . . have regard to the provisions of the development plan, so far as material to the application, and to any other material considerations.' Thus the development plan is only one factor to be taken into account by local planning authorities in exercising their discretion. In addition, and depending on the circumstances surrounding a particular development, the local planning authorities are required to take into account a range of other material considerations, some of which may not have been taken into account in the preparation of the plan, e.g. the suitability of the particular site for the development proposed; traffic flow conditions; the availability of infrastructure, and policy guidelines issued by the Secretary of State in the form of circulars, advice notes, appeal decisions, and research reports.

The development plan is defined by statute and regulation, and as a result there is little ambiguity as to what should or should not be considered under this heading by a local planning authority in determining an application for permission to develop. Unfortunately no definition exists for 'other material considerations', and whilst it is clear that such considerations must have a planning basis, there are difficulties in determining their precise boundaries, especially in cases which involve factors other than traditional land use factors, e.g. the adequacy of existing social infrastructure. However, over the years a body of case law has developed which provides basic guidelines as to what constitutes 'other material considerations'.

It is the aim of this chapter to examine the development control process and its inter-relationship with local planning. In particular, attention is focused on:

1 the range and scope of other material considerations, including policy guidance from central government;

2   officer–councillor relationships and bargaining in the decision-making
    process; and
3   the determination of Section 36 appeals.

Chapter 8 goes on to review the interaction between the development
process and local planning. However, it should be appreciated that this
structure, although reflecting the statutory planning framework, nevertheless represents an artificial differentiation between development control and
implementation. Development control is in reality an integral feature of the
implementation process.

7.1.1   Local plans and the development plan

The development plan provided for in Part II of the *Town and Country
Planning Act 1971*, as amended, consists of the structure plan and any local
plans. The structure plan is produced in England and Wales by the county
planning authority for its area and consists of a written statement which sets
out the main strategic planning policies for the area and the most important
proposals for change. The written statement is illustrated in general terms by
a key diagram; the policies and proposals relate directly to the development
and other use of land and they are approved with or without modification by
the Secretary of State. In Scotland the structure plan is produced by the
regional and island authorities, and approved by the Secretary of State for
Scotland.

   Chapters 3 and 4 have shown that local plans develop the policies and
general proposals of the structure plan and relate them to precise sites or
areas of land. They comprise a written statement, which sets out the planning authority's proposals for the development and other use of land in the
area covered by the plan, and a proposals map on an appropriate ordnance
survey map base. There is no requirement that local plans be prepared for
every area – rather, planning authorities are advised to produce them only
where they are considered to be necessary (DoE, 1984g, para. 3.11).
Although local plans can be produced by both county and district planning
authorities, there is a presumption in the legislation that they will normally
be prepared by the district authorities. Local plans must conform generally
to the structure plan, and are adopted by the authority responsible for their
preparation, after consultation, public participation and usually a local
inquiry or hearing.

   Over the whole country the 'new' development plan provided for by the
Town and Country Planning Act 1971 operates conjointly with the 'old'
development plan which was provided for initially by the *Town and Country
Planning Act 1947*, and latterly by the *Town and Country Planning Act
1962*, until such times as the 'old' system for a particular area is revoked.
Thus where a structure plan has been approved, but no local plan has been
adopted for the area, the old development plan continues in force so far as it

does not conflict with the structure plan. However, when a local plan is adopted or approved for an area then the old development plan ceases to have effect for that area. In cases where conflict arises within the provisions of the development plan, then the following priorities are observed:

1 A structure plan prevails over the old development plan.
2 A local plan prevails over the old development plan.
3 A local plan prevails over the structure plan, unless the County Council has announced that following modification or replacement of the structure plan the local plan no longer conforms; then the structure plan prevails.
4 The most recently adopted/approved local plan prevails over another local plan (DoE, 1984g, para. 1.8).

The importance of the structure plan in development control is emphasized by the provisions of the *Local Government Planning and Land Act 1980* (s.3) which states 'It shall be the duty of a local planning authority when exercising their functions under Section 29 of the *Town and Country Planning Act 1971* (determination of applications) to seek the achievement of the general objectives of the structure plan for the time being in force for their area'. This provision makes it clear that whilst the *Local Government Planning and Land Act 1980* gives greater development control powers to the district councils, they should not ignore the policies of the structure plan for their area. At the same time decisions in the courts have helped to clarify the extent to which the development plan should feature in the determination of applications. Thus '. . . the planning authority are to consider all the material considerations, of which the development plan is one' (Simpson *v.* Edinburgh Corporation, 1960, S.C.313; also Enfield London Borough Council *v.* Secretary of State for the Environment, 1974, Estates Gazette 53). However this discretion does not entitle the local planning authority or the Secretary of State to ignore the development plan in favour of some other consideration. For example, in Richmond-on-Thames London Borough Council *v.* Secretary of State for the Environment an Inspector was held to have erred in his judgement that the Report of the Property Advisory Group, *Planning Gain* (1981) took precedence over certain provisions contained in the adopted action area plan for the area.[1]

7.1.2 Other material considerations

The 1971 Act requires that in determining a planning application the local planning authority should have regard to all considerations which are material to the application and that to be material, considerations must be planning considerations. However as the Act offers little guidance as to what constitutes material planning considerations, the limits of discretion in this area and the weight to be given to other material considerations are set in a number of ways including central government policy and advice; precedent and court decisions.

Central government policy and advice

This is clearly capable of being taken into account as a material consideration in development control. The Secretary of State for the Environment is responsible for supervising the land use planning system and his policies are communicated to the local planning authorities in a variety of ways, e.g. circulars; ministerial statements, White Papers; Directions, or Orders issued under the Act; appeal decisions and the approval of forward planning matters. Directions and Orders are binding on the local planning authority but the other means of communicating policy are not. However a circular which for example sets out policy guidance on housing land availability is likely to be taken into account by the Secretary of State in determining Section 36 appeals relating to housing developments and the interpretation of that policy is liable to be reviewed in the courts. Such policy guidance vehicles are thus material considerations, and are dealt with in this book in Section 7.3. Circular 22/84 and the associated memorandum on structure and local plans is particularly relevant in this context as it makes it clear that emerging structure and local plans, and proposals for the alteration, repeal or replacement of such plans may be taken into account as a material consideration for development control purposes, while going through the statutory procedures leading to approval or adoption. The weight to be given to such plans or proposals increases as successive stages are reached in the statutory procedures. The relevant stages for this purpose are identified as:

1   the publicising of matters which it is proposed to include in the plan(s);
2   the submission of the structure plan or the placing on deposit of a local plan;
3   proposed alterations to the structure plan following an examination in public or proposed modifications to a local plan following a local inquiry;
4   the proposed adoption of a local plan.

These stages are all the subject of publicity or advertisement by the local planning authority under the 1971 Act or the *Structure and Local Plan Regulations 1982* (DoE, 1984g, para. 1.12). Supplementary planning guidance in the form of development control practice notes, development briefs or sketch layouts may also be taken into account as material considerations. The weight accorded such guidance also increases when it has been prepared in consultation with the public and been made the subject of a council resolution (DoE, 1984g, paras. 1.14–1.15.). In addition, experience has shown that despite the advice of central government informal or nonstatutory local plans and policy instruments are taken into account as other material considerations in the determination of Section 36 appeals. The evidence suggests that the nature and extent of the preparation and adoption procedures to which the policies and proposals have been subjected influences the weight given to these instruments in the same way as it influences the weighting given to emerging structure and local plans (Bruton and Nicholson, 1984c).

Precedent

Section 36 appeal decisions have, over a period of time, provided an indication of what the Secretary of State and Inspectors consider to be other material considerations. Thus third-party interests are a factor to be taken into account in determining planning applications, as are personal circumstances relating to a particular application, e.g. hardship.

Court decisions

A number of relevant decisions are fully reported in the *Encyclopedia of Planning Law*, Vol. 2, Section 2–875. In summary these decisions establish the following guidelines which are generally followed in practice. Thus, 'In principle . . . any consideration which relates to the use and development of land is capable of being a planning consideration. Whether a particular consideration falling within that broad class is material in any given case will depend on the circumstances' (Stringer *v.* Ministry of Housing and Local Government, 1971). In interpreting this somewhat broad definition a number of principles have emerged, e.g. the planning authority may refuse planning permission to protect private interests; it may also consider the value of existing use rights in determining an application; permission may be refused on the grounds of the possible precedent effect of the decision; an existing planning permission may be a material consideration; the relative planning advantages of a particular proposal (planning gain) are a material consideration; permission may not be refused to induce a developer to contribute an advantage in the form of planning gain which should be properly provided by the local authority or another authority; the personal circumstances of the applicant may play a part in influencing a decision.

7.2 The development control process[2]

Figure 7.1 shows the procedure followed in determining a planning application. The basic approach adopted towards the control of development is:

1 to establish that development requires permission;
2 to provide a broad definition of development; and
3 to exclude certain forms of development by exemption or exclusion.

Thus Section 22 of the *Town and Country Planning Act 1971* defines development as the carrying out of building, engineering, mining or other

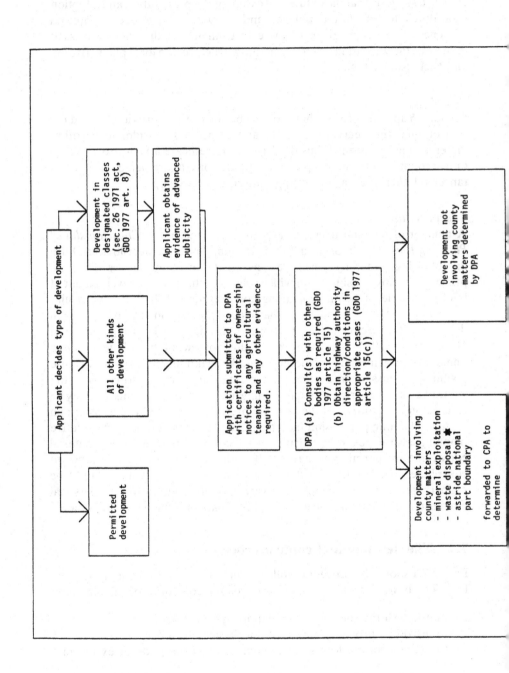

Figure 7.1 *The development control process*

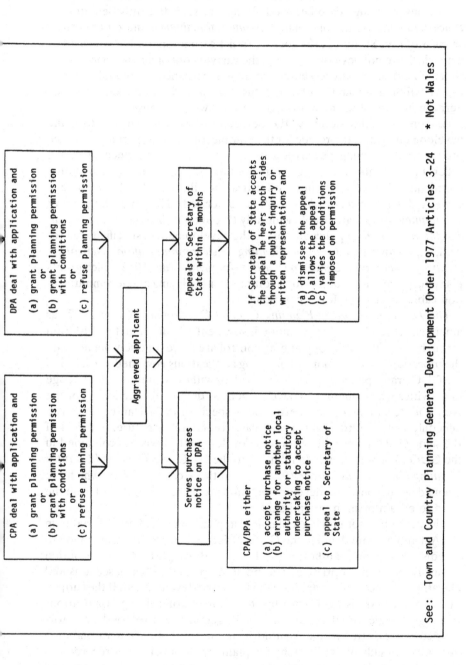

CPA deal with application and

(a) grant planning permission
    or
(b) grant planning permission
    with conditions
    or
(c) refuse planning permission

DPA deal with application and

(a) grant planning permission
    or
(b) grant planning permission
    with conditions
    or
(c) refuse planning permission

Aggrieved applicant

Serves purchases
notice on DPA

Appeals to Secretary of
State within 6 months

CPA/DPA either

(a) accept purchase notice
(b) arrange for another local
    authority or statutory
    undertaking to accept
    purchase notice

(c) appeal to Secretary of
    State

If Secretary of State accepts
the appeal he hears both sides
through a public inquiry or
written representations and

(a) dismisses the appeal
(b) allows the appeal
(c) varies the conditions
    imposed on permission

See: Town and Country Planning General Development Order 1977 Articles 3-24   * Not Wales operations in, on, over or under land or the making of any material change of use of any buildings or other land. It can be seen that this definition is concerned with, on the one hand, *operational activities*, and on the other, *changes of use*. Section 22(2) goes on to list certain operations and uses of land which are not development, e.g. the carrying out of maintenance works which do not materially affect the external appearance of the building; the use of buildings or land within the curtilage of a dwelling house for any purpose incidental to the enjoyment of the dwelling house.

Section 24 of the Act gives the Secretary of State the power to make development orders to provide for the granting of planning permission. Such orders grant planning permission for any developments specified in the order and may also lay down the procedure for the granting of planning permission by local planning authorities for all other development, e.g. *General Development Order 1977*, as amended. They may be *general development orders* applicable generally throughout the country, or *special development orders*, applicable only to the development of specified areas of land, e.g. *Town and Country Planning (Telecommunications Networks) (Railway Operational Land) Special Development Order 1982* which grants planning permission, subject to conditions, for the laying of telecommunication cables on British Rail operational land.

The Town and Country Planning General Development Order 1977, as amended, (the GDO) is the major instrument of subordinate legislation through which the limits of planning control are more precisely defined and the procedure for the making of planning applications governed. Article 3 of the GDO grants planning permission, with or without conditions, for a range of activities which amount to development as defined by Section 22 of the Act. The development thus permitted is specified in 21 clauses set out in Schedule 1 of the Order and in most cases is of a minor nature. Article 4 of the same Order allows the permitted development provisions contained in the GDO to be withdrawn. In addition to the GDO, the *Town and Country Planning (Use Classes) Order 1972* sets out those changes of use which are deemed not to constitute development, and are therefore excluded from the need to obtain planning permission.

The procedure for making a planning application is laid down broadly in the Act and in more detail in the GDO. Briefly, all applications are made to the district planning authority (or London Borough or City of London Council) on the appropriate forms and accompanied by plans, sections and elevations as necessary to describe the proposed development. If the subject of the application is dealt with by the County Council, e.g. applications relating to mineral extraction and, in England, waste disposal, it is forwarded by the district planning authority to the appropriate county planning authority. Details of applications for planning permission are recorded in the register of planning applications, which is kept by the district planning authority (except in National Parks, when it is recorded by the county), and

London Boroughs or the City of London. The register has to be available for public inspection.

Applications for certain types of 'bad neighbour' development, e.g. slaughter house; dance hall; must be advertised in the local press to allow neighbours and other interested parties to make representations to the local planning authority before the decision is made. In many cases the local planning authority will have to consult with other bodies in accordance with the requirements set out in the *Local Government Planning and Land Act 1980* and the GDO. Having complied with such requirements as may apply, the authority consider the application and decide whether it should be refused, allowed or allowed subject to conditions. A decision is required to be given within 8 weeks from the date of receipt of the application unless the applicant agrees to an extension. If no such extension is agreed to, or if the application is not determined within an agreed time extension, the applicant can appeal as if the application had been refused. If the proposed development involves a departure from the development plan and the authority wish to grant permission there is a special procedure to be followed including a programme of publicity and notification of the Secretary of State, so that s/he might intervene in the process if s/he does not want the matter to be decided locally. Similarly, if a proposed development is considered by the Secretary of State to be important or controversial s/he can call it in so that s/he can determine it personally. Applications may be made for outline permission where approval in principle is sought to the development, prior to drawing up detailed proposals. Such a permission if granted will be subject to the imposition of conditions reserving matters such as design, access and materials for further consideration.

If an application is refused, or granted subject to conditions, or deemed to have been refused because the local planning authority failed to give a decision within the statutory period, the applicant can appeal to the Secretary of State under Section 36 of the 1971 Act. The appeal must be made within 6 months of the date of issue of the notice of decision; must set out the grounds of appeal and indicate whether the applicant (now the appellant) is prepared to have the matter considered by written representations or public local inquiry. On receipt of the notice of appeal an Inspector from the Planning Inspectorate is appointed to consider the appeal. If the local planning authority and the appellant agree the matter may be dealt with by the *written representations procedure*, otherwise a *public local inquiry* will be held.[3] The procedure of the inquiry is governed by the two sets of *Town and Country Planning (Inquiries Procedure) Rules 1974* (one set for Secretary of State inquiries, the other for inquiries transferred to Inspectors).

The decision on an appeal is set out in a decision letter signed by the Inspector, if he or she has been given the delegated power to take the decision, or a representative of the Secretary of State if the latter has retained the right to determine the appeal. The decision letter gives the decision, i.e. to allow or dismiss the appeal, and the reasons for it. Once issued the decision as a planning matter is final and can only be challenged on legal grounds in the High Court (DoE, 1983).

7.3   Policy guidance on development control matters from central government

'Other material considerations' in the form of policy guidance is issued by the government of the day in a number of ways, e.g. Green Papers, White Papers, Ministerial statements, policy notes, research reports. However, circulars are by far the most frequent and influential form of policy guidance used by central government, and in development control they form a significant factor in defining 'other material considerations' and influencing the way in which the development control system is operated and managed at local authority level. Whilst policy guidance set out in this way does not bind local planning authorities directly, it influences their use of discretionary powers. Indeed because the provisions of circulars are capable of enforcement through the administrative machinery the advice offered is invariably followed (Nott and Morgan, 1984). Thus advice on the form and content of structure plans, if ignored, may nevertheless be implemented by the Secretary of State modifying the submitted plan on approval, while central government policy on development control matters may be similarly imposed in decisions on appeal, or by calling in applications. However, policy guidance is not the same thing as statutory obligation and there is often very little indication given of the status which should be given to advice set out in circulars which variously advise, warn, hope, urge, encourage, ask, and recommend. More recently (1979 to present) the emphasis which central government thinks should be accorded to circulars has been made more explicit. Nevertheless it is clear that the degree of importance of a circular is dependent to a large extent on the level of support accorded it in practice. 'If the Secretary of State and the Inspectorate are seen to be applying its terms rigorously, then the greater its influence on authorities is likely to be' (Grant, 1982, p. 64).

A recent (1985) High Court decision, which quashed a phasing policy in the Hertfordshire Structure Plan, as approved by the Secretary of State, clarifies the extent to which both the local planning authorities and the Secretary of State must adhere to advice given in circulars. Giving judgement on an application by the House Builders Federation to quash a phasing policy (Policy 11c) in the modifications to the structure plan, Justice Farquharson made it clear that while the Secretary of State was not bound by circulars and could depart from his own advice if he so wished, such departures must be supported by reasoned justification.[4]

Given the relationship between the development plan and development control, many of the circulars issued by central government to provide policy guidance impinge on both matters. The following section reviews the most important aspects of these circulars in so far as the inter-relationship between local plans, local planning and development control is concerned. In addition the advice given by central government in relation to development control matters through the series of Development Control Policy Notes (DCPNs) now issued by the Department of the Environment, and research reports and design guides, is also summarized. Notwithstanding the importance of circulars and DCPNs, it must be emphasized that each application for permission to develop is determined on the merits of the case.

7.3.1 Circulars and policy guidance in development control[5]

1 Development Control and the Development Plan
Circular 22/84 and the associated memorandum on *Structure and Local Plans* has already been referred to in the preceding section. It is important in the context of development control as it defines the inter-relationship between the development plan and development control, and provides guidance on the boundaries and weighting of other material considerations. Thus, 'Structure and local plans should set out policies and proposals for the scale and location of provision for all types of development. Sites for development . . . should be identified in local plans' (*ibid.*, para. 4.13). Three groups of development control policies are identified (para. 4.14):

(i)   policies to control particular types of development in the plan area;
(ii) policies to control particular aspects of development in the plan area, e.g. control of pollution or visual intrusion; and
(iii) policies to control development in particular parts of the plan area, e.g. Green Belt; conservation area.

It is suggested that local plans should set out the more detailed development control criteria adopted and where appropriate specify the policies which the authority proposes to implement through the imposition of planning conditions, Article 4 directions and planning gain (*ibid.*, para. 4.15).

The use of planning guidance to supplement the policies set out in the structure plan and local plan is recommended, particularly practice notes for development control (e.g. car parking standards, policy for flat conversion) and development briefs indicating how the local planning authority expect a site or group of sites to be developed. Such guidance should be consistent with approved or adopted statutory plans for the area; introduce no new land use policies; be prepared in consultation with the public; be adopted by resolution of the council; be published separately from the policies and proposals set out in the development plan; and be kept publicly available (*ibid.*, paras. 1.14–1.15). At the same time local planning authorities are warned off producing non-statutory plans or policies as the following statement shows. 'Any plan containing proposals for the development and other use of land which is not included in the development plan scheme as a local plan, or as proposals for the alteration of a local plan, and which has not been subject to any of the stages in the statutory procedures identified . . . as giving weight to plan proposals can have little weight for development control purposes. It cannot be treated as an emerging local plan' (*ibid.*, para. 1.13).

2 Development Control – Policy and Practice

Circular 22/80 *Development Control – Policy and Practice* and Circular 1/85 *The Use of Conditions in Planning Permission* provide basic guidance on how the development control system should operate. In addition to suggesting ways in which the system can become more efficient and publicly accountable *Circular 22/80* makes the following clear:

(i)   Priority should be given to the handling of applications which will contribute most to national and local economic activity (para. 11).

(ii) The government is particularly keen to encourage the formation and expansion of small-scale businesses (para. 12), and that '. . . when small-scale commercial and industrial activities are proposed particularly in existing buildings, in areas which are primarily residential or rural, permission should be granted unless there are specific and convincing objections such as intrusion into open countryside, noise, smell, safety, health or excessive traffic generation' (para. 13).

(iii) Enforcement and discontinuance powers against unauthorized development should only be used where planning reasons clearly warrant such action, and there is no alternative to enforcement proceedings (para. 15).

(iv) A five-year supply of housing land should be identified in accordance with structure plan policies. In the absence of such an identified supply there should be a presumption in favour of granting planning permission for housing except where there are clear planning objections (Annex A, paras. 2 and 3).

Circular 1/85 sets out the basis on which conditions should be imposed when planning permission is granted. The criteria establishing the validity of conditions are clearly articulated, i.e. they '. . . should only be imposed where they are

(a) necessary,
(b) relevant to planning,
(c) relevant to the development to be permitted,
(d) enforceable,
(e) precise and
(f) reasonable in all other respects' (Annex, para. 11).

Advice is given on the application of conditions to a range of different types of application. In particular, conditions restricting the occupation of commercial or industrial premises to local firms are identified as contributing to offices or factories lying empty for long periods. For this reason such conditions are regarded as undesirable in principle.

3 Development and Employment

Circular 14/85 *Development and Employment*, which was published as an annex to the White Paper *Lifting the Burden* (Cmnd. 9571), is a most important circular which deals with the implications of that White Paper for development control. Briefly, the White Paper'. . . sets out the case'for more freedom in the business sector and the need to deregulate' (para. 1.2). In relation to the town planning system it proposes, amongst other things:

(i)   to give strong guidance to local authorities on development and employment (para. 3.5);
(ii)   to introduce Simplified Planning Zones (para. 3.6(i));
(iii)   to extend the GDO to permit more exemptions from specific planning permission (para. 3.6(ii));
(iv)   to review the Use Classes Order (para. 3.6(iii));
(v)   to review the *Control of Advertisement Regulations* (para. 3.6(iv));
(vi)   to amend the *Town and Country Planning Act 1971* to include provision for the award of costs against authorities or others who act unreasonably (para. 3.6(v)); and
(vii)   to improve the system of major public inquiries (para. 3.6(vi)) and to review measures to speed up written planning appeals (para. 3.10).

The circular is short and to the point, as the following extracts show:

1   . . . This circular . . . deals with policy on development control under the *Town and Country Planning Acts.*
2   New development contributes to economic activity and to the provision of jobs. It is in the national interest to promote and encourage it. The planning system must respond positively and promptly to proposals for development. Delay adds to the costs of development.
3   Development proposals are not always acceptable. There are other important objectives to which the Government is firmly committed: the need to preserve our heritage, to improve the quality of the environment, to protect the Green Belts and conserve good agricultural land. The planning system, however, fails in its function whenever it prevents, inhibits or delays development which could reasonably have been permitted. *There is therefore always a presumption in favour of allowing applications for development, having regard to all material considerations, unless that development would cause demonstrable harm to interests of acknowledged importance.*
4   Authorities are obliged, under Article 7 of the *Town and Country Planning General Development Order 1977*, to give reasons whenever they refuse planning permission. Those reasons must be precise, specific and relevant to the

application: they must demonstrate clearly why, in the local planning authority's view, the proposed development cannot be permitted. Without such a clear demonstration the developer will not know whether or not his proposal can be made acceptable, or the grounds on which he can base an appeal against refusal. As a result, valuable investment and new jobs, in construction, in commerce and in industry, may be delayed or lost.

5   In dealing with applications for planning permission, Section 29(1) of the *Town and Country Planning Act 1971* requires that the authority *shall have regard to the provisions of the development plan, so far as material to the application, and to any other material considerations.* Development plans are therefore one, but only one, of the material considerations that must be taken into account in dealing with planning applications. Many development plans were approved or adopted several years ago, often several years after they had been prepared and based on even earlier information. The policies which they contain, and the assumptions on which they were based, may therefore be out of date and not well related to today's conditions. they cannot be adapted rapidly to changing conditions, and they cannot be expected to anticipate every need or opportunity for economic development that may arise. They should not be regarded as overriding other material considerations, especially where the plan does not deal adequately with new types of development or is no longer relevant to today's needs and conditions – particularly the need to encourage employment and to provide the right conditions for economic growth.

6   It is important that local planning authorities should have regard to the special needs of small firms and the self-employed. The planning system can present serious difficulties for those seeking to set up or expand their business, even on a very modest scale. Local authorities can do much to help both in assisting small firms to cope with the planning process and by avoiding unnecessarily onerous and complex controls. The Department of the Environment and the Welsh Office, in consultation with the Local Authority Associations, will be issuing a further circular on this subject, together with an explanatory booklet for small firms.

7   The Secretaries of State and their Inspectors will have regard to the terms of this circular in dealing with planning appeals and with any application that may be made to them for the award of costs.

In so far as the operation of the development control system is concerned, the circular emphasizes that there should always be a presumption in favour of allowing applications for development, having regard to all material considerations, unless that development would cause demonstable harm to interests of acknowledged importance (para. 3); and re-affirms that development plans are only one of the material considerations that must be taken into account in determining planning applications (para. 5). At the time of writing development control decisions based on or ignoring the advice offered in Circular 14/85 have not been tested in the courts. As a result any interpretation put on the implications of the circular must be seen as hypothetical. However, it would seem that the circular:

(i) positively shifts the emphasis away from restrictive local policies, such as the Warwickshire County Structure Plan local housing needs policies, *unless* the objective of such policies is to avoid '. . . demonstrable harm to interests of acknowledged importance.' (para. 3);

(ii) establishes 'the need to preserve our heritage, to improve the quality of the environment, to protect the Green Belts and conserve good agricultural land . . . .' (para. 3) as interests of acknowledged importance; which in turn

(iii) implies that physical factors are likely to be the main matters on which development control decisions are decided, e.g. unreasonable loss of agricultural land; noise or disturbance to local residents; traffic hazards; harm to the character or appearance of a conservation area or the countryside, thus effectively ignoring the complex inter-related nature of social, economic and physical issues.

4 Land for Housing

Circular 15/84 *Land for Housing* (which applies only to England) sets out the government's concern to bring home ownership within the reach of as many people as possible. It establishes that in areas where substantial housing development is proposed in the structure plan, local plans should be prepared and should ensure that suitable land is allocated to implement those proposals in ways which take account of local conditions.

Great emphasis is placed on the need to guarantee that sufficient land is practically available to enable approved housing policies and proposals to be implemented. Thus sites must be free from planning, physical and ownership constraints; capable of being economically developed; and located in areas where potential house buyers want to live. At the same time 'Local authorities should aim to ensure that at all times land is or will become available within the next 5 years which can be developed . . . within that period and which in total provides at least 5 years supply. . . . Within this context the aim should always be at least 2 years supply available on which development can start straightaway' (para. 12).

Land availability studies undertaken jointly by local planning authorities and housebuilders are seen as being particularly helpful in identifying land suitable for development, as such studies bring together the housebuilders' assessment of market demand and the development potential of particular sites, and the local planning authorities' assessment of planning objectives.

With regard to planning applications the circular is emphatic when it states '. . . applications for housing should be considered on their merits having regard to the provisions of the development plan *and* other material considerations. The results of land availability studies should continue to be treated as a material consideration in determining planning appeals' (para. 18).

Annex A of the circular (which is reprinted from Circular 22/80 in a slightly amended form) gives more detailed advice on planning permission for private sector housing, making it clear that

- development plan policies do not in themselves ensure that the house-building industry can produce the houses needed;
- planning authorities should identify specific sites which provide a five-year supply of housing land in accordance with structure plan policies;
- in the absence of an identified five-year supply there should be a presumption in favour of granting permission for housing, except where there are clear planning objections;
- where a five-year supply of land for housing has been identified this should not preclude residential development on other sites;
- where necessary infrastructure, e.g. sewerage, drainage, access, is not available, then the possibility of providing it through an agreement with the developer under Section 52 of the *Town and Country Planning Act 1971* should be explored, rather than refusing the application;
- more intensive development, including low cost starter houses, in appropriate locations should be encouraged;
- standards of design for footpaths and roads in residential areas which are higher than those set out in Design Bulletin No. 32 *Residential Roads and Footpaths* (DoE and Department of Transport, 1977) will not be supported by the Secretary of State;
- functional requirements within a development, e.g. provision of garages, internal space standards, size of gardens are primarily a matter for the developer to determine;
- isolated pockets or ribbons of development are to be avoided, especially when they intrude into surrounding countryside;
- modest expansion of villages and the development of infill sites is possible in many instances, and should be encouraged where it does not create isolated pockets or ribbons of development or intrude into open countryside.

5   Industrial Development

Circular 16/84 *Industrial Development* sets out the government's advice on applications relating to development concerned with the production of goods and services, including ancillary developments such as warehousing. It is seen as an expansion of the advice given in Circular 22/80 *Development Control — Policy and Practice*, which sets out government advice on the general aims and policies of development control. The stated aim of the Circular is '. . . to promote a sound and efficient balance between economic and environmental considerations in facilitating industrial development' (para. 20). To this end, local planning authorities are advised to '. . . see that the [structure and local] plans reflect an informed view of the scale, diversity

and distribution of sites which may be required for industrial development
. . . . When framing development plan policies . . . planning authorities
should aim to ensure that, within the constraints of national policies and in
line with the policies in structure plans, there is sufficient land available for
industry, and that the supply of sites allows developers to choose on the basis
of their individual needs between sites of different sizes with different
facilities' (*ibid.*, paras. 7 and 8).

Given that approved and adopted plans prepared some years ago may not
incorporate policies which meet the needs of modern industry, the circular
acknowledges the potential for conflict between such plans and proposals for
development (*ibid.*, para. 6). Thus local planning authorities are advised to
consider applications for industry on their merits, '. . . having regard to the
development plan *and* other material considerations. In the modern
economy it is not always possible to anticipate in the development plan all
the needs and opportunities which may arise. Thus where a developer
applies for permission for a development which is contrary to the policies
and proposals of an approved development plan this does not, in itself,
justify a refusal of permission . . . . While the decision will obviously be more
difficult than in cases which conform to development plan policies, the onus
nevertheless remains with the planning authority to examine the issues
raised by each specific application and where necessary to demonstrate that
a particular proposal is unacceptable on specific planning grounds' (para. 9).

High-technology industries are singled out for special mention in the
Circular which draws attention to their vital role in industrial regeneration;
the importance attached by the Secretary of State to the sympathetic treatment of such uses in development control; and the frequent compatibility of
high-technology industries with service industry, offices or housing (para. 18
and Annex).

6 Green Belts

The importance attached by successive governments to Green Belts is
strongly re-affirmed in Circular 14/84 *Green Belts*. Thus, 'There must continue to be a general presumption against inappropriate development within
Green Belts' . . . which '. . . have a broad and positive planning role in
checking the unrestricted sprawl of built-up areas, safeguarding the surrounding countryside from further encroachment and assisting in urban
regeneration' (para. 1). The essential characteristic of Green Belts is seen as
their permanence and the circular emphasizes that Green Belt boundaries
once approved should only be altered in exceptional circumstances; and that
where such boundaries have not yet been defined it is necessary to establish
boundaries that will endure. Thus they should not be drawn too tightly
around existing built-up areas and should not include land which it is
unnecessary to keep permanently open for the purpose of the Green Belt. In
following such a policy it is clear that in some cases land between the urban

area and the Green Belt, which may in due course be needed to accommodate the longer term development needs of the area, will need to be safeguarded.

Local planning authorities are encouraged to work together with land-owners and voluntary groups to enhance the countryside, especially those areas of land within the Green Belt that are close to existing urban development and are vulnerable to neglect or damage. 'The overall aim should be to develop and maintain a positive approach to land use management which *both* makes adequate provision for necessary development *and* ensures that the Green Belt serves its proper purpose' (para. 6).

7 Historic Buildings and Conservation Areas

Circular 23/77 *Historic Buildings and Conservation Areas – Policy and Procedure* consolidates advice issued in previous circulars about historic buildings and conservation areas. Circular 12/81, *Historic Buildings and Conservation Areas* re-affirms that advice in the light of legislative changes made in the *Local Government Planning and Land Act 1980*. This latter circular makes it plain that the Government is determined to implement policies to preserve the best of the nation's heritage. Indeed, the Secretary of State for the Environment '. . . will not be prepared to grant listed building consent for the demolition of a listed building unless he is satisfied that every possible effort has been made to continue the present use or to find a suitable alternative use for the building' (para. 3, 12/81). Thus the issue of finance in circumstances such as this becomes a material consideration.

In determining applications for development involving historic buildings local planning authorities are advised that new uses for old buildings often hold the key to preservation, and controls over land use allocation, density, plot ratio, day-lighting and other controls should be relaxed where it enables historic buildings to be given a new lease of life (para. 4, 12/81). In particular, changing patterns of farming and rural life mean that new uses must be found for buildings such as stables, coach houses, barns and oast houses that play an important part in the history and appearance of the countryside. Thus, 'If these buildings are used as workshops, craft studios or as holiday accommodation they can often make a contribution to the rural economy by providing employment. Local authorities should therefore be flexible in dealing with applications for changes of use of buildings of architectural or historic interest' (paras. 7 and 8, 12/81). In the same vein, Circular 23/77 advises that in determining applications in conservation areas special regard should be had to such matters as bulk, height, materials, colour, vertical or horizontal emphasis and design (para. 39, 23/77).

The establishment of conservation area advisory committees is recommended, consisting primarily of people who are not members of the authority. The advice of these committees should be sought in connection with applications which would, in the opinion of the authority, affect the character or appearance of the conservation area. They could also play a part in the general care and maintenance of conservation areas and make positive proposals for their enhancement (para. 44, 23/77).

Other current circulars offering policy guidance on historic buildings and conservation areas include Circular 8/84 *Establishment of the Historic Building and Monuments Commission for England*.

8 Wildlife and Conservation

Circular 108/77 *Nature Conservation and Planning* sets out the government's concern at the need to conserve the nation's heritage and advises local planning authorities to take this fully into account in determining planning applications. The role of the Nature Conservancy Council is outlined and the need to consult with it on policies for nature conservation and in considering the impact of development proposals is emphasized (paras. 3 and 4). The role of National Nature Reserves, Local Nature Reserves and Sites of Special Scientific Interest (SSSIs) is also reviewed (paras. 5 to 15). The benefits to be derived from involving voluntary conservation bodies in an advisory capacity, particularly in devising management plans for local nature reserves established by local authorities, are also promulgated (paras. 22 to 23).

Circular 32/81 *Wildlife and Countryside Act 1981* describes in general terms the provisions of that Act whilst Circular 24/82, *Wildlife and Countryside Act 1981, Commencement of Part I*, deals in more detail with the implications of Part I of the Act which became operative in 1982.

9 Agriculture

Government policy for the protection of agricultural land is set out in Circular 75/76 *Development Involving Agricultural Land*. As far as possible land of higher agricultural quality should not be taken for development where land of lower quality is available. The development plan should be used to safeguard agricultural land whilst the agricultural interest should be properly considered on individual proposals for development. The potential conflict between development and agriculture is acknowledged and the role of the Ministry of Agriculture, Fisheries and Food (MAFF) in the process of structure and local plan preparation is emphasized. Annex A to the circular establishes the agricultural factors to be considered in preparing local plans, including agricultural land quality; the location of development in relation to farms; farm size and structure; farm buildings and fixed equipment and the general effects of development on agriculture.

Where applications for permission to develop agricultural land accord with the provisions of the development plan there is not normally a need for further consultations with MAFF. Where applications for non-agricultural development do not accord with the provisions of the development plan, planning authorities are required by the GDO 1977 to refer to MAFF proposals which involve the development of 10 acres or more of land which is being, or was last, used for agricultural purposes. MAFF can object to the loss of agricultural land for development, but only for agricultural reasons, and local planning authorities are requested to give full consideration to the views expressed by MAFF and to make every effort to come to an agreement with MAFF.

The arrangements to be followed in connection with applications for development for agricultural purposes are set out in Circular 24/73 *Development for Agricultural Purposes*. The Annex to this circular establishes that:

(i)   the applicant should normally establish the need for additional dwellings on a farm before planning permission is granted;

(ii)   the presumption is that the need for additional agricultural dwellings will be met in a nearby village, hamlet or existing group of buildings; and

(iii) the factors to be considered in establishing need include the viability of the farm; its labour requirements; existing accommodation and the number of workers who need to live on the farm.

10   Other Development

In addition to the circulars referred to above the following circulars are important in varying degrees in establishing the inter-relationship between the development plan and development control. Development involving *opencast mining and mineral extraction* is dealt with in Circular 3/84, *Opencast Coal Mining*, which provides guidance on the issues to be considered in drawing up opencast policies and programmes, and sets out the arrangements for determining opencast coal applications during the period when responsibility for such applications is being transferred from the Secretary of State for Energy to local authorities. Circular 14/79, *Town and Country Planning (Minerals) Regulations 1971: Time-limited planning permissions* gives advice on the way in which local authorities should treat new applications made following the lapsing of existing permissions granted before 1st April 1969 which did not have a time limit imposed. Circular 58/78, *Report of the Committee on Planning Control over Mineral Working* draws attention to the conclusions of the Report (the Stevens Report) on the inter-relationship between mineral extraction and planning control. Following this report the *Town and Country Planning (Minerals) Act 1981* was introduced which distinguishes minerals development from all other forms of development to which planning controls extend, and confers a range of special new powers for the control of minerals developments. Thus the definition of development is extended to include the removal of material from certain types of deposit, including deposits of mineral waste; the capacity of local planning authorities to impose restoration and after-care conditions is increased and it establishes that compensation is not payable by the planning authority in respect of a refusal of planning permission or conditions imposed upon the grant of permission. Circular 50/78, *Report of the Advisory Committee on Aggregates* draws attention to the conclusions of the Report (the Verney Report) which deals primarily with the problems associated with the demand for aggregates for building and road construction, and the control of aggregate extraction. The implications for land use planning of the *Inner Urban Areas Act 1978* is dealt with in Circular 68/78. In particular the criteria for financial assistance for inner areas are set out, and the power to designate Improvement Areas amplified. The impact of large new stores is covered in Circular 96/77, *Large New Stores*; revised model standards for licensed caravan sites in England and Wales are established in Circular 23/83, *Caravan Sites and Control of Development Act 1960*; and issues concerning the accommodation of gypsies are covered in Circular 28/77, *Caravan Sites Act 1968 – Part II Gypsy Caravan Sites* and Circular 57/78, *Accommodation for Gypsies: Report by Sir John Cripps*.

Other important circulars include Circular 18/84, *Crown Land and Crown Development*; Circular 17/82, *Development in Flood Risk Areas*; Circular 9/84, *Planning Controls over Hazardous Development*; and Circular 39/81, *Safeguarding of Aerodromes, Technical Sites and Explosives Storage Areas*.

This brief review of the more important circulars concerned with the interaction between development control and the development plan highlights three main themes:

(i)   Clarification of the role of the development plan in the development control process.
(ii) The need for a flexible approach to be adopted in determining applications and considering other material considerations.
(iii) The indication that the land use planning system is concerned to manage change through the involvement of bodies other than the local planning authority in that process.

These themes are dealt with below. However, it is important to reiterate that each application is determined on its merits, taking account of policy guidance and any other material consideration as it applies to the case in hand.

7.3.2 The development plan in development control

The importance of the development plan in the determination of planning applications is clarified by almost all of the circulars reviewed. Thus Circular 22/84 makes it quite clear that the development plan and in particular local plans should set out the policies and proposals for the scale and location of all types of development, including development control policies and criteria. Supplementary planning guidance, subject to a number of conditions, is appropriate for more detailed planning control whilst informal or non-statutory plans '. . . can have little weight for development control purposes' (Circular 22/84, para. 1.13). Circular 15/84, *Land for Housing*, similarly re-inforces the position of the development plan by establishing that where substantial housing development is proposed in the structure plan, local plans should be prepared to ensure that the proposals are implemented in a way which reflects local conditions (para. 9).

Advice is also given on:

- Green Belts, where there is a presumption against inappropriate development (Circular 14/84, para. 1);
- historic buildings and conservation areas, where the Secretary of State will not agree to demolition of listed buildings unless he is satisfied that all alternative solutions have been explored (Circular 12/81, para. 3);
- agricultural developments, where there is a presumption that additional agricultural dwellings will be provided in existing settlements (Circular 24/73, Annex). This also reinforces the position of the development plan as policy guidance in development control.

However, the same circulars also encourage the local planning authorities to be flexible in interpreting other material considerations, and provide guidance on this interpretation, whilst Circular 14/85 *Development and Employment* gives point to this flexibility when it states 'Development plans are therefore one, but only one, of the material considerations that must be taken into account in dealing with planning applications' (para. 5).

7.3.3   Flexibility and other material considerations in development control

Circular 22/80, *Development Control — Policy and Practice*, whilst acknow-ledging the role of the development plan in development control, indicates that applications for small-scale businesses, even in residential or rural areas, should be granted planning permission unless there are over-riding planning objections. The same circular also suggests that enforcement and discontinuance powers should only be used against similar unauthorized developments where planning reasons merit it, and that there should be a presumption of granting planning permission for housing except where there are clear planning objections. The implications are that despite what is set out in the development plan, 'other material considerations' could determine the outcome of a large number of applications relating to com-mercial, industrial and housing developments. *Land for Housing* (Circular 15/84) re-inforces this point when it gives emphasis to the need to ensure a five-year supply of usable housing land, and the importance of land availabil-ity studies as other material considerations in determining planning appeals. Indeed the circular emphasizes the importance of other material considera-tions relative to the development plan and indicates a number of factors which constitute other material considerations, e.g. sites free from con-straints, capable of economic development, and located in areas where people want to live.

Circular 16/84, *Industrial Development*, goes even further when it explicitly acknowledges that in the modern economy it is not always possible to anticipate in the development plan all the needs of industry, and as a consequence advises local planning authorities to examine the issues raised by each specific application and to demonstrate where necessary that a particular proposal is unacceptable on specific planning grounds (para. 9). In short, a refusal only on the grounds that it does not accord with policy established in the development plan is not necessarily acceptable. High-technology industries are singled out as being suitably located in areas given over to residential, office or service industry uses whilst again emphasis is given to other material considerations relative to the development plan in determining applications (para. 9). Similarly, advice contained in Circular 12/81 on *Historic Buildings and Conservation Areas* encourages local planning authorities to be flexible in their attitude towards new uses in old buildings, especially where such uses contribute to the local economy. As has been pointed out, the most recent circular at the time of writing, Circular 14/85, *Development and Employment*, emphatically consolidates this advice. Now there is always a presumption in favour of allowing applications for development (para. 3); development plans are therefore only one of the material considerations which must be taken into account in dealing with planning applications (para. 5).

Thus it is clear that the circulars re-inforce the statutory requirement for the determination of planning applications, i.e. that the local planning authority 'should have regard to the provisions of the development plan, so far as material to the application, and to any other material considerations'. (Section 29(1), 1971 Act). The status of the development plan is clearly established as are the boundaries of 'other material considerations', especially in the circulars relating to housing land (Circular 15/84), industrial development (Circular 16/84), the policy and practice of development control (Circular 22/80), and development and employment (Circular 14/85). Indeed it can be argued that these four circulars in particular suggest that in certain situations 'other material considerations' should be given greater prominence than the provisions of the development plan. At the same time, however, it should be understood that the advice set out in circulars will invariably be taken into account in plan preparation. Thus whilst circulars may appear to be giving emphasis to 'other material considerations', over time the advice could well be incorporated in the development plan in the form of statutory policy.

7.3.4 Land use planning and the management of change

In Chapter 2 it was argued that the planner's role in practice is, or should be, primarily concerned with managing change in the environment by applying the basic managerial processes exercised by all managers in all types of organization, i.e. by

– *planning*, which involves the establishment of goals and objectives for the organization and determining ways of achieving these goals;

- *organizing*, which includes the provision of human and physical resources needed to accomplish the plan objectives; the definition of the tasks of individuals and groups within the organization and the establishment of relationships between groups and individuals;
- *controlling*, which is concerned to ensure that the implementation conforms to the plan requirement; and
- *leading*, which is the process of motivating and influencing others so that they contribute to the achievement of the goals set for the organization. The same chapter also suggested that the complex and 'wicked' problems with which planners have to deal can only be handled if the constraints which are contingent to the particular issue at hand are acknowledged and accepted.

There is sufficient evidence in the circulars reviewed earlier to suggest by implication that central government is encouraging practice to move towards a management of change approach which begins to acknowledge contingent factors peculiar to a particular issue or locality. Thus the circular on housing land availability (15/84) positively exhorts local planning authorities to 'organize' local groups concerned with the provision of housing in ways which 'controls' the outcome of housing policies in the plan, at the same time as taking account of local contingent factors which might affect the supply of and demand for housing. Circular 23/77 *Historic Buildings and Conservation Areas* adopts a similar stance in recommending the establishment of conservation area advisory committees, as does the *Nature Conservation and Planning* Circular 108/77 when it recommends the involvement of voluntary bodies in devising management plans for local nature reserves.

7.3.5 Development control policy notes as policy guidance[6]

In addition to the establishment of policy guidance through circulars, central government has issued a series of Development Control Policy Notes (DCPNs) which are intended to consolidate advice on development control. The notes were first published in 1969 and some have been amended from time to time. They are not binding on local planning authorities; do not claim to be comprehensive; and have to be read in conjunction with the circulars referred to earlier in this chapter and other government publications. Nevertheless they form an important body of advice which must be considered under the heading of 'other material considerations' in determining planning applications. In early 1985 the following Development Control Policy Notes were operative:

General Principles (DoE 1974), which summarizes the development control process.

Development in Residential Areas (DoE 1978), which deals with the issues of residential densities; piecemeal development; development of back land; water and sewerage; conversion of houses into flats; extensions to houses; garages and other out-buildings; garden walls and fences; car-parking and other development in residential areas. The note refers to Ministry of Housing and Local Government (MHLG) Planning Bulletin No. 2, *Residential Areas: Higher Densities* (MHLG 1962b) and Planning Bulletin No. 5, *Planning for Sunlight and Daylight* (MHLG 1964) for more detailed advice on these matters, which implies that these reports also constitute 'other material considerations'.

Industrial and Commercial Development (MHLG 1963), which gives general advice on factors to be taken into account in industrial and office development. Part of this DCPN has been superseded by DCPN 11 whilst sections relating to Industrial Development Certificates (IDCs) are now out of date.

Development in Rural Areas (DoE 1975), advises on ways in which development demands in rural areas can be met without detriment to the countryside or at the expense of high quality agricultural land. It deals with open country; National Parks and AONBs; villages; the availability of services; agricultural dwellings; farm amalgamations; and consultations with MAFF.

Development in Town Centres (MHLG 1969), advises on the complex processes of renewal and redevelopment in town centres. It covers issues such as piecemeal and comprehensive development; car-parking; and conservation. It refers to Planning Bulletin No. 1, *Town Centres: An Approach to Renewal* (MHLG 1962a); Planning Bulletin No. 3, *Town Centres: Cost and Control of Redevelopment* (MHLG, 1963); and Planning Bulletin No. 7, *Parking in Town Centres* (MHLG and Ministry of Transport, 1965) for more detailed advice. Again it must be assumed that these reports constitute 'other material considerations'.

Road Safety and Traffic Requirements (MHLG 1969), which advises on the ways in which road safety and traffic requirements can be met in development and redevelopment. Detailed guidance on visibility splays and access roads provided in the appendix to this note has been superseded by a series of publications including *Roads in Urban Areas* (Ministry of Transport, 1966); Design Bulletin No. 32, *Residential Roads and Footpaths* (DoE and Department of Transport, 1977); *Road Layout and Geometry: Highway Link Design* (Department of Transport, 1981a); and *Junctions and Accesses: The Layout of Major/Minor Junctions* (Department of Transport, 1981b).

Preservation of Historic Buildings and Areas (DoE 1976) is concerned to advise how the character of historic towns and villages can be preserved whilst at the same time not precluding development.

Caravan Sites (MHLG 1969) offers advice on caravans as homes; caravans for holidays and caravans for gypsies.

Petrol Filling Stations and Motels (MHLG 1969); the advice offered on motels is superseded by DCPN 12.

Design (MHLG 1969), which deals with the treatment of design issues in development control.

Service Uses in Shopping Areas (DoE 1985), which sets out government advice on issues such as the character of the shopping centre; advertisement control; building type; traffic generation and congestion; noise and disturbance; smell and litter; and opening hours. Particular reference is made to amusement centres; banks and building societies; betting offices; estate and employment agencies; restaurants; cafes and hot food take-away shops; launderettes and dry cleaners.

Hotels (DoE 1972), which deals with factors relating to hotel development including consultations; location; plot ratio; car-parking; hotels in London, rural areas and historic towns; and motels.

Large New Stores (DoE 1977), offers advice on the way in which large new retail stores should be handled in the development control process. It deals with issues such as changes in retailing; the mobility of shoppers and store location; assessment of demand; site considerations; highway considerations; and town centre and green field locations.

Warehouses – wholesale cash and carry, etc. (DoE 1974). This DCPN provides clarification on the meaning of 'warehouse' and advises on the approach to be adopted in dealing with applications for such developments.

Hostels and Homes (DoE 1975), which deals with the criteria to be considered in providing hostels and homes for persons in special need, e.g. young people in care; mentally handicapped.

Access for the Disabled (DoE 1985), which is concerned with the provision of suitable access for disabled people where buildings will be open to the public.

Although many of these DCPNs are now out of date they still represent formal current central government advice on the matters with which they deal. In conjunction with other policy guidance such as circulars, research reports, and planning and design bulletins, they constitute a central plank in establishing 'other material considerations'. They should thus be taken into account by the local planning authority in determining planning applications. The evidence indicates that where appropriate they are taken into account in determining Section 36 appeals.

7.4 Local plans, local planning and the determination of planning applications and Section 36 appeals

It is clear from the preceding sections that a wide range of matters must be considered by the local planning authority in determining planning applications. The Department of the Environment claims 'Structure and Local plans provide the necessary framework for development control ....' (DoE, 1984g, para. 1), and emphasizes that one of the main functions of local plans is '. . . to provide a detailed basis for development control' (DoE, 1984g, para. 3.1). At the same time it is clear from the earlier discussion that 'other material considerations' must be taken into account in determining planning applications. The local planning authority has the discretion to deal with each application on its merits provided it carries out the necessary consultations and considers all matters relevant to the application. It has been noted that it can refuse permission for the proposed development or grant permission for it with or without conditions. If the decision is a conditional permission or a refusal, then reasons for the imposition of the conditions or the refusal must be stated. They must be sound planning reasons. Despite the apparent simplicity of the process it is in reality highly complex with a wide range of factors influencing the eventual decision, including officer–councillor relationships, the development plan, other material considerations and bargaining skills.

7.4.1 Officer–councillor relationships in development control

In most district planning authorities the determination of planning applications is carried out by a sub-committee of elected members which meets on a three or four weekly cycle. This sub-committee is advised by professional officers who will generally prepare written reports on the more important applications and provide written recommendations, supported by verbal advice if necessary, on the less significant applications. Some authorities delegate the responsibility for this latter group of applications to the chief planning officer for determination. The relationship between the elected member and the professional planning officer is an important element in this process which has been criticised in recent years on a number of counts. At a fundamental level it is felt that the popular concept of representative local democracy is not always followed in practice, i.e. rather than councillors making decisions on the advice of officers, which the latter then implement, the technical expertise of officers is given pre-eminence in the process, with the councillors serving primarily as channels of communication between the electorate and the council, and as a means of legitimizing technocratic decisions (Dennis, 1972; Elkins, 1974; Collins *et al.*, 1978).

At a more practical level the interaction between councillors and professional officers in development control is criticized on the grounds that:

- councillors are often not competent in planning or development matters;
- local issues are often given disproportionate weight by councillors as 'other material considerations' in determining planning applications;
- councillors can, and do, refuse applications against the advice of their professional officers. For an applicant who has spent some time in discussing and amending his proposed application to a point where it has support from the professional officers, such an occurrence is both costly and galling (Pountney and Kingsbury, 1983b).

This last point suggests that in certain situations the planning officer may act as no more than a liaison point between the applicant and the committee. The planning officer's loyalty to his councillors (employing authority) may be matched by a potentially conflicting loyalty to his or her profession, a problem highlighted by Benyon (1982). Councillors too may be faced with a dual loyalty dilemma in determining an application which is opposed by the electorate of the ward which he or she represents but which is generally considered to be in the best interests of the local authority as a whole.

7.4.2   Development control decisions and local plans

Studies of the operation of the development control system have emphasized the importance herein of discretionary judgement and have shown that the development plan as a whole is only one of the information sources utilized by development control case workers (McLoughlin, 1973; Underwood, 1981). McLoughlin for instance points to the role of discussions and consultations, both with colleagues (e.g. about precedents) and applicants (often involving negotiations and bargaining, whereby initial stances on both sides are adjusted and proposals modified), as sources of information used by development control workers in framing their recommendations, in addition to statutory development plans, informal plans, site visits and other sources such as circulars. Notwithstanding the range of factors impinging on development control decisions, local plans can be expected to be important sources of information given their role as statements of local authority policy on the development and other use of land at a level of detail which identifies specific sites and precise areas of land. The inter-relationship between development control decisions and policies established in local plans has been examined in detail by a research project undertaken by the Building Research Establishment in 1979 (Pountney and Kingsbury, 1983a, b). The aims of the project were '. . . first to examine the framework provided by

Table 7.1   *Characteristics of the seven plans studied by Pountney and Kingsbury*

Inner London Borough
   Plan area – borough-wide district plan excluding three small action areas
   Plan adopted January 1979

Outer London Borough
   Plan area – borough-wide district plan
   Plan adopted April 1980

Midlands City
   Plan area – inner-city action area
   Plan adopted February 1977

Small county town in southern England
   Plan area – town-wide district plan
   Plan adopted August 1979

West Midlands Authority
   (a) *Market town:*
      Plan area – town-wide district plan
      Plan adopted September 1980
   (b) *Rural area:*
      District plan for the area
      Plan adopted March 1980

Green Belt in Midlands
   Subject plan for area
   Plan adopted December 1977

Source: Pountney and Kingsbury, 1983a.

local plans for development control . . . and secondly to consider the extent to which various types of development control information are used in the preparation and review of local plans' (Pountney and Kingsbury 1983a, p. 140). The outcome of this project provided the first systematic evidence of the significance of local plans in development control decisions, although more recent work undertaken by Davies, Edwards and Rowley (1986) for the DoE has generally reinforced the findings of this earlier project.

Basically Pountney and Kingsbury selected seven local plans for detailed study representing a wide variety of environment and administrative arrangements (see Table 7.1); sampled planning applications for each plan area; and noted the decision taken and related it to policies and proposals contained in the relevant plan. In an attempt to secure consistency of analysis each application was examined under the following headings:

– the type of development; i.e. change of use or operational development;
– the existing and proposed use of the site;

- the decision taken, i.e. permitted with or without conditions, or refused with reasons;
- whether the decision was a committee decision or delegated to chief officer;
- the relationship of the decision to the local plan, i.e. developments conforming with the plan; developments that contradict the plan policies; and developments for which the plan has no specific provisions. Figure 7.2 summarizes the results of the analysis.

For the seven plans studied, the following results emerge:

(i) Between 24 per cent (inner London) and 65 per cent (Green Belt) of the applications/decisions sampled made no specific reference to policies or proposals contained in the written statement or proposals map, i.e. they were determined on 'other material considerations'. However, a more detailed study of the nature of these applications shows that they are primarily concerned with minor developments, '. . . which have little impact environmentally, economically, or socially' (Pountney and Kingsbury 1983a, p. 143), e.g. advertisements, minor alterations, extensions. This wide range is not surprising given that the nature of the plans varies. Thus for example the Green Belt plan, which is primarily concerned to constrain and control the location of new development, includes development control policies to achieve this end rather than policies concerned with the detailed form of new development. By contrast the county town plan includes detailed policies concerning advertisements in the town centre. In effect the proportion of decisions falling into this unspecific category '. . . varies from plan to plan reflecting both the degree of detail required by the circumstances and the policies associated with the different plans' (Pountney and Kingsbury 1983a, p. 144).

(ii) Only 7 per cent of decisions overall were judged as being in conflict with policies and proposals in the plan, and that in most cases special local factors or personal circumstances explained the decision.

(iii) Between 45 per cent (market town) and 65 per cent (approximately) (outer and inner London, and rural area) decisions were in accordance with plan policies and proposals.

(iv) Refusals ranged from 10 per cent of all applications in the inner London borough plan to 27 per cent (approximately) in the Green Belt subject plan, and the rural area plan. This variation is explained on two counts – first that in the inner London borough, where the pressures for development are varied and complex, pre-application negotiations are a central feature of the authority's approach to development control as a means of ensuring that the applicant is made fully aware of the authority's attitude toward the proposed development so that it can be modified accordingly prior to submission. This also saves the applicant

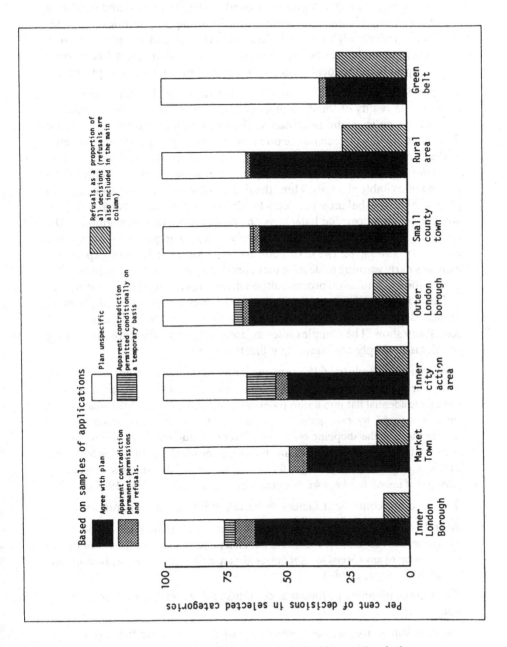

Figure 7.2 Relationship of development control decisions to local plans
Source: Pountney, M.T. and Kingsbury, P.W. (1983), 'Aspects of development control, Part 1',
Town Planning Review, Volume **54**, No. 2.

unnecessary expense which can be associated with the refusal of a major and complicated development. Second, in the Green Belt and rural area plans the major aim is to defend the area from inappropriate development. The relatively high rate of refusals reflects this and the pressure from developers who periodically test what is and what is not likely to be acceptable to the local authority by submitting a planning application.

The main conclusions drawn by the researchers are that the interactions between the reality of the development control process and the generalizations of the policies and proposals in the local plan are complex, and that there is a conflict between on the one hand the need for specificity and detail in the plan to ensure consistency in development control decisions, and on the other the need to retain flexibility in order to accommodate unforeseeable but inevitable changes. Thus 'the degree of specificity introduced into a plan affects the balance between flexibility and rigidity which must be achieved in practice. The balance is one of the important intangibles in the planning process which is not formally acknowledged' (Pountney and Kingsbury 1983a, p. 148). In addition, development control experience provides early warning of development pressures and is fed back into the plan preparation or alteration process, although much of this feedback is informal.

These findings illustrate the theoretical perspectives for local planning discussed in Chapter 2, as the following quotations from Pountney and Kingsbury show. The complex inter-related and 'wicked' nature of planning problems is simply and amusingly illustrated by:

Now even minor and straightforward applications may overlap a number of policy goals. For example an application to expand a specialist shop by change of use of an upstairs residential flat into a storeroom in the shopping centre of the small county town is covered by two quite distinct policy statements. One encourages the development of the shopping centre, specifically mentioning specialist goods . . . . Yet, because there is pressure for more housing in the town but little scope for further development, it is not surprising to find in another part of the written statement that 'changes of use of dwellings will not generally be allowed' . . . (*ibid.*, 1983a, p. 148).

The need for contingent factors to be taken into account is reflected in:

Apart from being in accord with the long-term aims of a specific plan, decisions have to take account of relevant changes in national policies and, ideally, should try and take account of any special local or personal circumstances affecting a given applications (*ibid.*, 1983a, p. 148).

The view of planning problems as complex and 'wicked' is reinforced by the conclusion that:

the variations in the decisions reflect the fact that within the framework of the development control process the abstractions of the plan preparation stage are put to the test of having to react with reality. The interaction is often complicated (*ibid.*, 1983a, p. 147).

This also emphasizes the need to vary one's approach and decisions in the light of contingent factors and the need for requisite variety in the control system to match the variety being controlled. Finally development control is quite clearly presented as a bargaining process:

The development control system, in the processing of applications and the associated appeals procedure in effect provides a very sophisticated 'forum' where individuals, relevant interest groups and organizations, elected representatives, professional planners and other officials and . . . the developers can all have their say and argue their case (1983a, p. 150).

In 1983–85, Davies, Edwards and Rowley, on behalf of the DoE, examined the relationship between development plans and development control in an attempt to determine (a) whether or not development plans provide an adequate, up-to-date framework for development control, and (b) the extent to which development control and appeals have regard to the development plan. Their study looked at this relationship in 12 district councils in England and Wales '. . . in terms of its substantive content, the planning issues actually raised by planning and control. It looks at the relationship in terms of the general purposes of planning common to all districts and types of development, rather than the specific issues in individual districts. And it looks at the relationship from the perspective of development control, in order that the range of issues need not be arbitrarily cut off by the precise content of individual development plans' (Davies, Edwards and Rowley, 1986, p. 5). In reviewing the general relationship between the development plan and development control, they establish that there are elements of the development plan which do not relate to development control, e.g. the allocation of resources, the promotion of development. Equally, they show that there are aspects of development control which have to rely on sources of policy other than the development plan. They present this relationship in the form of a diagram (Figure 7.3). Using two rectangles to represent development plan and development control, the diagram shows quite clearly the limited area of overlap between the development plan as a source of policy and development control, and the area of development control relying on other sources of policy.

The research reviews the relationship between the development plan and development control against two sets of questions: the first set relating to those operating the local authority development control system, the second set relating to applicants and other interested parties. The questions relating to development controllers were:

- What considerations are the development controllers required to take into account by the development plan?
- Where does the authority lie for the considerations other than those specified in the development plan which are also taken into account by the development controllers?

Sources of Policy

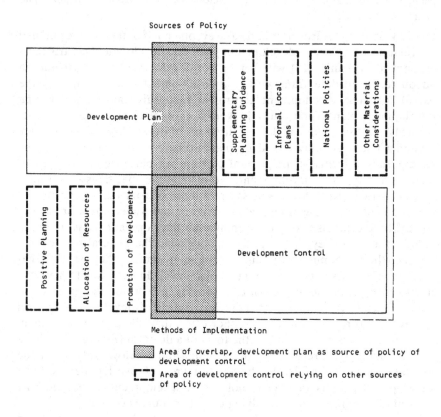

Figure 7.3   *Policy and implementation: the development plan and development control*
Source: Davies, H. W. E., Edwards, D. and Rowley, A. R. (1986). 'The Relationship between Development Plans, Development Control and Appeals'. Department of Local Management and Development, University of Reading, Figure 1.1. Reproduced with the permission of the Controller of HMSO.

For the applicants and the other interested parties, the questions were:

- What considerations are the development controllers likely to take into account and where is their authority?
- What are the criteria or standards attached to these considerations and how firm or negotiable is their status?

To overcome the problems associated with establishing a precise relationship between development plan and development control, given that the policies contained in the development plan are not comprehensive and do not cover every issue that is likely to be raised in development control, an attempt was made to link policy and control by the use of what the researchers term 'planning considerations'. These they defined as '. . . the characteristics of development proposals that are examined by local planning authorities when assessing the merits of a planning application' (Davies, Edwards and Rowley 1986, p. 6). The list of planning considerations was produced by an iterative process which tested and revised the considerations in the light of practical experience of operating the development control system. The final list used in the research contained 87 considerations, ranging from detailed matters such as car-parking and density standards to more problematic issues such as financial implications. The list is presented in Table 7.2, and in effect it indicates the potential scope of development control and the factors which could be taken into account in determining applications for planning permission.

The researchers then used this checklist to assess the extent to which the range of considerations was used in the determination of planning applications; reviewed the extent to which the considerations so used were derived from written policy either in the development plan or other non-statutory plans or policies; and established the extent to which development plan policy and the development control system overlapped.

The study establishes that: (a) the scope of development control is potentially very wide, ranging from detailed matters such as physical layout and issues of amenity, to broad issues such as the volume and location of development (p. 65); (b) there is a wide measure of agreement that the core of development control is concerned more with the practical considerations of design, form and layout of development, than with the strategic issues of volume, location or phasing of development (p. 65); (c) there is much less agreement about the use in development control of strategic considerations concerned with quantity, distribution and co-ordination of development, and newer issues such as planning gain or the financial viability of development (p. 66); (d) in relation to Figure 7.3 illustrating the relationship between the development plan and development control 'First the area of overlap between the two rectangles is quite narrow, and varies for different kinds of development, in different places and at different times. Second, the

Table 7.2   *Checklist of planning considerations used in the determination of planning applications*

1st Tier	2nd Tier	3rd Tier	
101 Amenity (including appearances)	201 Site characteristics	301	Topography
		302	Landscape features: vegetation, water, etc.
		303	Impact on historic etc. buildings
		304	Archaeological style
	202 Design (visual quality)	305	Issues of architectural style
		306	Intrinsic architectural merit: scale, mass, etc.
		307	Relationship with surroundings
		308	Treatment of external spaces
		309	Density/plot ratio
	203 Physical impact/ quality	310	Daylight
		311	Sunlight
		312	Protection from noise
		313	Visual privacy
		314	Orientation
		315	Outlook
		316	Noxious/hazardous uses
	204 Operation/amenity (effect on amenity)	317	Hours of operation
		318	Effects of construction
		319	Litter
		320	Obstruction (i.e. 'comings and goings')
	205 Relationships to surroundings (i.e. off-site)	321	Impact on historic etc. buildings
		322	Impact on protected land – amenity
		323	Non-conforming use – removal
102 Arrangement	206 On-site layout	324	Roads – layout
		325	Roads – capacity
		326	Parking – layout
		327	Parking – capacity
		328	Pedestrian movement: gradients, etc.
		329	Creation of new pedestrian routes
		330	Cyclists
		331	Disabled persons
		332	Refuse collection

1st Tier	2nd Tier	3rd Tier	
		333	Children's play space – layout
		334	Children's play space – space provision
		335	Residential internal accommodation – layout
		336	Residential internal accommodation – provision/facilities
		337	Residential private open space – layout
		338	Residential private open space – provision
		339	Residential amenity/open space – layout
		340	Residential amenity/open space – provision
		341	Non-residential ancillary accommodation – layout
		342	Non-residential ancillary accommodation – provision
		343	Backland development
		344	Security: e.g., defensible spaces
	207 Off-site relationships	345	Compatible/related uses and activities – proximity
		346	Compatible/related uses and activities – capacity
		347	Proximity to incompatible uses and activities
		348	Highway network – proximity
		349	Highway network – capacity
		350	Public transport – proximity/ access
		351	Public transport – capacity
		352	Utilities – proximity
		353	Utilities – capacity
103 Efficiency	208 Resources on-site	354	Sub-surface conditions
		355	Condition of buildings
		356	Conversion potential of buildings
		357	Vacant land/buildings
		358	Loss of natural resources: agricultural land, etc.

1st Tier	2nd Tier	3rd Tier	
	209 Off-site	359	Blight – physical
104 Co-ordination	210 Phasing: on-site	360 361	Linkages in mixed use schemes Interim measures

	211
	Phasing: off-site
		362
		
		363
		
		364
		365
			Other linked proposals:
			  physical
			Other linked proposals:
			  socio-economic
			Phasing – by quantity
			Phasing – by areas

	212 Operation/time	366	Temporary uses

105
Quantity and
distribution
	213
	Quantity
		367
		368
		369
		
		370
		371
		
		372
		
		373
		
		374
		375
			Loss of existing use
			Addition or increase in use
			Residential mix/non-
			  residential unit
			Ancillary uses
			Special categories
			  (residential)
			Special categories
			  (non-residential)
			Expansion of existing
			  premises
			Employment generation
			Impact on existing uses –
			  off-site

	214 Distribution/location	376 377	By sub-areas By groups

106
Other
considerations
	No 2nd tier	378
		379
		380
		
		381
		382
		
		383
		384
		385
		
		386
		
		387
			Planning gain
			Agencies
			Prejudicial to another
			  (preferred) scheme
			Competition
			Applicants' needs (including
			  personal occupancy)
			Site assembly
			Financial viability
			Impact on existing occupier –
			  on-site
			Impact on existing occupier –
			  off-site
			Precedent

Source: Davies, Edwards and Rowley, (1986), Figure 2.1, p. 15.

definition of the area of overlap is made fuzzy by the flexibility and discretion which characterize policy statements in development plans. Third, the area of development control outside the area of overlap is very wide and the issues there raised also vary for different kinds of development, in different places and at different times' (p. 86).

In relation to sources of policy, the 12 districts studied relied on an average of 22 plan or policy documents in 1982, including current and/or emerging structure plans, supplementary planning guidance, and non-statutory local plans and policy frameworks. However, analysis of these documents showed that between them, they covered only a small part of the checklist of development control considerations, e.g. on average only 46 per cent of the considerations relating to residential development were covered by the structure and local plan together. Put another way, in the 12 districts studied other material considerations, including policy guidance from central government and physical factors accounted for 54 per cent of the checklist of development control considerations in relation to residential development. This is not surprising, given that statements in plan or policy documents may be imprecise in their wording and flexible in their application. Indeed, many of the statements in development plans do not necessarily amount to a planning consideration in so far as development control is concerned.

The conclusions drawn are that '. . . development plans can provide an adequate up-to-date framework for development control, and development control does have regard to the development plan. However, this is not always the case: development plans do not always provide the framework in the fullest sense; development control does not only have regard to the development plan . . . The situation is that the subject matter of development control . . . varies too widely and too much of it lies beyond the usual limits of development plans, which themselves are very diverse, for the relationship between plan and control to be universal and comprehensive' (Davies, Edwards and Rowley, 1986, p. 82).

These findings support the arguments set out earlier in this chapter concerning the role of other material considerations in development control and indeed quantify in general terms the relative use of the development plan and other material considerations in development control. In addition, the conclusions reached by Davies, Edwards and Rowley (1986) support the theoretical perspectives relating to complexity and contingency discussed in Chapter 2, as the following quotations show:

'Uniformity would miss the point about the system's chief virtue, its ability to enable a sensitive response to local conditions. The so-called strategic considerations raised in development control inevitably are specific to a particular locality, responding to its particular settlement patterns and social economic conditions and to the key issues identified as being of local concern' (p. 83);

'The shifting and varied pattern of the relationship between plan and control . . . is an inevitable response to the . . . delegation of policy formulation to local authorities, and discretion in the exercise of control' (p. 82);

'The range of considerations taken into account by development control has evolved through the varied response to changing local circumstances and policy stances' (p. 82);

'The relationship between development plans and development control . . . is properly complex and varied. Any attempt to impose greater simplicity, uniformity and rigidity would be counterproductive whilst the planning system retains its responsibility for the use and development of land, widely interpreted' (p. 83).

7.4.3 Development control decisions and non-statutory local plans

The DoE is emphatic in its advice that local plans and policies which are not included in the development plan scheme and have not been subject to any of the statutory procedures leading to adoption '. . . can have little weight for development control purposes' (Circular 22/84, para. 1.13). Yet despite this stricture it is clear that local planning authorities make extensive use of non-statutory local planning instruments in the development control process as 'other material considerations'. A survey in 1983 of the use made of such instruments by local planning authorities in England indicates that:

- 6500 (approximately) informal or non-statutory local planning instruments were then in use;
- 40 per cent (approximately) of these instruments were acceptable as supplementary planning guidance, although a significant number of these instruments which are acceptable to the DoE did not accord with the criteria or procedures set out for their production in the then current circular (Circular 4/79 now replaced by Circular 22/84);
- for the authorities who responded to this particular question 98 per cent (approximately) of non-statutory local planning instruments were being used daily in development control, generally within a context set by the structure plan, or in conjunction with adopted local plans or 'old style' development plans (Bruton and Nicholson 1984a, c).

Table 7.3 illustrates the statutory context and the different types of non-statutory instrument used in the development control process. It can be seen from the table that informal local plans; single-topic frameworks (similar to a statutory subject plan) and other policy frameworks are normally set within a statutory context set by the structure plan. Indeed statutory local plans are rarely used to set the context for these types of instrument, which suggests that for development control purposes they are seen as substitutes for the statutory plan.

Despite the widespread use of such non-statutory instruments the local authorities acknowledge that their most serious potential disadvantage is the

Table 7.3 *Statutory context for the use of non-statutory local planning instruments in development control by type of instrument, England, March 1983*

Type of Instrument	Statutory context									Total
	Local plan	Structure plan	Structure and local plan	Town map	Town map and structure plan	Town map, structure and local plan	Other statutory context	Not used in development control	No response	
Development control note (%)*	59** (7.9)	112 (17.0)	31 (4.7)	21 (3.2)	63 (9.5)	3 (0.5)	56 (8.5)	7 (1.1)	308 (46.7)	660 (14.7)
Development brief (%)	105 (9.1)	156 (13.6)	156 (13.6)	94 (8.2)	69 (6.0)	0 (–)	52 (4.5)	8 (0.7)	507 (44.2)	1147 (25.6)
Informal local plan (%)	33 (2.8)	428 (36.0)	29 (2.4)	47 (3.9)	170 (14.3)	6 (0.5)	25 (2.1)	12 (1.0)	440 (37.0)	1190 (26.5)
Single-topic framework (%)	4 (1.1)	97 (27.3)	4 (1.1)	9 (2.5)	21 (5.9)	3 (0.9)	18 (5.2)	8 (2.3)	191 (53.8)	355 (7.9)
Other policy framework (%)	7 (3.7)	80 (41.9)	5 (2.6)	7 (3.7)	18 (9.4)	4 (2.1)	11 (5.8)	3 (1.6)	56 (29.3)	191 (4.3)
Local plan brief (%)	1 (0.5)	62 (34.1)	2 (1.1)	0 (–)	31 (17.0)	0 (–)	0 (–)	11 (6.0)	75 (41.2)	182 (4.1)
Informal joint study (%)	3 (4.8)	23 (37.1)	2 (3.2)	1 (1.6)	4 (6.5)	4 (6.5)	1 (1.6)	0 (–)	24 (38.7)	62 (1.4)
Special area work (%)	3 (0.7)	39 (8.9)	7 (1.6)	60 (13.8)	18 (4.1)	0 (–)	68 (15.6)	38 (8.7)	203 (46.6)	436 (9.7)
Design guide (%)	20 (10.2)	33 (16.8)	11 (5.6)	15 (7.7)	36 (18.4)	0 (–)	10 (5.1)	0 (–)	71 (36.2)	196 (4.4)
Unspecified (%)	0 (–)	2 (2.9)	0 (–)	0 (–)	30 (43.5)	0 (–)	0 (–)	0 (–)	37 (53.6)	69 (1.5)
Total (%)	235 (5.2)	1032 (23.0)	247 (5.5)	254 (5.7)	460 (10.2)	20 (0.4)	241 (5.4)	87 (1.9)	1912 (42.6)	4488 (100.0)

Source: Bruton and Nicholson, 1984a.

*Row percentages based on the number of instruments of each type in use.

**Figures refer to number of instruments.

possible lack of status in determining development control applications and any subsequent Section 36 appeals. Nevertheless under the terms of the Act they constitute 'other material consideration' and contribute to the establishment of policy guidance.

7.4.4   Development briefs and development control

Development briefs are seen by the DoE as a means of assisting developers in the preparation of planning applications for sites or groups of sites, and local planning authorities certainly use them in this way. They are also used as a means of co-ordinating the implementation of development. Indeed Bruton and Nicholson (1984a) found that in England in 1983, 1147 development briefs were in use by the 70 per cent of local planning authorities who responded to their questionnaire survey on the use of non-statutory local planning material. Put another way, 25 per cent of all informal local planning instruments were found to be development briefs. This section examines the role of development briefs in development control by reviewing a number of situations where they have been used in practice.

Development briefs may be defined '. . . as documents seeking to provide detailed guidance on the form a development should take on a particular site or sites . . .' (Bruton and Nicholson, 1983a, p. 434). An analysis of a range of development briefs in use shows that they can be used to establish a framework for development, including the location of desired land uses and disposal terms where the authority is land owner, or to respond to individual planning applications as appraisal documents, setting out in full the planning authority's stance to the proposed development. Their content includes factors such as site-specific details, e.g. location, area, ownership; physical and service constraints on development, and design criteria, e.g. form, massing, height, materials. Indeed the development briefs that have been prepared and used by Cardiff City Council over a number of years, particularly in connection with the development of large peripheral housing estates, are typical of the type of brief encouraged by the DoE, e.g. the development briefs for Thornhill, a large peripheral housing estate (8000–9000 persons) on the north of Cardiff, and St. Mellons and Trowbridge Mawr, an even larger peripheral housing estate (18,000 persons) on the east of Cardiff. In both cases, given the following, informal local plans were produced for these areas (Cardiff City Council, 1975 and 1976):

- The pressures then being experienced (mid-1970s) for residential development in the Cardiff area.
- Central government advice on land availability for housing, which recommended the use of informal interim local plans.
- The fact that the structure plan had not been approved for South Glamorgan.

It was the City Council's stated intention at that time to progress these plans to formal adoption as soon as possible, but given the formal inclusion of these areas for residential development in the structure plan and the speed with which development has taken place since, this has not been necessary. Both informal local plans, which provide a framework for the development briefs, set out the policy and proposals to be adopted in relation to population, housing, transport, schools and social facilities, shopping; recreation and open space, employment and industry, and landscape. Plans setting out the main structure for the areas are included, and although produced on an Ordnance Survey base, the proposals are represented diagrammatically. The St. Mellons plan makes no specific reference to the use of development briefs in guiding the form of the proposed development, although in practice a number of briefs variously referred to as 'planning briefs' and 'design briefs' were prepared and used for different phases of residential development, the provision of open space, the design of the district centre and the location of employment opportunities. By contrast the Thornhill informal local plan refers explicitly to the use of 'Detailed design briefs . . . for stages of the housing development in order to maximize the potential and character inherent in each of the varied sites and to set development guidelines for development control purposes' (Cardiff City Council, 1976, para. 5.5). This statement is reinforced by the section dealing with development control policies which states:

15.1 All proposals for development in the development area shall be in accord with the provisions of this District Plan.
15.2 Land currently used for farming for which no development is proposed, shall be retained in agricultural use.
15.3 Design briefs will be prepared to guide the development and permissions will be conditioned accordingly . . . .

The plan then goes on to indicate the priority for the production of the design briefs (Cardiff City Council, 1976, para. 15).

In all, nine planning briefs have been prepared and used for housing developments in conjunction with the Thornhill informal local plan; six briefs for housing development; two briefs for employment; and a design brief for the district centre. The housing briefs are similar in form and content and set out the objectives of the brief; the existing features to be retained and incorporated in the development; the density of development; access and circulation requirements; car-parking standards; open-space provision; service requirements and constraints; and design criteria including matters relating to internal privacy, sound insulation, sunlight and daylight, and materials.

The briefs relating to the provision of local employment follow a similar format dealing in particular with site constraints, pedestrian and vehicular circulation, and design policies including layout, materials, and landscaping. The brief for the district centre is again of similar form and content although it deals with the provision of community facilities associated with a district shopping centre and gives fuller attention to design matters, including a location map showing physical characteristics and points of access into the site of the centre, and the relationship between it and the adjoining residential development.

The impression created by these briefs is one of repetition, and in many cases the differentiation between the housing areas dealt with by numerous briefs becomes hazy. However the development on the ground, whilst unmistakably speculative and mass-produced, does not give the impression of dull repetition. Given the speed with which both Thornhill and St. Mellons are being developed the absence of a statutory local plan and the low incidence of Section 36 appeals against refusals for planning permission, then the use of development briefs in this way in development control can be considered successful. This is especially so when account is taken of the multiplicity of building firms and architects involved in the process.

The Heathcote example

As has been shown, development briefs are used widely throughout the country. Many of these briefs are similar in form and content to those used by Cardiff City Council. However it would be unwise to assume that they all operate in the straightforward way suggested by the Cardiff experience. Indeed the experience of Warwick District Council with a wider range of development briefs suggests that, as with all other aspects of planning, the issues involved are complex and inter-related and much influenced by contingent factors (Bruton and Nicholson, 1984b). For example, the development brief produced by Warwick District Council (1980) for the development of an industrial estate – *Heathcote* – in the joint ownership of the County and District Councils is also typical of the type of brief envisaged by the DoE and used in Cardiff. It sub-divided the estate into phases of development to relate the provision of infrastructure to the income received from site disposal; it sub-divided phases 1 and 2 of the development into individual single-user sites for disposal; it established guidelines for the local planning authority to deal with planning applications relating to individual sites, including matters of form, materials, landscaping, parking standards, fencing and security, access and circulation and the storage of liquids. The brief was concerned to promote a unified and harmonious estate of sufficiently high environmental quality within which high-technology industry could be encouraged to locate. In many respects the brief could be quoted as a model of the sort encouraged by the DoE. Unfortunately in the period 1979–83 it had very little impact on the control of development which took place on the estate for a number of reasons, the most important of which were (1) the basis on which the servicing of the development was to be funded; and (2) the decline in the market for industrial land and premises during this period.

Briefly, the two authorities owning the site agreed that the provision of basic infrastructure should be financed out of sales of serviced sites, and that any development costs incurred should be in line with demand for land released by the provision of infrastructure. The development was phased with this in mind. However with the decline in the market for such land, especially high-quality single-user prestige sites, the authorities soon modified their outlook to recoup as quickly as possible capital laid out to provide the initial servicing, by responding to the then market demand for small serviced industrial units. Financially this proved to be a sound policy in that by March 1982 on 6.2 hectares of land, sale receipts exceeded costs by £300,000 (approximately). However, the impact of the change in policy on the physical appearance of the development anticipated in the development brief was significant – small serviced plots rather than larger sites for prestige users now dominate the estate, whilst effective planning control over matters of detailed design has been weakened by the need to achieve sales, since in a slack market restrictions on development imposed through conditions on planning permission or by stringent design requirements could mean either a reduced price or no sale at all. The pattern and scale of development has determined the character and physical appearance of the estate, so that it is now unlikely to prove attractive to high-technology industry which typically requires high-quality landscaped settings.

Whilst it is not possible to draw general conclusions from one isolated case it is clear that the process of implementation and the process of development control are highly inter-related. Given the basis on which the authorities chose to fund the Heathcote venture it was inevitable that once market conditions changed the approach to developing the site also had to change. Thus despite a firm land use designation in the urban structure plan, despite a clear vision of what form the estate would take which was articulated in a development brief, the end product is very different from that original vision. Uncertainty and contingent factors have influenced the outcome; the development brief has been at best peripheral in guiding the development although the particular approach adopted by the authorities has been sufficiently robust to respond to the complex inter-relationships involved in a way that has allowed the site to be developed, and at a profit.

Further examples

Two further examples taken from the same District can be used to illustrate other facets of the use of development briefs in development control – the first relates to an area in Warwick referred to as West Rock, the second relates to the proposed development of a new shopping centre in Leamington Town Centre.

West Rock: This is an area on the western fringe of Warwick town centre. Following the abandonment in 1976 of a long-standing proposal for an inner relief road and pending the production of the proposed Warwick Town Centre local plan, the District Council established by council decision a policy framework within which two sites which had been acquired by the authority and cleared for the now abandoned inner relief road, could be developed to positively enhance the street scene and improve traffic movements in the West Rock area. The need to achieve a development of high visual quality, to develop the area for housing or offices depending on market conditions, and to rationalize access and car-parking was given emphasis.[7]

In an attempt to meet these guidelines the District Council produced a planning brief for the larger of the two sites in its ownership (Site A) and invited prospective developers to submit sketch schemes on the basis of the brief. The planning brief identified the areas to be developed for housing and car-parking; established the means of access; and identified buildings to be demolished and the broad pattern of landscaping expected. From the submissions received a shortlist of schemes which were considered to be acceptable *on design grounds alone* was selected and the developers were invited to submit a financial bid for the site. It was made clear to all developers that the District Council was not obliged to accept the highest bid, and that as a condition of the eventual sale a detailed planning application would have to be made in accordance with the winning scheme and within a specified timescale. It was similarly specified that the development would have to be substantially completed within two years of detailed approval being granted. The selection of the shortlist and the winning scheme involved members of the committee which would ultimately be responsible for granting planning permission, as well as members of the Warwick Town Council, which is normally consulted by the District Council on development control matters in Warwick.

The initial sketch layouts were received in March 1978; a shortlist of four was selected on design grounds alone in May 1978; the winning entry was chosen on the basis of both design and finance in August 1978. The disposal of the site was concluded in September 1978; a detailed application was submitted and planning permission was granted in November 1978. The development of the site was completed by the end of 1980.

Following successful disposal of Site A, the District Council turned its attention to the second site (Site B). Given that this site was less prominent than Site A, and that the pattern and form of development in the area was largely fixed by the proposed development on Site A, the District Council decided to prepare a detailed planning brief for Site B and to dispose of it on the open market with the benefit of an outline planning permission based on the brief. Tenders were invited which consisted of sketch schemes for the development and a financial bid. The most suitable tender was then chosen on the twin basis of design and finance. The brief established access into the site, specifying kerb radii and vision requirements; and emphasized the high-quality design required which should be set within the context of the proposed development on Site A and which had been subject to a competition. The form and timing of the development was controlled after disposal through conditions in the sale contract, which were similar to those specified on Site A.

This example again emphasizes the point that the process of implementation and development control are inter-related. The initial use of a competition in conjunction with a development brief to bring forward good-quality design schemes which were judged solely on design grounds, acknowledges that high-quality design is expensive and will be reflected in total site development costs and the price the developer can afford to pay for the land. At the same time the approach adopted by Warwick District Council illustrates quite clearly the significance of contingent factors.

The development of the first site was tightly controlled throughout, although it must be acknowledged that much of the control stemmed from the District Council's powers as land owner which allowed conditions to be imposed on the developer through the sale of the site which could not have been imposed under town and country planning legislation. By contrast, the control over the development of the second site was more relaxed, in that a joint design and price mechanism was used. In this latter case the design criteria were enforced and achieved through both land-owning and planning powers possessed by the District Council. To have released Site A simply with the benefit of an outline planning permission would not have guaranteed that a scheme of sufficient quality would have been forthcoming. Instead the design competition was used as a mechanism to bring forward acceptable schemes. In this case, the brief for Site A set the competition rules at the same time as promoting the development potential of the site. In turn, the successful disposal and development of Site A meant that the progression of proposals for Site B was made easier as there was proven developer interest in the site and the design context had been set. Thus it is clear from this example that development briefs, when used in conjunction with land-owning powers, can be a powerful force for controlling and achieving the type of development sought. Whether such achievements are possible through the use of development briefs alone is another matter. In so far as development control is concerned, the inter-relationship between the development briefs, the competition and the selection process of the successful tenders by elected members determined what was eventually built. There is no doubt that the development briefs were influential in moulding the framework. Equally it is evident that contingent factors such as quality of the surrounding environment; the quality of design aimed for; the cost of the development and the associated return to the district council; and the perceptions of the elected members, their professional officers and the local residents; were also factors which influenced the form of the development.

Leamington Spa Town centre: The circumstances surrounding the use of development briefs in relation to Leamington Spa Town Centre differ from those applying in the case of Heathcote and West Rock, in that the statutory context for development control in the area is provided by a recently adopted (1983) town centre local plan (Warwick District Council, 1983). Although the emphasis in the plan is on the management and maintenance of the existing urban fabric, new developments are proposed on key sites, the most significant of which is for new shopping development to be provided through a joint public–private venture on the area referred to as Site D. The policy relating to this development was set out in the adopted local plan as follows:

Proposal S.6: A major shopping development comprising at least one large store devoted to the sale of durable goods, with other small individual shop units, will be provided on Key Site D . . . . (Warwick District Council, 1983, p. 40)

The site is largely in the ownership of the District Council and the method of establishing a relationship with a private developer was as follows:

(i)   A general invitation was issued to interested parties to register an interest in participating in the proposed shopping development.

(ii)   A preliminary development brief was issued in June 1982 to all bodies who had registered an interest. This brief gave emphasis to the need to provide an additional 85,000 square feet of shopping floor space comprising a single major user of 50,000 square feet, together with a selection of shop units of varying size.

(iii)   Thirty organizations responded to the brief by August 1982; a shortlist of 12 were interviewed and a final shortlist of three were selected in October 1982.

(iv)   A more detailed brief in the form of a design and finance guide produced by consultants was issued to the final three companies in October 1982. This brief amplified the requirements of the local plan policy in a way which allowed the three selected developers to submit detailed schemes. The needs of shoppers, design quality and servicing requirements were emphasized in the brief.

(v)   The three schemes submitted were placed on public display in February 1983. The winning scheme was selected in March 1983 taking into account the public response to the submissions (including concern about the scale of the proposed shopping provision), the criteria specified in the more detailed second brief, and the strength and quality of each developer's team.

(vi)   Planning permission for the winning submission was granted in June 1983 by the District Council and a partnership agreement was entered into whereby the developer was required to acquire all the land on Site D not currently owned by the District Council, assisted where necessary by the use of District Council powers of compulsory acquisition. The freehold of all properties thus acquired will eventually be transferred to the District Council, who in turn will lease the entire site to the developer for 150 years.

(vii) The developer will then construct the proposed development in accordance with the previously approved plans, including a car-parking element which will be sub-let to the District Council.

Whilst the above process is relatively clear and straightforward, it was complicated by the involvement of a second development company. This company had been unsuccessful in its tender in the early stages of the process but had proceeded to acquire sites on the open market which formed part of Site D, and had submitted a planning application to develop Site D as a shopping centre. The District Council refused this application on the grounds that by proposing 120,000 square feet of additional floor space, including two large retail stores, it was contrary to the policy of the local plan. The development company appealed against this decision and objected to the compulsory purchase notice served on it in respect of the land it owned which formed part of Site D. The appeal and objections were heard conjointly and were dismissed. The developer selected by the District Council to develop the site is currently working towards the implementation of the shopping development which has detailed planning permission

Although the circumstances surrounding the proposed shopping development in Leamington town centre differ in scale and complexity from those which existed in West Rock, nevertheless the two-stage process whereby the developer was selected is similar. In both cases the District Council was concerned to produce good design quickly and cheaply; to give emphasis to the importance of design; and to create a range of innovative design solutions. The combination of competition and design briefs was certainly effective in achieving this and resulted in the granting of detailed planning permission for a complex shopping development in a minimum of time and in a way which satisfied both the developer, the District Council and the general public. However, it should be noted that the Council's position as the major land owner of the sites involved was almost certainly the prime factor in preventing unacceptable development from taking place on the site. Planning control, although featuring in the process, almost certainly occupied a secondary position.

It is interesting to note that the involvement of the second and unsuccessful development company in the process exhibits some of the characteristics of distributional bargaining outlined in Chapter 2. For example, by acquiring land which fell within Site D and submitting an application for permission to develop a shopping centre on the site in competition with the District Council's ideas, it adopted a commitment and communicated that commitment to the District Council:

- The District Council by refusing that application and serving a compulsory purchase order on land within Site D owned by the development company, adopted an equal and opposite commitment and communicated it to the developer.
- The acquisition of land within Site D by the developer immediately introduced intersecting negotiations into the process, in that if the District Council approved the second developer's proposals then the problem of acquiring the land owned by that company disappeared.
- By appealing against the planning refusal and objecting to the compulsory purchase order the would-be developer resorted to arbitration to resolve the impasse between himself and the District Council.

It can be argued that power and authority in the form of a statutory development plan, supplemented by development briefs amplifying that plan, were the deciding factors in this issue. However the attempts made by the District Council to take account of contingent factors in respect of design quality, the level of provision of excessive floorspace which was strongly opposed locally, and the quality of the environment, by involving elite groups like the Chamber of Trade and the Leamington Society, and the public in the decision, also made a contribution to the resolution of this issue. Despite these uncertainties, and in the light of the experience reviewed above it is clear that development briefs can be used to progress development proposals towards implementation on the ground within a policy framework set by the following:

(i) A statutory plan, e.g. Heathcote industrial estate within the framework set by the urban structure plan; the shopping centre proposals within the framework set by the Leamington Town Centre District Plan.
(ii) A non-statutory local plan, e.g. the Thornhill development in Cardiff.
(iii) Committee decision, e.g. the policy guidelines put forward in connection with the West Rock area.

In terms of development control the examples reviewed suggest that development briefs fulfil two main functions. The first reflects the role envisaged for briefs by the DoE – that is to assist developers in the preparation of planning applications. Here briefs are 'internalized' within the development control process, where their function is to provide statements of the criteria that will be applied in determining planning applications on the site(s) in question. This articulation helps to strengthen subsequent decisions since it shows they have been made on a consistent basis. The Thornhill, St. Mellons and Heathcote briefs most clearly fill this role. Although in the case of the Heathcote development, conflicts of interest arose over quality of design and development, the indications are that where the development control process can operate unhindered from such difficulties, briefs offer a useful means of formalizing the criteria that the process intends to apply to the sites in question. Indeed the Cardiff experience of successfully co-ordinating the development and design aspirations of a number of speculative house-building firms and their associated architects in a way which gives coherence and interest to the large developments at St. Mellons and Thornhill, is testimony to the efficacy of development briefs in development control.

The second role for briefs in development control, where they are used in a positive way to stimulate the desired development, is slightly more complicated. This entrepreneurial role can rest solely on planning powers, where the brief offers planning permission to interested developers, but is more powerful and is therefore likely to be more effective if the local authority is also land-owner. In this case, the brief can offer guaranteed land availability, with no site-assembly problems, backed if necessary by compulsory purchase powers, together with planning permission. The West Rock and Leamington Site D briefs provide examples of the latter; these documents rested largely on the available land-owning powers, with the submission of planning applications and the granting of permission being relegated to a minor technical role after developer and scheme selection. Detailed discussions over design, for instance, were held within the context of the competitions rather than as part of the processing of individual planning applications.

However, the development control process does not have to take a subsidiary role; the unsuccessful second development company recognized this on Site D in Leamington by not only seeking to challenge the District Council as land-owner by buying into the site but also as local planning authority by submitting a planning application and subsequent appeal. Moreover, at West Rock the possibility that a winning scheme could subsequently have been refused planning permission was recognized and guarded against by involving in the selection process those who would eventually consider the planning application. Authorities trying to bring forward schemes in this way should remember that though a scheme may agree closely with the terms of a brief this does not in any sense replace the need for planning permission. The use of such informal documents as development briefs must always be related to the statutory requirements of the land use planning system. Finally it is interesting to note the following:

(i) The use of the brief for the *Heathcote* development provides a good illustration of how changing circumstances can render invalid a planning process which was designed to cope with certainty to achieve agreed ends by the application of known technology. To achieve what the District Council regarded as the most important end (the development of the site on a self-financing basis) the means were adjusted and the goal of a high-quality industrial estate sacrificed. In Christensen's terminology the process changed gradually and probably not consciously from a professional optimizing approach to a pragmatic adjustment to changing circumstances, i.e. a planning process for agreed ends but using 'means' which are not proven.

(ii) The *West Rock* and *Leamington Spa Town Centre* briefs, by contrast, show a marked concern to apply from the outset a planning process where the ends were agreed but there were doubts as to how best to achieve these ends. The overriding 'end' to achieve a high-quality design, led to the use of a pragmatic, experimental approach to establishing the most appropriate 'means' – the architectural competition combined with a financial agreement proving most effective in coping with the particular uncertainties and other contingencies.

7.4.5 The determination of Section 36 appeals

Appeals against the refusal of planning permission, the refusal to approve details following an outline permission, the imposition of conditions or the failure to issue a decision within the specified time period can be made under Section 36 of the *Town and Country Planning Act 1971*. Such appeals can be dealt with by written representations, if both parties agree, or through a public inquiry. The Secretary of State for the Environment appoints an Inspector to hear such appeals, which may be transferred to the Inspector for determination under the provisions of Schedule 9 of the 1971 Act (para. 1) or retained by the Secretary of State if he so decides. In the former situation, the Inspector sends the decision letter determining the appeal, under his or her signature. In the latter situation the Inspector will prepare a full report on the appeal, including recommendations for the Secretary of State.

In 1984, 9793 appeals were determined through written representations of which 31 per cent were allowed, whilst 1551 were heard at inquiry, with 40.6 per cent being allowed. In the same year a total of 480 appeals were determined by the Secretary of State (37.3 per cent allowed) and a total of 11,163 were determined by Inspectors under transferred authority (32.2 per cent allowed) (DoE 1985b, Tables 3.2 and 3.3). The advantages of having an appeal heard by written representations are claimed to be that the method is cheaper and quicker (DoE 1983, para. 2.7). In 1984 in England the average time taken to determine written representations was 22 weeks if authority was transferred to the Inspector, and 42 weeks if the decision was retained by the Secretary of State. The average time to determine appeals through a public inquiry was 28 weeks by Inspectors and 53 weeks by the Secretary of State (DoE 1985b, Tables 2.1 and 2.2).

Regardless of whether appeals are determined by the Inspector or the Secretary of State, the decision must be taken in the full knowledge of the policies involved and all the facts of the case. Thus the Inspector in hearing the appeal must '. . . have regard to the provisions of the development plan, so far as material to the application, and to any other material considerations' (s.29(1), 1971 Act). As the previous discussion suggests the development plan might on some occasions quote all the policies which bear on the case, although it is more likely that other policies at central and local government level, by virtue of being relevant, are also material considerations. Equally the facts relating to a particular appeal, in so far as they are relevant, also constitute other material considerations. In determining the appeal, or recommending to the Secretary of State, the Inspector needs to take account of all relevant factors which can generally be considered under three main headings:

Government policies

These are articulated in White Papers, departmental circulars and bulletins, Ministerial statements, appeal decisions and press notices. Established government policies must be accepted by Inspectors without question and the DoE circulars referred to in Section 7.3 *must* form a central part of the Inspector's consideration of appeals. There is evidence to suggest that as such circulars generally deal with the broad objectives underlying the development control process there are many occasions where the particular circumstances of an appeal justify departing from established policy, i.e. the proposed development does not conflict markedly with the policy and is unlikely to obstruct the government's aim. However, in recent years, as circulars have become more explicit, and with the increase in the number of appeal cases transferred to Inspectors from the Secretary of State for determination, it has become apparent that circulars are now something more than a matter to which regard should be paid.

Local authority policies

Like central government, local authorities formulate policies to achieve their objectives and these may be incorporated in approved development plans, draft statutory plans, non-statutory plans, resolutions of council or local codes of practice. Taken together, the framework of accepted planning policies stemming from both central and local government provides the Inspector with a means 'by which to orient his judgement and align it objectively with that of his colleagues' (Allan, 1974, p. 4). Earlier sections in this chapter have dealt with the role of statutory and non-statutory plans in establishing local authority policies for the purposes of development control, and these are by far the most important means of establishing local authority policy. However, council resolution can, and is used to, establish development control policy, generally in response to an emerging planning problem evidenced by a series of applications which require a policy to be determined. Such policies are thus a response to contingent factors; e.g. Cardiff City Council has a large number of development control policies established solely on the basis of council approval and which are recorded only in the minutes of the appropriate Council meeting and the minds of officers and councillors operating the development control system. These policies relate to issues such as building societies and banks in central area shopping streets, activities and displays in pedestrianized areas, flat conversions and office development (Bruton and Nicholson, 1983b, p. 21). Such policies are quoted in support of development control decisions and in any subsequent appeal. If relevant to the case in hand they constitute 'other material considerations'.

Non-statutory local planning instruments are widely used in development control (see p. 314) and despite the advice of the DoE that such plans can have little weight for these purposes (DoE, 1984g, para. 1.13), there is ample evidence to show that they are considered by Inspectors as 'other material considerations' in determining appeals. Indeed Bruton and Nicholson (1984c) found that where relevant, all categories of non-statutory local planning instruments are considered by Inspectors in appeals, not only those defined as acceptable supplementary planning guidance. Moreover, the level of significance attached to such instruments appears to be related directly to the nature and extent of the preparation and adoption procedures used, i.e. if the instrument has been subject to participation, consultation, adoption by Council and is freely available to the public, then it will be given greater weight than if it is only at the early stages of preparation.

The 'facts' of the application

Factors under this heading, which may be taken into account if material, include details of the application, such as ownership, and site boundaries; the existence of previous planning permissions, existing use rights, third-party representations and the personal circumstances of the applicant. In addition to establishing the 'facts', the Inspector is also required to exercise his or her judgement on a range of factors which include precedent and the consequences of the proposed development on the locality; the desirability of retaining the existing use and the consideration of financial matters in influencing a particular development, e.g. permission for the office use of upper floors above shops in listed buildings could enable the listed buildings to be maintained satisfactorily as a result of the relatively high income derived from rental, which a residential use would not. The Inspector must judge to what extent these circumstances should influence the determination. Allan (1974) makes the point forcibly that the determination of an appeal focuses on the issue(s) to be resolved. By implication the establishment of the issue(s) is a subjective judgement of the Inspector based on an analysis of the relevant policies, fact, and other material considerations. Indeed, Allan sees the establishment of the issue(s) as being a critical step in an appeal, for these not only provide objective criteria for the assembly of relevant factors, but also define the considerations on which the Inspector's decision will turn. Inevitably, issues at appeal vary from physical issues such as visual intrusion on the one hand to issues such as the amount of retail shopping floorspace an area can support on the other. In this latter case, the

Inspector will need to rely heavily on written policies and supporting evidence in determining the appeal. These policies and evidence may only be available in a local plan, be it statutory or non-statutory. The physical type of issue, by contrast, can be much influenced by the particular location of a proposed development and the developments which adjoin it. In such situations the Inspector can quite validly determine appeals on the basis of 'other material considerations', provided that they are referred to in the appeal statements or raised at the inquiry.

From this outline of the appeals process, it is clear that the development plan is only one factor to be taken into account in determining an appeal under Section 36 of the 1971 Act. Indeed, appeal decision letters indicate that a wide range of factors are taken into account, including the development plan, non-statutory plans and other material considerations, and that appeals are determined on their merits, with no one set of factors automatically being given greater emphasis. As far as the role of policy is concerned, research for the DoE by Davies, Edwards and Rowley (1986, p. i) has investigated the relationship between development plans and appeals by assessing whether development plans are providing an adequate up-to-date framework for appeals and to what extent appeals are having regard to development plans. They analysed a 10 per cent random sample of appeal decision letters issued in 1982; checked the validity of this sample by an in-depth analysis of appeal files in 12 district councils in England and Wales, and interviewed a variety of participants in the appeal process, including members of the Inspectorate, local authority planning officers and agents acting on behalf of appellants. Their findings in relation to the role of policy in appeals can be summarized as follows:

(i) Policy may not always be regarded as one of the principal considerations in an appeal but it is part of the background against which the decision is made.

(ii) In the national 10 per cent sample of 1286 appeals analysed, one-third (approximately) of the decision letters made no reference to policy of any kind. Rather, these decisions were determined by the facts of the case, the arguments put forward and the Inspector's judgement. Indeed, in 41 per cent of all appeals in which policy was mentioned, it was not regarded as a determining factor.

(iii) Policy is important in determining appeals where strategic considerations are important, e.g. major developments, where policy determined 81 per cent of all such appeals.

(iv) Policy is less important in determining householder appeals (37 per cent of such appeals), where amenity and design considerations tend to predominate.

(v) In appeals where policy was the determining factor 88 per cent of such appeals were dismissed, compared with a 58 per cent dismissal rate where policy is not a factor. Thus '. . . the presence of policy, in particular a structure plan, or, in the few appeals where it is mentioned, a local plan, usually becomes a determining factor' (*ibid.*, 1986, p. 67).

(vi) Central government policy as set out in circulars and DCPNs, is frequently used by appellants against the original decision of the local authority. Indeed it was mentioned in one-third of all appeals containing any reference to policy across the whole range of development. Once mentioned and accepted as a determining factor, a relatively low proportion of dismissals resulted.

Thus the study demonstrates that the value of the development plan in appeals is most obvious where strategic considerations are the determinants of the decision. Conversely, in appeals where issues of amenity, design and socio-economic considerations are significant the development plan does not provide an adequate framework. Indeed on those occasions where policy on these matters is available it would seem that it is usually to be found in non-statutory local documents and central government policy guidance (*ibid.*, 1986, Table 6.7, p. 59).

7.5 Bargaining in development control[8] (Bruton 1983b)

In reviewing the theoretical perspectives which impinge on local planning, it is suggested in Chapter 2 that distributional bargaining is an integral part of the local planning process. In an attempt to determine the extent to which development control is characterized by bargaining, two cases of proposed major developments were analysed against an analytical framework developed from the work of Schelling (1960), and Bacharach and Lawler (1981). The findings suggest that bargaining is indeed an essential ingredient in development control, at least in so far as major development proposals are concerned. The proposed developments were (1) for industrial and commercial developments at Miskin, some 8 km north-west of Cardiff and adjacent to the M4; and (2) for a superstore development at Baginton, some 2 km south of Coventry city centre.

Four local planning authorities were directly involved in each case (South Glamorgan CC and Mid Glamorgan CC, Taff-Ely BC and the Vale of Glamorgan BC in the case of Miskin; the West Midlands MCC, Warwickshire CC, Warwick DC and Coventry City Council in the case of Baginton), whilst the applicants were corporate bodies (a charitable trust and the Land Authority for Wales in Miskin, and Sainsbury's and British Home Stores in Baginton). Both sites were within urban fringe areas for which structure plan policies precluding the form of development proposed had recently been approved.

Table 7.4 summarizes the unique and common features of the theories of Schelling, and of Bacharach and Lawler, and presents the analytical framework against which:

- the parties involved in the bargaining, their resources, the roles they play and their actions were identified;
- their relationships and interactions were defined;
- their values and interests were established;
- the relative distribution of power amongst the parties, and the tactics adopted in the process of bargaining, was reviewed;
- constraints acting on the parties involved in the bargaining were assessed.

Whilst this framework clearly identifies the separate elements involved in a bargaining situation and suggests that they are independent, this is far from the case. The relationship between these elements is vitally important for they are interdependent and individually almost meaningless.

Evidence from the analysis suggests that development control is a highly complex decision-taking process whicn exhibits many of the characteristics associated with bargaining. In particular, it is clear that at least in the two instances examined the structural characteristics of the decision-taking framework, the distribution of power between the elements of that framework and the way those elements interacted are important factors in the process. The dependency relationship established between the major participants, and the way those relationships were changed through the application of aspects of the distributional bargaining process appear to have been significant factors in influencing decisions, e.g. in the case of Baginton the dependency relationships established amongst the four local authorities involved in opposing the proposed development left little scope for the applicant to negotiate or bargain. By contrast, the dependency relationships between the four local authorities involved in the proposed Miskin development, which split the authorities into two groups – those supporting and those opposing the development – provided scope for the applicant to bargain, with some success.

Authority; influence; the adoption of a commitment; the use of bargaining agents, intersecting negotiations and arbitration; are all evident in the two case studies. The proposed superstore at Baginton was refused planning permission and the subsequent appeal dismissed by the Secretary of State. After two major Section 36 inquiries which both recommended approval of some form of industrial and commercial development at Miskin, the Secretary of State for Wales eventually granted planning permission subject to the implementation of a Section 52 agreement between the applicant and the Vale of Glamorgan Borough Council.

In these cases, the development plan may be seen as an essential and integral element in the bargaining process which is development control – it is the element of the process which establishes the boundaries of authority of planning policy and legitimizes the role of the local planning authorities. It is the element of the process against which the applicant (and appellant) use their counter arguments, power, influence, and resources in an attempt to

Table 7.4 *An analytical framework for examining the nature and characteristics of any bargaining undertaken as part of the development control process after Schelling, and Bacharach and Lawler*

The structural characteristics of the decision-making system

1 Who are **the main actors** involved?
2 What is **the relationship** between the actors?
    (a) Is it **formal, structured** and bounded by rules or
           informal and/or ambiguous?
    (b) Is the relationship based on:
           conflict – indicating distributional bargaining; or
           co-operation –indicating integrative bargaining?

Patterns of interaction between the actors

1 Is the interaction:
    direct, and possibly involving the use of bargaining agents; or
    indirect, implying a tacit approach to the bargaining?
2 Is the interaction characterized by:
    – **perfect information flow** between all actors?
    – **incomplete information flow** or non-communication?
    – **continuous** two-way interaction? or
    – **discontinuous** and one way?

Values and interests of the actors

1 Do the values and interests represent:
    conflict or **consensus** (suggesting distributional or integrated bargaining)?
2 Are the **values and interests**: sectional; individual or general?
3 Do the actors represent whom they claim to represent?
4 Are those being represented aware of this representation?

Power distribution and the behaviour of actors

1 Is power **unevenly distributed**?
2 **Who has power** and from what does it **derive**?
    – **commitment and bargaining skill**?
    – **dependence**?
    – **authority**?
3 **What tactics are used** by actors to achieve their objectives?
    (a) **an explicit approach** to bargaining involving commitment; bargaining agents; restrictive agenda, intersecting negotiations, compensation and/or arbitration?
    (b) **a tacit approach** which includes in addition to the tactics outlined in the explicit approach, the manipulation of information; communication through a third party; and exclusion of certain groups which are designed to give an impression of power; to reduce one's dependence and to increase the dependence of others?

Power distribution and the behaviour of actors – cont.

4 How is the **final agreement** reached?
  – **by bargaining with concessions** on each side until an acceptable bargain is struck?
  – **by compensation** or **arbitration**?
  –by those with power **imposing** it on others?

Constraining characteristics

1 *Of the system*
  – resources, including time, manpower, money
  – legal and statutory requirements
  – desire to avoid mutually damaging behaviour
2 *External to the system*
  – other related issues (intersecting negotiations)
  – pressure from outside on various actors
  – the need for public bodies to be seen to be acting legitimately

---

implement development which would otherwise not be permitted. On the basis of the two case studies, it would seem that as long as those holding power and authority in any development control situation are united, then the provisions of the development plan will dictate the decision at local authority level. Appeal to a higher and more powerful authority could well change that – but again only through the exercise of authority, which might be influenced by bargaining.

7.6 Conclusions

By statute the development control process is seen as a somewhat mechanistic and technical process. Section 29 of the *Town and Country Planning Act 1971* (as amended) requires that in determining an application for permission to develop, the local planning authority must have regard to the provisions of the development plan and any other material consideration. In reality, development control is a highly complex process with the complexity deriving from:

– the establishment of exactly what constitutes other material considerations for each application; and
– the implicit and sometimes explicit bargaining which forms an integral part of the process.

The importance of other material considerations and consultations in the development control process means that development control does not merely have a narrow concern with implementing the development plan, but is also a mechanism for the management of change in the environment. In other words, development control thus 'moves away from a fundamentally straightforward administrative process into a more complex negotiating process between the various interests concerned in development' (Davies, 1980, p. 12). The development plan forms one element in the bargaining process, and in many instances can be the predominant factor in the determination of an application. However:

- *policy guidance* from central government in the form of circulars; Development Control Policy Notes; Green Papers, White Papers and ministerial statements;
- *local authority policy* reflected in instruments other than statutory development plans;
- *officer–councillor relationships* in the development control decision-taking process;
- *bargaining*; and
- the 'facts' relating to a particular development on a particular site, e.g. traffic, amenity, topography;

these are all factors which will influence the determination of applications for permission to develop, or appeals against such determinations. Certainly in Section 36 appeals this group of factors appears to have a greater influence in determining the outcome of the appeal than the development plan.

This chapter has attempted to examine the relative contribution of statutory local plan policies and other material considerations in the determination of planning applications. It is clear from the work of Pountney and Kingsbury (1983a, b) and Davies, Edwards and Rowley (1986) that statutory local plans are only one factor in this complex decision-making process, and that their importance in the process is at times marginal. This is not altogether surprising given the range of other material considerations; the limited number of statutory local plans that have been adopted or placed on deposit and the complex inter-actions between the applicant, the officers and councillors of the local authority, third-party interests, the inspectorate and the Secretary of State, which often characterize the progress of planning applications.

Notes

1   This case was reported in *The Times*, 16 May 1983.
2   Grant M., *Urban Planning Law*, Sweet and Maxwell, London, 1982, pp. 145–645; and Alder, J., *Development Control*, Sweet and Maxwell, London, 1979, cover the development control process in detail.
3   Carnworth, R., *Blundell and Dobry's Planning Appeals and Inquiries*; 3rd edn., Sweet and Maxwell, London, 1982, deals in detail with the appeals procedure.

4 See 'Court out on phasing policy', *Planning*, No. 620, 31 May 1985, p. 1.
5 All circulars referred to in this section are DoE circulars unless otherwise stated and are published by HMSO, London. The second element of the circular number indicates the year of publication, e.g. Circular 22/84 now published in 1984. They are not included in the references given for this chapter.
6 The Development Control Policy Notes referred to in this section are published by HMSO, London. The early DCPNs were produced by the Ministry of Housing and Local Government. Since 1970 they have been produced by the DoE. They are not included in the references given for this chapter.
7 The planning brief for Site A, West Rock, Warwick, was approved by Warwick District Council on 18 January 1978. The brief for Site B was approved in March 1979.
8 This section is based on work undertaken by Lucy Carlyle, reported in an unpublished M.Sc. dissertation, Department of Town Planning, UWIST, 1982; details of the bargaining process undertaken in connection with development control may be found in Bruton, M. J. (1983b), *Bargaining and the Development Control Process*, Papers in Planning Research, No. 60, Department of Town Planning, UWIST, Cardiff.

8 Local planning and implementation
II: The development process

8.1 Introduction

In the British planning system, local planning policy on land use and development is set out and progressed by a wide variety of plans and other material. Local planning policies can be established by or derived from the following:

1 The statutory local plan, prepared by local planning authorities under town and country planning legislation (Chapters 3 and 4).
2 Additional statutory plans, programmes and initiatives with a bearing on land use and development issues, produced and/or progressed by a variety of public agencies (including the local authority) under a range of statutes or central government policy initiatives (Chapter 5).
3 The range of non-statutory local planning material prepared by local planning authorities either instead of or as well as statutory local plans (Chapter 6).

As manifested in local plans and other statements of policy judged to be 'other material considerations', local planning policy forms one input to the process of development control and the appeals system as administered by local authorities and central government (see Chapter 7). Local planning policy, however, is not primarily concerned to provide a more or less specific aid to decision-making in development control; rather it is concerned to intervene in wider processes of public and private development and land use change so as to influence or manage events in pursuit of policy aims. Development control is an important means to this end. Hence the activity of local planning can only be fully assessed by locating local planning and the development control system that it informs within some understanding of these wider processes of development. This entails a shift in focus, away from considerations of how planning policies are established towards a concern with how far these policies as expressed are or have been successfully carried through to implementation.

This chapter: (1) considers some of the methodological issues involved in assessing the extent to which local planning policies have been implemented; (2) reviews the nature of the development process and the role therein of local planning policy; and (3) within this framework considers how public sector agencies and private sector development interests operate in carrying out land and property development, focusing on the ways in which local planning policies have influenced this activity.

8.2 Policy, plans and implementation

Little attention has been given to exploring and assessing the effects (both intended and unintended) of local planning in general and of statutory local plans in particular. The study by Hall *et al*. (1973a, b; Clawson and Hall, 1983b) is the only comprehensive investigation at the national level of the operation of the British land use planning system. However, this deals with the development plan system as established in 1947 rather than with the present two-tier structure and local plan arrangements, while the main concern is with the overall national consequences of the system rather than with its specific effects at detailed local level. The study thus concludes that the central policy objective of urban containment was achieved, but gives little attention to how this overall policy was expressed at local level or detailed in development plans. Foley (1972) is similarly concerned with the implications for the South East of national and regional policies and strategies for containment and dispersal, dealing with their progression in the context of the 1965 reform of local government in London. A limited number of other writers have focused, often via case studies, on particular substantive policy themes such as Green Belt (Mandelker, 1962; Thomas, 1970; Gregory, 1977; Munton, 1982; Elson, 1986); AONBS (Blacksell and Gilg, 1977; Anderson, 1981), housing land availability (Pearce and Tricker, 1977) and urban renewal (Dennis, 1970, 1972; Davies, 1972; Ambrose and Colenutt, 1975). These accounts have addressed the role of plans where such formal statements of policy were available, e.g. Anderson's study of the Sussex Downs AONB was concerned to establish the extent to which county structure plan policies had been implemented through the development control process. However, the bulk of these studies refer to 'old-style' development plans, town maps and CDAs. Research for the DoE by Healey *et al*. (1982, 1985) is the first study to explore directly (rather than as part of a specific policy interest) the implementation of planning policies contained in structure and local plans; other work is now emerging (Blyth, 1984).

The relative lack of attention that has been given to the implementation of policies set out in and progressed by the 'new' development plan system is partly due to the fact that from the point of view of assessing how policies in plans have been carried through, structure and local plans have become available only relatively recently. It also reflects the initial concern of planning practice with structure and local plan preparation, a concern reflected in practice-oriented research. A further factor is that questions of policy implementation are inherently difficult to research.

The hierarchical model of policy control which has been seen to underpin the development plan system offers one perspective on the relationship between policy, plans and implementation. Policies as expressed in planning statements are successively and sequentially refined down through the levels of the hierarchy. Ultimately these policies are 'applied' via local plans through the process of development control; at this point (or shortly afterwards) implementation 'occurs'. In this view policy is assumed to flow from the top (or centre) of the hierarchy, and implementing agencies are seen as compliant agents for those making policy. Implementation is regarded as an unproblematic technical or administrative process, following on from the setting of policy. It thus appears as a relatively unimportant topic for research when compared to policy formulation and plan preparation, activities located in higher levels of the hierarchy.

An alternative perspective on the relationship between policy and implementation rejects these sequential and hierarchical assumptions as unrealistic, referring instead to an interactive process of negotiation or bargaining between policy-makers and implementing agencies. Implementation is here seen as an essentially political process of bargaining and negotiation which is undertaken within the context of the power relationships between the agencies and organizations involved in order to reach an acceptable compromise or agreement (see Chapter 2). From this perspective, policies lose their previously fixed and absolute characteristics as 'given' inputs handed down from higher levels, and can be seen instead as 'a series of intentions around which bargaining takes place and which may be modified as each set of actors attempts to negotiate to maximize its own interests and priorities' (Barrett and Fudge, 1981, p. 25). Policy is thus flexible and may be adjusted as required in the bargaining process. Policy statements (i.e., plans) may be seen as 'snapshots' of an agency's position on the policies in question at one point in time. As statements of policy positions, plans can become outdated and redundant as an authority's stance on the topics and issues involved shifts. This creates a problem for the hierarchically ordered statutory development plan system, where local plans for instance represent both a means of establishing local land use planning policies *and* a mechanism for implementation as applied in development control. To some extent the tensions that this creates between *current* policy and *former* policy as still extant in plans can be resolved by virtue of the discretionary nature of development control. A further implication is that local planning authorities concerned to maintain an up-to-date framework of plans featuring current policies may favour a non-statutory approach to local planning given the relative procedural complications of statutory local plan review and modification.

Plans and other formal policy statements have been ascribed a variety of roles in the negotiative processes which characterize policy formulation and implementation. Leach and Moore (1979, p. 168) argue in the context of county–district relationships that formal legal or quasi-legal documents and agreements (e.g. structure plans, development plan schemes) have crucial roles to play, both in providing a focus for potential sources of conflict over policies, and as weapons or 'bargaining counters' to be used in playing out the conflict that results. This is not to suggest that such documents inevitably create disagreement, but is to argue that formal policy statements are not neutral documents in the evolution of a county–district relationship. An approved structure plan for instance gives a county a degree of influence over the policy content of district-prepared local plans, difficulties over enforcing conformity notwithstanding. Development plan schemes can be seen as the agreed outcome of a negotiative process determining the number, timing and priority of local plans to be prepared by each tier. In this process, districts usually see themselves as the appropriate authority to carry out the bulk of local planning work. For example, the preparation of an informal local plan by Warwick DC during negotiations with the county over the development plan scheme was clearly intended to demonstrate the district's ability and commitment towards producing local plans for its area (Bruton, 1980). Development plan scheme preparation and agreement is important, since schemes (through their assignment of proposed local plans to local planning authorities) will affect *subsequent* power relationships; 'as much as being about substance, each plan is staking a claim for the power and influence of the authority who prepares the plan' (McAuslan and Bevan, 1977, p. 10). In the context of local plans being used to guide private sector development interests, Healey concludes that policies are primarily conveyed to such interests by negotiation. Policies as expressed in plans may then be used selectively by both planning authority and developer as appropriate e.g. as justification for a particular decision or standpoint. Plans thus 'provide a formal base line for negotiations, while statements within them may be used as counters in bargaining situations'; hence plans as such can hardly be said to 'guide' development (Healey, 1983, p. 228).

The view of plans as policy snapshots highlights the important distinction between policy as embodied in plans and policies as actually being pursued. The two do not necessarily coincide, so it cannot be assumed for instance that statutory development plans represent a full and accurate statement of local planning policies in operation. For studies of the operation of the land use planning system, this raises the problem of identifying policy objectives. As Hall *et al.* (1973b, p. 35) note in connection with the 1947 Act development plans, 'the objectives were by no means clearly spelt out in the plans themselves; it is often necessary to imply them from what is said . . . there may well have been other objectives, which were not written out anywhere . . . we cannot rely wholly, or even principally, on statements in the plans themselves'. Moreover, policies stated in plans may not be seriously followed, while events which have occurred apparently in line with policies or plans might have happened anyway without intervention; also, it is difficult to separate the effects of a policy from those of the plan in which it is expressed.

Further difficulties are created by the way in which local planning policies and proposals are formulated and expressed. Statutory local plans for example often lack site-specific proposals over much of the plan area, favouring instead a variety of plan-wide and area-specific policies (see Chapter 4). In terms of the determination of planning applications, statutory local plans more usually provide sets of performance criteria and other policies rather than a series of proposed site-related uses. A further complication is that policies in local plans are often characterized by deliberate ambiguity or lack of specificity. This may be in order to preserve a certain amount of flexibility so as to cope with uncertainty; to reflect in some 'form of words' the compromise necessary to reach agreement over policies; or to retain a degree of discretion in development control decision-making. Local plan policy ambiguity thus leaves scope for a negotiative relationship between development control officers and applicants. However, where plans do not provide many specific proposals to which development could or could not conform, the relationship between development and policies in plans cannot be considered in a straightforward manner. Studies of the implementation of plans have perforce had to consider 'the extent to which the patterns of development conforms with the policy principles and detailed development proposals expressed in plans, and how far the plans themselves as tools have been instrumental in producing this result' (Healey *et al.*, 1982, p. 80). Outside of relatively simple situations, such as the phasing of land release where both sites and timing are specified, it may be difficult to determine precisely how far or how successfully a local plan is being implemented.

Assessing the effects of plans and of the policies expressed within them is thus sometimes a matter of judgement as to whether a plan's policies have been followed, while there is no guarantee that these policies do actually represent those pursued by the agency in question. However, it should not be assumed that policies in plans and planning statements are *invariably* or *necessarily* different from those being followed by the agency in question. Statutory local plans for instance are widely publicized statements of local planning authority policy; it could be politically difficult for an authority to diverge openly from policies endorsed by public participation and usually by a public local inquiry. Here is a further role for ambiguity in the framing of plan policies, i.e. policies which can bear several different interpretations subsequently allow an authority some freedom of manoeuvre within the boundaries of policy as legitimized in the plan. Additionally, if planning applications for development that conformed to the local plan were to be refused (e.g. following a shift in policy), stated local plan policies could conceivably be used at appeal against the authority's case. Local planning authorities will wish to avoid this and seek to project a consistent policy position. Nevertheless, statutory local plans and other formal statements of local planning policy cannot readily be taken as clear and unambiguous summaries of the policy position of the authority or organization responsible for their preparation. If policy is a series of intentions modified through negotiation, and plans represent 'snapshots' of such intentions, it is more appropriate to focus on the negotiations and bargaining positions undertaken and adopted by the organizations and interests involved. These positions define policies which are occasionally expressed in plans. In turn, these interactions and negotiations occur within processes of development producing modifications and additions to the built environment. It is these processes that local planning policy may be very broadly considered as seeking to influence or manage, with a view to obtaining desired adjustments to the pattern, location and other characteristics (e.g. density, location, design) of the resultant development. Hence, this assessment of the effects of local planning policy on land use and development begins by establishing in general terms the nature of the development process. The ways in which local planning policies have influenced the development activities of public sector agencies and private sector interests are then considered within this context. As a consequence, this chapter has a broad area of concern, in that it deals not so much with the expression of local planning policies or with their application in development control, but with the influence of such policies on a range of different development interests. In particular, in order to reach conclusions about the effects of local planning policies, the chapter necessarily reviews how and why different public and private agencies and interests participate in the development process. The analysis summarizes the various roles of these different agencies and interests within the development process, focusing on the impact of local planning policies and the significance of plans.

8.3 The development process

There are a number of different approaches towards modelling and conceptualizing the process of land and property development. These include: the ecological and economic perspectives proposed by Burgess (1925), Hurd (1924), Haig (1926), Ratcliff (1949) and Alonso (1964), whereby urban land use is seen to be determined by a sorting process involving competitive bidding between activities for desirable locations; behavioural approaches focusing on the decisions and motives of agents in the development process (Rogers, 1967; Denman and Prodano, 1972; Drewett, 1973a), particularly in terms of residential development and the role therein of the pre-development landowner (Clawson, 1962, 1971; Kaiser and Weiss, 1970; Goodchild, 1979; Bryant, Russwurm and McLellan, 1982); and critical analyses of the land development process which situate this activity within a wider account of the economic, social and political organization of society (Lamarche, 1976; Roweis and Scott, 1978; Scott, 1980; Boddy, 1981). At a more descriptive level, a number of case studies of residential (Craven, 1969) and commercial development (Ambrose and Colenutt, 1975) have emerged, while others have worked up classifications of development agents in terms of the objectives of development (Cadman and Austin-Crowe, 1978) or the ways in which interests in land change during the development process (McNamara, 1983). A number of schematic representations of the process have also been suggested in the form of flow charts linking specific events in the development of land (Ratcliffe, 1978; Joint Unit for Research on the Urban Environment, 1977). Variants of this approach incorporate an element of flexibility regarding the sequence of and relationships between events (Barrett, Stewart and Underwood, 1978) and the roles of the agencies involved (Goodchild and Munton, 1985), and thus avoid making assumptions about the precise sequence of events or about the roles of different agencies in the various activities involved in bringing resources together to achieve development. These schemas provide a useful if simplified approximation of the complex interactions which constitute the development process. The process of land and property development subsumes numerous agents and agencies – individuals, groups, organizations and institutions in both the public and private sectors – with widely differing (and sometimes conflicting) attitudes and interests. These include developers and builders (who are dependent on the process), finance institutions (searching for investment yields), and statutory undertakers (whose main function is to provide a service); in short, 'the development process is characterized by ambiguity of interest, by inter-organizational overlap, and by complexity' (Barrett *et al.*, 1978, p. 37). There is no necessary correspondence between role and agency, so that 'the roles played by each actor can vary greatly in the development of land' (Goodchild and Munton, 1985, p. 89). In relation to the industrial development process, for instance, Boddy and Barrett (1979) show that the relationship between roles and agencies is complex, with no simple correspondence between the two e.g. local authority roles in industrial development variously encompass planning functions, infrastructure provision, land purchase and site assembly, provision of buildings, related financial assistance, and development partnerships. Barrett and Whitting (1983) suggest that the relationship between an agency and the roles it performs is not static but evolves over time, so that roles in the development process cannot be definitively assigned to the public or private sectors – public agencies may be carrying out any or all of the functions necessary to achieve development. Thus, while the development process can be seen as a series of activities or roles serving to bring together various resources (including public sector powers to intervene in the process as well as finance, land, materials, labour and professional skills), the agencies involved here are to a greater or lesser extent substitutable for one another in terms of the roles they pursue. At the same time, since resources necessary for the achievement of development are controlled or possessed by different agencies, a situation of mutual dependence arises between agencies (Barrett and Whitting, 1983). The relative degree of dependency on particular agencies is determined by the extent to which the resources they possess or control can be substituted from elsewhere. Hence the local planning authority is in a relatively strong position given its control over development rights afforded by town and country planning legislation, while the dependence of a development company on a particular landowner is influenced by the availability of similar (and therefore substitutable) land in the 'area of search'. The way in which the relationship between a given agency and roles in the development process works out in practice is related to: (1) *scope for action*, represented by control or possession of resources; here, central government prescribes certain roles to public agencies while effectively limiting and regulating the activities of public agencies; (2) *attitudes to risk and reward*; and (3) *agency interests and objectives*, defining the extent to which it needs or desires to become involved in the development process.

These factors determine *who* does *what* in the development process, so that different groups of agencies may be classified on this basis (Table 8.1). To summarize, 'involvement in the development process comes about either because an agency seeks, or is given, a role, or because the development process offers one way (amongst several) of furthering its own objectives' (Barrett and Whitting, 1983, p. 15). It must be stressed that there is no *necessary* correspondence between role and agency or sector, so that for example the important 'developer' role can be undertaken by private development companies, local authorities or other public agencies established for the purpose. The *objectives* providing the rationale for intervention or involvement, however, do vary between agencies *and* between sectors. The 'developer' role is carried out by private sector interests seeking profit maximization, by local authorities in order to meet their own development needs or in pursuit of wider local policy aims, and by public development agencies in fulfilment of responsibilities set by statute and national policy. Development (that is, the *promotion* of development) is also undertaken by industrial and commercial companies seeking to provide property for their own occupation, by construction companies extending their contracting role so as to include the risk of development in pursuit of profit and possibly other objectives (e.g. continued employment for labour force), and by financial institutions moving directly into property development in order to meet at least partly their investment objectives. An associated point here is that the *identities* of agencies (particularly those in the private sector) can overlap; for example, the financial institutions have acquired substantial shareholdings in property companies, while property companies may take on the investor role and seek to combine within their portfolios both profits from development and the creation and holding of property investments (Cadman and Austin-Crowe, 1978). Taken as a whole, the development process incorporates a complex range of positions,

Table 8.1   *Agency roles, interests and objectives in the development process*

Agencies	Role in process	Agency interests and objectives
1. Private		
Property companies and housebuilders	Entrepreneurs seeking to co-ordinate or undertake development	Maximize profits: dependent on process in long-term but typically only short-term interest in product: various strategies for dealing with risk
Financial institutions*	Supply development and investment funding for property development	Liabilities determine investment requirements (yield, liquidity, security): aim to match liabilities with appropriate investments
Construction companies, contractors	Construction	Dependent on process in long-term but otherwise short-term provision of service: financial return; diversification to reduce risks
2. Public		
Statutory undertakers	Infrastructure provision (highways, gas, electricity, drainage, water supply and sewerage)	Fulfil agency duties and functions as given in legislation: servicing of *new* development may be relatively low priority: also undertake development to meet own requirements (as do the nationalized industries)
Public development agencies**	Undertake/facilitate development in accordance with policy initiatives	Again aim to fulfil duties and functions as set by statute: more direct interest than statutory undertakers in progressing development
Local planning authorities	Regulation of development and other use of land	Conformity of product of development process to planning policy
Local authorities as developers	Undertake or facilitate development	In relation to wider statutory roles (e.g. as housing authority) leading to development needs/ requirements, or in pursuit of other policy aims (e.g. economic development) rather than financial return *per se*

Source: Adapted from Barrett and Whitting (1983).
*Insurance companies, pensions funds and banks.
**For example new town development corporations, UDCs, Land Authority for Wales, Scottish and Welsh Development Agencies.

interests and roles, with considerable variety in the objectives and roles of agencies *within* the public and private sectors as well as *between* these sectors.

This view of the development process, as comprising a wide variety of public and private agencies each concerned to achieve or further its own objectives and interests within the legislation, rules and requirements prescribed by central government, begs the further question of how and on what basis inter-agency relationships are established. This is an important issue given the degree of dependency between agencies arising from their differential access to or possession of the necessary resources to allow development to proceed. Barrett and Whitting (1983, p. 16) suggest that the actual development process can be seen as a series of interactions and transactions taking place between agencies, a process defined as:

negotiations taking place crucially over the sharing of *rewards* (or benefits) to be obtained from the outcome of development in relation to the contribution of *resources*. The incidence of risks and costs will be crucially determined by the relative balance of bargaining power ... those upon whose resources the process most depends will be in the position of greatest bargaining strength.

This interpretation of the development process follows the general arguments of Bacharach and Lawler (1981, p. 60) that bargaining power (which is rooted in the resource context) is based on an opponent's dependence; in particular, 'an increase in the dependence of bargainer A on opponent B increases opponent B's bargaining power'. For Bacharach and Lawler, bargaining is a process of exchange in which the parties need resources controlled by each other; the bargaining power of an agency is grounded in the other's dependence on that agency for resources. In turn, the other's dependence derives from the availability of similar (substitutable) outcomes from other relationships, and from the value (or commitment) given to a particular outcome. In the development process, an agency's control of resources confers bargaining power to the degree that these resources are not readily substitutable from elsewhere, and this will be reflected in the end result of the bargaining process. For example, the *financial institutions* possess substantial bargaining power as a consequence of their control over much development and investment funding and the fact that they may take such finance elsewhere (other development projects or other investment sectors) if the terms that are available (rates of return, levels of risk) are considered unsuitable when judged against their investment requirements and the other investment opportunities that are available. In the public sector, the *statutory undertakers* have control of an important development resource, infrastructure, on which others are heavily dependent; yet they themselves are dependent on other agencies to only a relatively limited degree in so far as meeting their own objectives is concerned. The consequent bargaining strength of these agencies in the development process means that the availability of services (particularly drainage/sewerage) is of critical importance to the viability of individual schemes and proposals. Barrett and Whitting, drawing on Strauss (1978), emphasize the importance of the contexts within which negotiations take place – these limits are set by such contingent factors as the scope of an agency's operation, the economic context for development, the agency's own objectives and interests, available bargaining power, and negotiating experience and precedent (the results of previous negotiations). These limits vary over time and are also subject to negotiation at a 'higher' level, e.g. the legitimate scope of a public agency's operation can be extended or reduced through a change in legislation or in the more detailed rules and requirements prescribed by central government.

If this view of the development process – as a complex series of transactions between agencies pursuing a range of objectives and interests, regulated by negotiations and bargaining over access to a variety of resources – is accepted, what are the implications for local planning authorities seeking the implementation of local planning policies? Resource ownership and issues of substitutability determine bargaining strength, which in turn is reflected in the relative incidence of risks, costs and rewards and in the extent to which an agency's objectives can be achieved. Local planning authorities possess a considerable degree of bargaining power, derived from their control over development rights conferred by the *Town and Country Planning Acts*. This power is partly contingent upon the support their development control decisions to refuse planning applications receive at appeal. The blunt veto power of refusing a planning application is supplemented in practice by informal negotiations over proposals, in which the planning authority (with the sanction of refusal in the background) seeks to influence schemes in pursuit of its policy objectives; by the use of conditions on planning permissions; and by formal planning agreements under Section 52 of the *Town and Country Planning Act 1971*. In addition, local authorities have become more actively involved in the development process, carrying out development either themselves or in partnership with private developers (see Section 8.4.3 of this book). This role implies accepting a degree of risk, which must be measured not simply against financial return but also against the importance attached to the policy objectives involved and the requirements of public accountability. Further constraints on the local authority developer role are the availability of powers and finance; successive reductions in public expenditure have effectively increased local authority dependence on other agencies in so far as they are interested in the active promotion or control of development. This suggests that if local authorities are to continue to intervene in the development process, they will increasingly need to bargain and negotiate with the resources that are still available to them (i.e. regulatory powers of development control, but also including powers of compulsory purchase) as they pursue the implementation of policy objectives (Barrett and Whitting, 1983). An additional effect will be to emphasize the basic (low risk) statutory aspects of local authority involvement in the development process (e.g. development control) at the expense of other (more costly) activities such as promotion or development.

Given this view, many traditional planning values (consensus, rationality and comprehensiveness) and practice (such as the production of plans) must be recognized as being inappropriate in some circumstances. Local planning authorities may on the one hand be seeking to prepare a logical and comprehensive framework for the control of change while on the other they actively intervene in development processes in an entreprenurial fashion. This is to recognize once again that as well as being producers of plans, planners are also managers of change. As such, planners seek the resolution, in furtherance of local planning policies, of the conflicts of interest inherent in the development process. This role necessarily involves negotiation, bargaining and compromise, as well as plan preparation to support policy positions. The various plans and other statements of intent such as planning applications produced during the development process represent formal statements of the bargaining positions adopted by the organizations and agencies involved. Development plans additionally represent a *resource* for the local planning authority in so far as they are used to support development control decisions; they may also be used selectively by applicants. Bargaining positions however are not immutable, e.g. they can be adjusted in order to allow an acceptable compromise to be reached. Formal planning statements recording such positions cannot therefore be regarded as full and accurate accounts of the policies that are or will be pursued by the organizations concerned.

8.4 The public sector

8.4.1 Public sector roles in the development process

Public sector agencies in Britain are engaged in the provision of a large number of services, goods and facilities. These include the provision of public housing for rent and of other community and social facilities; energy production and distribution; the provision of transport services; and the maintenance and extension of basic infrastructure networks. In turn, these activities contribute to change in the local environment and have a bearing on the development and other use of land, and so interact with local planning policies formulated by the local planning authority. The significance of this relationship for the implementation of local planning policies varies according to the importance of an agency's activities in the development process, particularly in terms of the degree of control they possess over key resources. For the public sector as a whole, these 'development' oriented roles are diverse, including land assembly, infrastructure and service provision, the regulation of development, and direct development in order to meet service, administration, production and distribution needs. Agencies may pursue a single role (e.g. the Land Authority for Wales: land assembly and supply), carry out a variety of divergent activities (e.g. the local authorities are involved in the formulation and progression of planning policy as well as direct development to provide for instance highway, education and social services facilities), or be responsible for a number of more closely related functions (e.g. the Welsh Development Agency: the development of industrial property, industrial investment and promotion, derelict land reclamation). The scale of operation varies from local (district and county councils) to regional (the Land Authority for Wales and the Welsh Development Agency) and national (statutory undertakers and nationalized industries, although these often have significant regional level 'area' organizations). As far as the influence of the activities of public organizations on land use and development are concerned, two broad sets of roles can be distinguished: (1) *the provision and maintenance of infrastructure*; and (2) *land and property development and land holding practices*, related to the carrying out of statutory responsibilities and policy decisions.

It must be emphasized that these two categories relate to development- or land-related activities rather than to individual organizations, in that agencies can, and do, carry out roles falling within both groups. The first category, infrastructure provision, is important in achieving many local planning policies, particularly those allocating (unserviced) greenfield sites for urban growth and policies for urban regeneration. The need here is for co-ordination of objectives and implementation programmes. The second category is a broad grouping embracing significant variations in the nature and purpose of involvement in the development process. Such development activities are carried out, for example, by local authorities and health authorities in order to meet community needs through the provision of housing, health, education and other facilities; by the nationalized industries and statutory undertakers so as to meet their land and property requirements arising from production, distribution and administration; and by bodies such as the Welsh and Scottish Development Agencies with an explicit concern for carrying out land and property development as the term might be understood in the private sector, e.g. the provision of advance factory units. In local planning terms, roles within the second category raise the issue of securing effective integration between land use policies formulated by the local planning authority and other public development schemes and land holding practices. Public inquiries have been used in an attempt to clarify and reconcile conflicts between large-scale service development proposals and land use planning objectives, e.g. between energy supply and landscape conservation in National Coal Board proposals to carry out coal-mining beneath the Vale of Belvoir. While public inquiries into specific schemes may adequately deal with such matters as environmental impact, it is unclear how they can satisfactorily address the wider policy questions which are often raised, particularly as these commonly involve the 'national interest', however defined. Land use issues also arise through the rationalization of public sector services leading to redundant or surplus land and property; the availability of land in this case depends on its release by the public agency concerned.

This raises a further aspect of the involvement of public agencies in the development process, namely the public ownership of land, necessarily required in the course of the provision of goods, services and facilities. Nationally, between 12 and 18.6 per cent of the land surface of Great Britain is estimated to be in the freehold ownership of public authorities (Dowrick, 1974; Massey and Catalano, 1978), although recent pressures from central government encouraging the disposal of public land and property assets[1] are likely to lead to a reduction in these aggregate figures. For individual cities, the proportion of land owned by the local authority is reported as being between one-third (e.g. Plymouth, Birmingham) and two-thirds (Manchester) of all land within the municipal boundaries (Dowrick, 1974; Manchester and Salford Inner Area Partnership Research Group, 1978); much of this publicly-held land undoubtedly relates to slum clearance and the provision of council housing. Other public agencies, especially the nationalized industries and the statutory undertakers, are also extensive landholders in urban areas and particularly in the inner cities (Royal Town Planning Institute, 1979). As a result, inner city land transactions usually involve the public sector as buyer or seller (or both). These points are important to the extent that public agencies (including local authorities) control sites allocated for development in local plans. It must not be assumed that the public ownership *per se* of allocated sites readily ensures that such sites are available – for example, apparently unused land held by British Rail may be part of a long-term operational reserve and therefore not surplus to requirements. Moreover, even if the agency concerned is willing to dispose of sites for development (or to participate in such developments) the approach taken to disposal or partnership arrangements will not necessarily be radically different from that of a private sector owner. Thus, public bodies are required to dispose of their holdings at the best price that can be obtained, and many regard surplus land as an asset which can be used to generate capital funds with which to support their main activity.

8.4.2 Infrastructure provision

Infrastructure can be considered as the networks of physical services upon which new development is dependent – gas, electricity, telecommunications, water and sewerage, and highways. Different public agencies are responsible for each of the physical services, their duties encompassing not only the extension of existing networks to meet the needs and demands of new developments, but also network maintenance. The expansion of the provision of services may be of only relatively low priority in the context of their overall operations. For instance, many of the investment projects of water authorities are concerned with the improvement or renewal of existing water and sewerage networks rather than with the servicing of new development. Development proposals in structure and local plans, particularly those allocating greenfield sites for urban expansion, have obvious implications for future infrastructure requirements. Land use planning allows the control of development so as to utilize existing infrastructure more efficiently (e.g. avoid ribbon development) and so as to co-ordinate new construction with the provision of new infrastructure. Land release for new development must perforce have regard to the adequacy of existing infrastructure and to the ease with which it can be extended, the absence of services being a valid reason for the refusal of planning applications on grounds of prematurity (Grant, 1982, p. 378). However, the lack of any formal links between local planning authorities and the major infrastructure agencies raises problems of integration. As Grant notes, either planning decisions may come to be effectively dictated by the capital programmes of other public agencies, particularly the water authorities, over which the planning authority has no control, or planning choices may have unanticipated implications for the expenditure programmes of other agencies. In practice these relationships remain ambiguous, essentially dependent upon informal co-operation. Organizational theories have been used to describe the interdependency between planning authorities and infrastructure agencies (Friend, Power and Yewlett, 1974; Payne, 1978; Hickling, Friend and Luckman, 1979).

Of particular importance among the infrastructure agencies are the water and highway authorities, for the provision of these services requires relatively substantial physical works which in both cases are essential for development to proceed. By contrast, electricity for example can be supplied comparatively cheaply, and alternative sources of energy supply are available. In England and Wales there are ten regional water authorities who are responsible for water supply and the provision of sewerage and sewage disposal facilities. Water authority objectives give priority to the servicing of new housing and industrial development above for example pollution control for the improvement of water quality (Payne, 1978), but these are only broad guidelines. Water authorities must be financially self-supporting; close financial control by central government has in recent years led to substantial reductions in capital investment, and hence in the capacity of water authorities to meet the demands of urban growth. However, water authorities can be requisitioned by both local authorities and private developers to provide sewerage and water supply for domestic purposes. This is not to say that this is the normal procedure for the servicing of new development. The courts have limited the use of requisitioning as applied to sewerage by insisting that the power may only be exercised once planning permission for the development has been granted; planning authorities may thus continue to take existing capacity and capital expenditure plans into account, via consultations with the water authority during the preparation of plans proposing land release and on particular planning applications (Grant, 1982, p. 380). Significantly, however, water authorities have no power of veto as regards the granting of planning permission. In this respect the bargaining power of water authorities in the development process, stemming from their control of necessary infrastructure, is circumscribed. As a consequence, water authorities may be concerned to ensure that local planning authorities do not release land that will be expensive or impracticable to service, and this may lead to what Grant terms an 'unofficial embargo' (*ibid.*, 1982, p. 73) on new development in areas where existing capacity is inadequate.

Concern has been expressed that the provision of water supplies and sewerage by water authorities and the planning policies of local planning authorities are not fully integrated, leading to the danger that water authority priorities (based on financial considerations) will prejudice land use planning decisions based on wider social considerations (Payne, 1978; Parker and Penning-Rowsell, 1980). It is likely that this fear initially reflected the disruptive consequences of local government reorganization in 1974 and the creation at the same time of the regional water authorities from some 1600 different agencies, including local authorities, water companies, sewerage boards and river and drainage boards. More recently, the relationship between water authorities and local planning authorities appears to have stabilized, presumably because mutual inter-dependence means that both sets of authorities can derive some advantage from co-operation in anticipating development and consequent servicing requirements. The issue of integration however continues to be raised (Penning-Rowsell, 1982) in the context of potential discrepancies between the development aims of county council structure plans and the local plans of the constituent districts. Water authorities have a statutory duty to have regard to structure plans in drawing up their investment plans, and undoubtedly use local plans as one way among others (e.g. informal contacts with district councils) of building up a picture of where future development is likely to take place. The use of emerging and published development plans as a source of information in this way emphasizes not so much the availability of plans as such but of access to information about the current state of commitments and the policy intentions of planning authorities.

County councils in England and Wales are the designated highway authorities, having responsibility for capital investment in all public roads other than those dealt with at national level by the Department of Transport (e.g. motorways). Like the water authorities, highway authorities are consulted during the preparation of development plans and on certain planning applications. The response to particular development proposals usually deals with relatively detailed questions of site access, safety and layout, where technical standards of highway design and parking requirements can be applied. Highway requirements can usually be accommodated in a scheme through negotiation. However, there will inevitably be occasions when development proposals necessitate substantial off-site capital investment such as the provision of new roads or highway improvement. This is a matter for the highway authority, there being no power for the requisitioning of new roads, although developers can make financial contributions towards the cost of off-site works undertaken by the authority in connection with new development. As with developer contributions to the cost of the provision of water supply however it is the initial developer, whose proposals and contributions unlock the area, who bears this burden – rather than all developers benefiting from the works (Grant, 1982, p. 381).

Structure plans appear to have been less than wholly successful in providing a wider context for highway capital investment given the timescale of plan preparation and the tendency to avoid explicit statements of priorities and hence controversial political questions. However, Hickling *et al*. (1979) consider that there is an emerging role for local plans as policy documents which reduce uncertainties about relative priorities for local road improvements and bypass schemes (rather than providing information on new claims for investment). Here, the emphasis is on continuing access to relevant data at all stages during the preparation and subsequent use of development plans, 'as opposed to whatever "snapshots" of land use commitments and predictions may be available for inclusion in the formal documents at the moment of their publication' (Hickling *et al*., 1979, para. C.6.4). In other words, highway authorities require information on current land use planning policies as actually being pursued rather than policies as given in plans. For the local planners, the production of plans as 'snapshots' is difficult where the highway authority refuses to commit itself to the timing or, in the case of the provision of new roads, to the location of such investment. Here the highway authority is no doubt attempting to avoid premature commitments or a reduction in programme flexibility at a time of reduced public expenditure. Where the precise route of a firm road proposal is not known, one solution for local plans is to identify by means of a site-specific proposal an area of land which will be safeguarded as an 'area of search' (DoE, 1984g, para. 4.25). The key problem is that highway authorities (and indeed water authorities) are responsible for the infrastructure needs and demands of a number of local planning authorities, so that individual districts and (in the case of water) counties must convince the infrastructure agencies of the legitimacy of their claims for investment. Statements in plans offer local planning authorities one way of attempting to attract scarce public resources to their areas, but in total this does not amount to a rational method for distributing capital infrastructure investments.

Research into the relationships between the development plan system and investment programmes emphasizes the basic interdependencies involved (Hickling *et al*, 1979). Infrastructure agencies see current planning policy intentions as a source of information to use in making decisions about resource allocation. Local planning authorities require current information about the progress of investment projects in order to make more confident policy decisions about the future development and use of land. Both sets of agencies are attempting to use information from the other in order to reduce uncertainty. For the infrastructure agencies, this uncertainty is primarily economic in nature, while planning authorities face political risks in that all relevant interests must be given an opportunity to speak prior to the formal establishment of planning policy. A view of these relationships between infrastructure and land use planning agencies as continuous processes of management is more appropriate than the linear view of the plan-making process, where the authoritative 'master plan' is implemented through a series of public and private investments. Published plans are thus not to be seen as comprehensive statements of future intentions or development patterns, but according to Hickling *et al.* (1979, p. iii) as:

only one form of output, among many other tangible and intangible products, from a process in which many kinds of activity are intertwined – activities of monitoring, policy review, issue identification, work scheduling and day-to-day operations. It is a process which generates a continuously evolving body of commitment to policies and actions, and which exercises influence over events in a variety of subtle and often gradual ways.

One particular form that this management process may take is illustrated by the Berkshire Development Programme, which operates informally as a liaison and information service for the agencies involved in order to improve response to unforeseen circumstances in the development of land (Patterson, 1978). The Programme explicity avoids interfering in the decisions of each agency about spending priorities, and does not attempt to formulate a written list of priorities. Rather, the 'output' of the Programme is knowledge as to the intentions of the other participants: 'its results are judged in terms of increased perceptions of problems and opportunities and the better co-ordination of timing capital projects' (Patterson, 1978, p. 58). The Programme seeks to improve available information on the timing and phasing of development in the County as a whole, to assist capital programming and infrastructure co-ordination, and to help local planning authorities in the implementation of structure plan strategies and the avoidance of unplanned land releases; participants in the exercise are reported to find the Programme and the informal contacts which result very helpful (Stoddart, 1983).

To summarize, this discussion of the relationships between local planning authorities and infrastructure agencies emphasizes the following:

1   That these links are founded on a situation of mutual inter-dependency between agencies as to the provision of information which can reduce the uncertainties involved in the operating context of each organization.

2   That relationships are typically not based on plans but take the form of continuous interaction, both formally and informally, through a variety of channels; local plans may record or formalize agreements reached, or represent promotional statements.

3   That local planning is as much about the management of change as the production of plans or the control of development; here, the key roles are organizing and providing leadership and a forum for integration as well as planning and controlling in the widest sense.

The complexity of providing infrastructure to allow land development to proceed in accordance with local planning policy has been managed in practice by the evolution of relationships between the agencies involved which are based on continuous interaction. These relationships have been set within the guiding framework of the policy hierarchy, which leaves considerable scope for such negotiations to proceed. Local planning is able to perform a useful function here in co-ordinating and managing these relationships and the resources for change that are involved. However, this optimistic conclusion as regards the roles of local plans and planning in integrating public sector infrastructure provision must be tempered by considering the extent to which other aspects of public sector land- and property-development activities have been influenced, controlled or in any sense managed by local planning authorities.

8.4.3   Public sector land and development interests

As indicated above, the land holding and property development activities of public sector organizations are diverse, including the following:

1   The provision of social and welfare facilities such as health care, education, social services and housing to supplement private sector forms of these services.

2   The necessary holding, development and disposal of land by the statutory undertakers and the nationalized industries.

3   The land and property development roles of various public development agencies, where involvement in the development process is a central concern rather than a by-product (however important) of some other activity.

8.4.3.1   *The provision of social and welfare facilities*

There are potential roles for local planning and local plans in allocating sites for the future development of social and welfare facilities, and in proposing new uses for redundant or surplus land and property. In practice such integration is inhibited by the sectoral structure of both central and local government, which encourages public agencies to plan and establish their own priorities for expenditure; local planning authorities have little effective influence over these processes.

Health authority forward planning, for example, reviews existing services and resources in establishing the need for additional premises. Resources include existing National Health Service landholdings, which are kept under review having regard to both changing circumstances (for instance, the rationalization of holdings made possible by the completion of new capital developments) and economic factors (in particular, as a means of raising additional capital funds as well as reducing estate management costs). This latter point is of increased significance in times of reductions in public expenditure, recent statements by Ministers making it clear that further growth of the National Health Service is dependent entirely on greater efficiency and capital receipts from the sale of redundant land and buildings.[2] Internal Department of Health and Social Security/Welsh Office memoranda[3] require health authorities to follow three main principles in land transactions:

1 To assess the potential of existing stocks so as to secure maximum return, including capital income from sale proceeds.
2 Not to retain land against remote contingencies, but to release land surplus to current or planned service needs.
3 To keep land purchases to a minimum consistent with operational needs.

Before acquiring land, health authorities should ensure that planning clearance has either been obtained or will be forthcoming for the proposed development. Government departments are entitled to Crown exemption from the provisions of the *Town and Country Planning Acts* in respect of development carried out for their functions, but a consultative mechanism exists whereby such bodies consult local planning authorities about proposed developments (DoE, 1984f). In undertaking land disposals, health authorities consider certain priority interests (alternative NHS uses, complementary health care services, other government departments, local authority social services departments, and other public interests), while the interests of the former owners of the land are also taken into account under the 'Crichel Down' procedure.[4] In any event, the best possible price must be obtained. An incentive to securing disposal is of course provided by the ability to use such receipts in boosting funding of the service. Overall, it is health authority forward planning strategies (incorporating service development, economic and 'commercial' factors) which determine land acquisition, development and disposal, rather than the planning policies progressed by the local planning authority. Local plans can include proposals identifying sites for future health service uses where a need has been expressed, but this merely signifies in advance that the consultative procedure for obtaining planning clearance will proceed smoothly on these sites so far as the planning authority is concerned. Such allocations say little about the timing of land acquisition or development, which will be dependent upon the health authority's assessments of priorities for service development.

Local authorities in England and Wales are largely responsible for the provision of public housing, education and social service facilities, these functions being undertaken by the district authorities except for education and social services, which outside the metropolitan areas are county responsibilities. Sectoral organization is evident in that different departments within local authorities are usually responsible for these different services, although local authority land and property holdings are commonly overseen and administered by estates departments. The forward-planning of service development is usually undertaken within the department concerned, with other departments (estates, planning) becoming involved when the land use implications are sufficiently clear to permit site search and land acquisition and disposal. Land and development requirements are usually met through 'the interaction of local political priorities among councillors, professional views of the way the services should be provided and the availability of central government loans and grants' (Healey, 1983, p. 221). Again, local plans have a peripheral rather than central role to play in this process. Reductions in local authority expenditure have made it difficult for local plans to produce programmed summaries of future municipal land and property requirements, either because the spending departments involved have not yet produced programmes for inclusion in land use plans, or because they wish to retain flexibility by avoiding commitments in publicly available documents. A local plan can attempt to safeguard a site for future school development by allocating it for educational use, but risks being left with undeveloped allocated land if for some service reason (falling school rolls, capital cutbacks) the development is not implemented. Planning departments have typically only limited influence over the intentions of other departments as to site acquisition, development or disposal, reflecting their lack of control over resources for 'implementation' and the political and financial strength of the main service departments.

Some of the most trenchant criticisms of local authority involvement in the development process concern slum clearance and comprehensive redevelopment, which have been undertaken in many major British cities since the second world war. Here, the major land use involved was housing, with ancillary social and highway development also typically being proposed. Bedevilled by financial restrictions, procedural delays and difficulties of inter-agency co-ordination, the ideal of carefully phased programmes of housing acquisition, clearance and redevelopment was hard to achieve, resulting in large tracts of publicly held vacant land in the inner cities and long timescales for the completion of the process (DoE, 1975d; Milligan, 1972; Muchnick, 1970; Nicholson, 1979; Norman, 1972; Stones, 1972, 1977). Muchnick's pluralist analysis, for example, emphasizes the fragmentation of agencies involved in comprehensive redevelopment in Liverpool, which as a consequence encompassed a variety of interests, priorities and proposed courses of action, the sectoral committee and department structure of local government militating against effective co-ordination. Attempts to develop a comprehensive approach by the planning department were constrained by competing interests, most notably those of the housing department and the Public Health Inspector, so that in practice key aspects of renewal took place largely independently of planning proposals (Norman, 1971; Paris, 1974). For this reason, charges that land use planning was solely responsible for the problems of comprehensive redevelopment are somewhat wide of the mark. The planner's role in other aspects of post-war urban development in Britain can also be seen as marginal, e.g. high-rise housing provided by local authorities to rehouse people from slum dwellings reflected the demands of the building industry and the support given to those demands by central government through the structure of subsidies paid to local authorities, rather than a series of land use planning decisions made at the local level (Pickvance, 1977; Dunleavy, 1977).

Away from large-scale comprehensive redevelopment, which is a special if significant case, problems of integrating local authority social and community services remain. In the context of the inner areas, these difficulties have led to calls for a 'total approach' to co-ordinate the policies of central government and local authority departments with private sector and voluntary interests (DoE, 1977a). On a more limited basis, advocates of area management propose decentralized administration of services, together with an attempt by the local authority to consider (and, if necessary, to improve) 'the degree of fit between local needs and priorities and the local effect of diverse policies and procedures' (Hambleton, 1976, p. 177). Hambleton suggests that local planning and area management could be combined in a single area-based approach encompassing both the planning and implementation of local authority policies to the area in question, or at least that there could be close collaboration between the two. An obvious limitation on this approach is the long-standing restriction on the content of the statutory local plan written statement to 'proposals for the development and other use of land for the plan area' (DoE, 1984g, para. 3.22). Pointing to growing concern about the limitations and relevance of the official view of statutory local planning, Hambleton argued in 1976 (p. 176) that 'many of the most promising experiments in area-based policy-making and management seem to be happening despite local planning and not because of it'. For instance, Haringey LB's experiments with area-based approaches to local planning through a 'district planning process' have made use of non-statutory documents capable of (*a*) encompassing a concern with both land use and development and an interest in co-ordinated local authority management, and (*b*) responding to rapidly changing events and taking into account changes in policy (Frith, 1976; Field, 1984). In Haringey area-specific non-statutory 'district plans' were envisaged as vehicles for 'across-the-board' local authority policy-making, but in the event their preparation served to highlight the need for an intermediate strategic framework to supplement the general policies of the GLDP. This has been provided in the form of a statutory borough plan, which as originally conceived was nonetheless 'designed to provide a framework for later district reviews which would not only be comprehensive area management tools extending beyond the scope of the statutory procedures, but also serve as a basis for monitoring the borough plan' (Field, 1984, p. 77). Field suggests that the Haringey approach can be considered to be a success given the completion of the statutory borough plan; the establishment of a machinery for local planning/area management, and the meeting of professional ideals for a wider role. In general however, the sectoral organization of central and local government, backed up by bids for resources on an intra- rather than inter-sector basis (e.g. TPPs, HIPs), must militate against the integration of service provision for local areas – whatever innovations in statutory or informal local planning the local planning authority can offer.

8.4.3.2   *The statutory undertakers and the nationalized industries*
The relationship between local planning proposals and the activities of the statutory undertakers and the nationalized industries in the development process is also likely to present problems of integration. These organizations, created by statute to operate major industries or services, are large single-purpose undertakings operating on a national or regional, rather than local, basis. They have broadly similar powers to acquire, hold and dispose of land for purposes connected with the discharge of their functions. However, these land and property requirements vary according to wider factors such as the functions and responsibilities set by statute; the operating environment, including demand for goods and services; various technical production or distribution factors, typically unique to each agency or industry; a range of financial considerations; and the policy contexts set by central government. Each of these factors is of varying significance for the landholding and development activities of individual agencies, so that their influence cannot easily be generalized. Moreover, the landholdings of different organizations are not of equal significance in terms of their role within the overall enterprise, while the management and disposal policies adopted with regard to surplus landholdings vary, as does the manner in which these surpluses arise. On the other hand, development proposals arising from statutory roles are often large-scale and pose special problems (such as environmental impact) for local authorities, particularly where the service agency concerned is able to claim that the national interest is involved. For local planning authorities, the key problem is that the statutory undertakers and the nationalized industries primarily hold and develop land in order to carry out a statutorily set activity, and define their priorities for land management accordingly rather than with regard to relatively small area local planning concerns.

For the National Coal Board, for instance, the activity of mineral extraction means that land must be held in reserve pending mining operations, which continually create worked-out surplus. Land becomes redundant not only because of the closure of mines in response to wider economic circumstances or government policy, but also because the land is simply worked out. The aim with regard to such redundant land is freehold disposal in order to generate capital funds with which to support the main activity. A similar concern underlies the operation of the British Railways Property Board, where surpluses arising from the rationalization of the rail network provide both capital and revenue returns through sale, letting or leasing for development in partnership with private developers (Dashwood, 1980). The British Gas Corporation maintain a gas supply network for the distribution of North Sea gas; growing demand has led to network expansion and a continuing programme of land acquisitions, typically in the form of relatively small sites located both within, and on, the outskirts of urban areas. In the long term, the inevitable depletion of North Sea gas reserves will mean a change in gas sources; current plans envisage the production of substitute natural gas as a suitable replacement. This will enable the existing distribution network to be used, but may lead to further substantial acquisitions for the siting of large-scale gas production plants. British Gas has been accused of retaining former gaswork sites (used for the local production of gas prior to the advent of supplies from the North Sea) to meet these future needs at low cost;[5] certainly current profit levels mean that the incentive to dispose of surplus holdings is reduced. Finally, the landholding policy of the British Steel Corporation similarly cannot be separated from the overall financial position of the industry, but with different consequences. In the long term for instance the expectation is of continued contraction in the face of international recession and import penetration (Cooke and Rees, 1981), while short-term losses have encouraged the disposal of land and property on a considerable scale e.g. such disposals realised £22 million in 1980–81 (British Steel Corporation, 1981). The closure of steel plants in particular localities poses a considerable challenge to the local planning authorities concerned; other public agencies may become involved as new landowners and as providers of infrastructure to allow redevelopment to proceed, and there is a role here for statutory local plans in providing a land use framework e.g. Cardiff City Council's East Moors Local Plan.

For the statutory undertakers and the nationalized industries, landholdings are typically viewed either on technical criteria if classed as operational or on commercial criteria if regarded as redundant. Land does not have to be in use to be part of a genuine operational reserve, but the drawing up of registers of vacant or underused public sector land under the *Local Government, Planning and Land Act 1980* will tend to discourage the retention of land against remote contingencies. Viewing landholdings as commercial assets can be expected to have a similar effect. However, neither technical nor commercial criteria for land management necessarily imply a concern for the implementation of local planning policies and proposals, while the scale of new development projects and the local economic significance of plant closures and service rationalizations mean that it is difficult for local planning authorities to devise an equivalent and appropriate response in the time available.

8.4.3.3 *Other public agencies*

Finally, there are a few examples of public sector agencies which carry out land and property development not as a necessary pre-condition for the production, supply or provision of goods and services (e.g. site acquisition and plant construction in order to produce steel; power station development in order to produce electricity) but in pursuit of wider policy objectives or statutory roles. The clearest instance here is the new town development corporations, which have actively participated in the process of development within their designated areas, providing residential, industrial, retail, commercial, and community facilities. However, this development role is carried through largely independently of the local planning authority, and is co-ordinated by the non-statutory master plans prepared by corporations rather than by local planning proposals. Similar comments apply to the activities of Urban Development Corporations (see Chapter 5). Local authorities may also act as developers, having been extensively involved in town centre retail and commercial redevelopment schemes, site assembly and land supply, and the production of factory units, as well as their provision of social and community facilities discussed above (for general reviews of local authority involvement in the development process, see Boddy and Barrett, 1979; Barrett and Whitting, 1983). As Ratcliffe (1978) notes, the extent to which authorities become involved in different forms of development naturally varies between authorities according to such factors as geographical location, local problems, resources and opportunities, and political composition. Such activities, which typically require the public ownership of land, represent an interventionist 'positive planning' by which authorities can supplement their regulatory powers of development control. The technical arguments for undertaking positive planning are powerful, but the issue remains controversial for political and financial reasons, while it is unclear how far local authorities have the necessary ability and suitability to act as entrepreneurs, or to what extent the commercial ends of positive planning can be reconciled with an authority's planning objectives (see Grant, 1982, p. 500). It must be noted that within policy areas the provision of land and/or property is only one of the ways open to authorities to achieve their aims. Local authority involvement in the industrial development process under the banner of 'local economic development', for instance, includes various forms of intervention (Boddy and Barrett, 1979): facilitation and promotion (policy statements, information and advice, promotion, advertising and advocacy); physical provision (land supply, premises, and improvements to the immediate physical environment e.g. through the designation of Industrial Improvement Areas); financial provision (loans and grants, sale and leaseback arrangements and equity finance); and labour market management (keyworker housing, job creation and training). Local plans have a role here in establishing a sympathetic policy context for the determination of planning applications for industrial development, while the allocation of land for industrial use can be supplemented by various activities aimed at bringing such land forward, e.g. infrastructure provision, land assembly, the provision of serviced plots in industrial estates. Explicit statements of planning policies in local plans can indicate and support an authority's direct development intentions or set out planning requirements, particularly where private developers are to be involved on a partnership basis (e.g. see Bruton and Nicholson, 1984b). However, local plans are neither necessary nor sufficient factors in local authority land-development activities.

Direct development by central government – as opposed to the establishment of legislative and policy frameworks within which development by both the public and private sectors occurs – is relatively limited, the land and property requirements of government departments being met by the Property Services Agency. However, the provision and management of industrial estates in the depressed areas is a traditional and continuing role of government. This function is presently administered by the English Industrial Estates Corporation (which also manages the rural estate and factory functions of the Development Commission and those of the Council for Small Industries in Rural Areas), and by the Welsh and Scottish Development Agencies, together with regional development boards such as the Highland and Islands Development Board and the Development Board for Rural Wales. These agencies undertake the provision of industrial property in order to promote economic development. The Welsh and Scottish Development Agencies also carry out programmes of land reclamation using central government grants, and therefore have important roles in both land use and property development. The Welsh Development Agency (WDA) for example oversees and funds a rolling programme of land reclamation, which is largely carried out by the local authorities (land acquisition, design, supervision). On the other hand, WDA involvement in factory construction is much more direct; such developments have been concentrated on existing WDA estates, e.g. at Treforest, or in localities severely affected by the contraction of the steel industry in Wales, e.g. at Llanwern, Port Talbot, Shotton, East Moors (Cardiff), and Ebbw Vale. Sites and land for these special programmes of factory building have been made available by local authorities and the British Steel Corporation. The WDA aims to 'complement and supplement, but not to compete with or duplicate, the activities of other public and private agencies, in the same field' (WDA, 1977, p. 6), with existing structure and local plans providing a framework for factory construction and land reclamation. Funds allocated by the Agency to these two activities are thus an important source of finance for the implementation of relevant local planning policies. Co-operation between agencies may be a further factor in progressing particularly large development projects; the sudden closure of the East Moors steelworks for instance resulted in a co-ordinated response by public agencies in clearing and redeveloping the site. This response involved the production of a statutory local plan for the area by Cardiff City in order to provide a co-ordinated framework for redevelopment (see Fortune, 1981), land assembly by the Land Authority for Wales, infrastructure provision by the county council, the development of workshops by BSC Industry, and advance factory construction by the WDA. Proposals for the redevelopment of redundant and derelict dockland adjacent to the East Moors site have also been progressed through the involvement of a range of authorities and agencies (Sant, 1985). Another example from Scotland is the collaborative Glasgow Eastern Area Renewal (GEAR) project, where a number of agencies, including the local authorities and the Scottish Development Agency, are co-operating in the comprehensive regeneration of the area.

Finally, the Land Authority for Wales (LAW) has a unique role within the development process in that it is the only public authority within the United Kingdom that is responsible for the supply of land for development by others. The Authority carries out land assembly (including the provision of services) in anticipation of demand, or intervenes when private development is obstructed by problems of ownership or service constraints. The Authority was created by the *Community Land Act 1975* in order to operate the community land scheme in Wales (in England and Scotland local authorities were given this function). A central function for LAW under the *Community Land Act* was to facilitate community control over land development in accordance with its needs and priorities, i.e. to intervene in the private development land market in order to ensure an adequate supply of available land to enable the construction industry to meet community needs within the framework of adopted planning policies. This introduced an element of positive implementation into the land use planning process; LAW actively participated in the private market for development land as assembler and dealer. LAW also became involved in structure and local plan preparation as far as the 'implementability' of plans was concerned (Hollingsworth and Cuddy, 1979). In practice, the community land legislation proved less than successful, largely because of an absence of financial support from central government, so that the scheme was finance- rather than planning-led from the start (Barrett, 1980). In any case, following the May 1979 change of administration the *Community Land Act 1975* was repealed by the *Local Government, Planning and Land Act 1980*; Barrett and Whitting (1980) provide a useful post-mortem on the scheme's demise. The LAW was retained due to its perceived success as an agency contributing to the implementation of policy, but with different operating priorities. LAW's status within the land scheme had placed the Authority at the interface between those responsible for allocating land for development (the local planning authorities) and those 'responsible' for achieving development (builders and developers). The 1980 legislative changes effectively resolved the tensions inherent in reconciling these two sets of interests in favour of the latter. The new role is to make 'development land available as quickly as possible where the private sector finds it difficult to complete transactions' (LAW, 1981, para. 1.2), land use planning taking second place to commercial considerations; 'land for which there is a "need" as the term is used in structure plan strategies, will only be given priority by LAW if there is a foreseeable market demand' (para. 2.1.1). The effect of the new terms of reference and of associated changes in financial arrangements has been to emphasize the commercial aspects of the Authority's role, where it responds to effective demand rather than need. The Authority now has a residual facilitating function for private sector development interests rather than an implementation or positive planning function for the land use planning priorities of local authorities (Barrett and Whitting, 1983). In carrying out this new role, the Authority is required to be self-financing on a short-term basis, yet is still expected to successfully intervene where the private sector finds it difficult to complete land assembly.

8.4.4 Local planning, local plans and the public sector
The evidence reviewed here suggests that it is difficult for local planning authorities to assess accurately the scale, location and timing of public sector land- and development-oriented activities in order for these to be integrated with local planning policies and proposals. The potential for co-ordination appears to be greatest between planning authorities and infrastructure and development agencies. Where planning policies and programmes of physical infrastructure have proceeded in parallel, however, development plan documents as such are not necessarily important (Healey *et al.*, 1982, p. 78). Co-ordination may relate to land use policies but be expressed and developed through informal or semi-formal channels of consultation and communication, i.e. derived from continuous interaction rather than being plan-based. This emphasizes local planning tasks of monitoring and the integration and management of relationships between public and private agencies involved in land development, rather than the production of land use plans. The value of plans to infrastructure agencies depends on the extent to which they contain firm information on patterns of future development, but this is difficult to predict accurately in structure and local plans, particularly for proposals dependent on the private sector for implementation. The activities of public development agencies, largely concentrated on the provision of industrial premises, are usually located within the adopted framework of planning policies and therefore are important for

implementation. There is a degree of overlap here in that local authorities also carry out local economic development strategies, but this is more likely to result in co-operation rather than competition, especially where large sites/development projects are involved. Local plans, statutory or informal, may be prepared to guide such schemes, but are not necessary. In Wales, the significance of LAW as an agency contributing to 'positive planning' is now limited and the Authority is restricted in the main to dealing in profitable sites on a predominantly commercial basis. Despite this, LAW's facilitating role with regard to the supply of land for private development still operates within the framework of approved and adopted planning policies.

Local planning authorities are faced with more intractable problems when other public organizations such as the nationalized industries become involved in landholding and development. These agencies are large and powerful undertakings operating to supply goods and services on a national basis in pursuit of a clearly defined statutory role. Operational land and property requirements, both current and expected, are defined on technical criteria; alternatively, landholdings may be viewed in commercial terms, as assets whose disposal can realize useful capital receipts, by agencies being pressed by government expenditure limits and poor profit levels. Hence, the land acquisition, management and disposal policies of such organizations are influenced by operational requirements, financial considerations and government social and economic policy. The concerns of local planning authorities, whether expressed in plans or not, are unlikely to feature significantly in determining for instance, policies on land release or the location of new plant. Finally, the provision of social and community facilities (public housing, schools, hospitals) is organized on a sectoral basis within both central and local government; in the latter for instance departments and committees typically have functional responsibilities for particular policy areas. Sectoral provision of services is both reflected in and strengthened by the emergence of policy vehicles (e.g. HIPs, health authority forward-planning, and TPPs for the provision of transport and highway infrastructure) linked to expenditure statements and bids for resources from the relevant central government department. Since these programmes are intended to plan particular policy areas, they are fundamentally different from local planning authority land use plans, for these are spatially bounded and seek to co-ordinate the activities of public and private agencies within the plan area in pursuit of strategic policy objectives. The difficulties of linking small-area local planning concerns with vertically oriented policy vehicles and investment programmes is a continuing problem for local plans and local planners; Bowen and Yates in 1974 (p. 502) pointed out that:

local planning departments or sections remain a relatively insignificant part of the total complex of agencies making day-to-day decisions relating to the destiny of our urban environment. It is arrogant to suppose that solutions lie within the profession or even within local government as a whole . . . [the powers of local planners] are strictly defined and our control over the allocation of resources is minimal . . . the local planner is left merely to co-ordinate these allocations in order to minimize the possible resultant confusions.

Plans may summarize or act as a guide to future public development activity where such information is available from the local authority department or agency concerned, and may as a result for instance allocate sites. Inclusion of such proposals in a plan however does not constrain the agency involved, and so says little about the timing of development, which will be determined by service forward-planning considerations rather than land use policy. It would seem that in this context the role of the local planner must focus on managing change in the environment.

The involvement of public sector organizations within the development process is characterized by both fragmentation and variety of roles and functions. Agencies have statutory responsibilities which involve them either directly or indirectly in the development process. Those with a central concern for land and development issues – the provision of infrastructure, the facilitation of development by others or direct provision – have an associated interest in local planning policies, which either provide a framework for their operations (here, agency powers and resources are important for the implementation of planning policies) or represent one source of information about likely future infrastructure requirements. Integration and co-operation in this case reflects common interests and a mutual dependency in carrying through land and development roles. On the other hand, those whose interest in the development process essentially arises as a result of some other activity bring operational (technical, service development) or commercial criteria to bear on their land acquisition, management, development and disposal practices. Local planning policies are of only peripheral concern in such cases. The organized complexity of public sector land and development interests constitutes a key 'wicked problem' faced by the development plan system in its efforts to translate general policies for change into development on the ground. Informal interaction between local planning authorities and the agencies involved is a pragmatic response to this complexity. Nonetheless, if the general socio-economic and associated physical development policies being followed by the public sector are not to be obscured, negotiative relationships forged at the local level must be set within the context of some form of guiding framework, such as that provided by the idealized hierarchy for strategic planning and implementation outlined in Chapter 2.

8.5 The private sector

8.5.1 Introduction: complexity and variety
A simple view of the relationship between the land use planning system and the development activities of the private sector sees national and regional

policies being successively refined in structure and local plans and then 'applied' through the process of development control, which provides the interface between development interests and land use planning policy. Controlled by the sanction or veto of publicly-held development rights, actual development subsequently occurs consistently with higher level policies. However, although the British planning system does have an obvious hierarchical structure, the manner in which this system operates in practice is considerably more complex than the hierarchical top-down model of policy control allows. This is partly because of the flexibility and ambiguity which characterizes the various components of the system. Land use allocations, policies and proposals in structure and local plans do not confer permission to develop, but form just one consideration in the discretionary development control process. Here, decisions on planning applications are taken with regard to the provisions of the development plan, so far as material, and to any other material considerations, and will thus reflect a range of other factors as well as development plan policies and proposals. Similarly, the output of development control represents no guarantee that the actual development involved will occur; local planning authorities cannot force the implementation of planning permissions, so development granted permission may never take place. In any case, policies in development plans may be only a subset of those being pursued by the local planning authority or may have been superseded to a greater or lesser degree. Moreover, generalizations about the extent to which development control decisions and actual development activity can be said to represent 'implementation' of development plan policies are further hampered by a series of variations in terms of development pressure and activity and the local authority response:

1   The composition of private sector development interests varies between areas due to differences in investment opportunities and the nature of demand for built space (Healey and Elson, 1981; Healey, 1983). However, private sector property development interests all share a concern to undertake or co-ordinate development with a view to maximizing profit, the process of development being essentially similar to other manufacturing processes producing commodities for sale.

2   Planning policies vary, from in crude terms development promotion to development restraint, as a consequence of local authority assessments of (and attempts to balance) the local relationships between need, demand and supply for any given policy concern.

3   The form in which such policies are expressed varies as the local planning authority considers appropriate, e.g. statutory local plan, informal plan, policy document, promotional briefs, or in the light of the policy aims and the nature of development interests and pressures.

4   Finally, policy formulation and plan preparation does not take place as a

technical process in a 'contextless' environment, essentially representing instead the outcome of interplay and negotiation between different public and private groups and interests.

These variations clearly pose a problem for examining the relationship between local planning policies and private sector development interests. Aside from the statutory, legal and administrative framework within which both are established and operate, neither offers a consistent profile which could as it were be held constant so as to view changes in the other. Considerable spatial variation in the mix of development interests and planning policies can be expected, with consequent variations in the outcome of the interplay between the two hindering general conclusions about the effectiveness of local planning policies and plans. No doubt for these reasons, recent studies concerned with the effects and consequences of planning policies have taken a case study approach, focusing on particular local situations and investigating therein the interplay between development interests and planning policies, as a prelude to the establishment of more general relationships.

8.5.2 The Implementation of development plans (Healey *et al.*, 1982 and 1985)
This work was carried out by Oxford Polytechnic for the DoE in the period 1981–85, and examined the implementation of planning policies and plans in two stages:

1 An exploratory study focusing on three areas in the outer South East, each of which represented only part of the relevant administrative district: High Wycombe (planning policies seeking restraint of development); Wokingham (development management); and Banbury (development promotion) (Healey *et al.*, 1982).
2 A wider study placing more emphasis on structure plan strategic planning policies for selected districts in Greater Manchester and the West Midlands (Healey *et al.*, 1985).

In the first stage, policies concerned with new residential development and employment-generating uses were chosen for study, as these policy areas were common to each of the three case studies, of central significance for land use planning in these localities, and clearly within the province of town and country planning powers. Briefly, the characteristic features of the three study areas selected for the first stage are as follows:

(i) *High Wycombe (Bucks).* This is within the metropolitan Green Belt around London, and faces considerable pressure from both housing and employment-generating development. Planning policy seeks to limit such development to marginal additions undertaken to meet local needs.

(ii) *Wokingham (Berks)*. In this area, local planning policies have acceded to development pressures following a regional strategy which identified the area as one where growth would be concentrated. This growth has been managed by the allocation of a limited number of large greenfield sites for housing and industrial development. Public sector investment in infrastructure is co-ordinated by a Development Programme (see above, Section 8.4.2), with the local planning authorities co-operating with developers in planning and servicing development. This strategy of 'growth with restraint' seeks to protect the remaining open areas from development pressure.

(iii) *Banbury (Oxfordshire)*. Here, town expansion schemes have been used to promote private housing and employment-generating development, with current policies seeking to consolidate that development.

In these three areas the research focused on the following key issues:

1   The pattern of development that had occurred in the study areas 1974–1980 and how far this related to planning policies and proposals as expressed in development plans and other planning guidance.
2   The processes responsible for producing this development and the role therein of planning authorities.
3   The specific role of development plans and other forms of planning guidance in the processes and outcomes analysed in (1) and (2).

The conclusions of the research on these issues are as follows.

1   Planning policies and the pattern of development
Policies for *new residential development* were successfully implemented where these sought to limit development to certain areas, zones or sites, despite considerable pressure for development in High Wycombe and Wokingham. The amount of development in the study areas 1974–80 was judged to be consistent with structure plan policies. However, 'policies which have proved more difficult to implement were those which sought to limit the amount and rate of development to definable local needs, those which sought to phase development in a "balanced" way, or in relation to particular infrastructure investments' (Healey *et al*., 1982, p. 36). For example, in Wokingham the county and district councils sought to manage development pressures and large-scale land release through negotiative partnership arrangements. Here, the aim of the local planning authorities was to resist pressure for land release elsewhere. These partnership arrangements involved public sector agencies (provision of phasing and layout frameworks, infrastructure and community facilities) and voluntary consortia of developers (sites and financial contributions), but proved problematic, initially in the face of problems of infrastructure provision and subsequently as a result of a downturn in the market which led to slow progress on the part of developers. Policies for *employment-generating*

development were also successfully achieved, particularly in terms of the location of development. In High Wycombe, the main policy concern was not with location but with limiting permitted development to that which could be justified on the grounds of local needs and demand; this was successfully implemented through local user conditions on planning permissions. The overall pattern in respect of both sets of policies is thus one of a substantial match between policy and development. This finding probably reflects the fact that the policies under consideration were achievable via the central planning power of restricting development/land release. Significantly, where policies were not capable of implementation through land release, considerable additional efforts were found to be required (e.g. provision of serviced land, keyworker housing and direct financial assistance by Cherwell DC in Banbury).

2 Development processes and the role of planning authorities

In all three areas the local planning authorities were judged to be able to exercise considerable influence over the pattern of development (Healey *et al.*, 1982, p. 63):

their policy stance provided a framework for the operations of smaller builders and developers and a baseline for negotiations with larger development interests. There can be little doubt that some potential proposals which the planning authorities would not accept have either been deflected from the areas, or have not been worked up. Some are lying in wait for a change in policy . . . the larger regional and national house builders and property companies were most likely to be deflected.

Policy was challenged only where alternative interpretations were possible. Again, the local authorities had the most influence on the timing and location of development when policy implementation involved limiting land release in areas of high demand; this is partly attributed to policy *consistency* (between levels of government) and *stability* (over time) as to the way in which land use planning powers over development have been exercised. Large-scale allocations nevertheless gave rise to difficulties in achieving the desired rate of development. Various development interests (particularly agents and larger and more sophisticated locally-based developers) enjoyed a close negotiative relationship with the planning authorities in High Wycombe and Wokingham, which gave them a key role in establishing the principle of development on sites and the detailed requirements for acceptable proposals. It is suggested (Healey *et al.*, 1982, p. 65) that both local authority policy intentions and

the local planning authority's role itself has reflected Both the mix of development interests operating in an area and the nature of the political interests represented there. Both of these have in turn been affected by central government policy and, in the case of the former, national economic investment trends and the structure of demand.

Different local authority approaches to implementation result from these varying pressures and factors. For example, in High · Wycombe a 'regulative–negotiative' approach reflected the accommodation reached between sophisticated and locally-based development and conservation interests, while in Wokingham the strategy of development with restraint led to the local planning authority adopting a distinctive co-ordination and facilitation role for the management of development. Finally, little evidence was found of substantial conflicts or difficulties of co-ordination within the public sector; public agencies have been reasonably effective in ensuring that land made available for development has been serviced.

3 The role of development plans and planning guidance

In relation to the two policy areas under consideration (new residential and employment-generating development) a considerable range of planning guidance was found within the study areas. Statutory guidance was seen to be most important, although only in Banbury was a formal local plan prepared; uncertainties over structure plan modifications have hindered local plan production in High Wycombe and Wokingham. In the three study areas, not all policies in use were expressed formally. Internal work practices and member attitudes were as important as formal policy statements given room for their interpretation as to meaning and relative priority. Because of these factors, larger development interests appreciated the need to interpret policy statements in the light of an authority's current policies. Central government attitudes, indicating the likely extent of support for district planning policies, were also important to such interests. Direct consultation was valued, both in order to determine current policy (development plans represented part of the relevant material here) and to discuss detailed design/layout matters. Plans were used selectively by both local planning authorities and development interests when it suited their own policy concerns and development objectives, and as a result such documents as town maps continued to be used even though generally considered to be outdated. *Statutory* plans could be regarded as indicating central government's attitude to appeals, but their significance depended on their *currency* in relation to present policies and on the standing given to local planning policies by the Inspectorate. The study offers the following conclusions about the role and value of development plans:

(i) Statutory plans gave the local authority an important tool for implementing the policies within them where these related to land release; if supported by the local authority and central government, such plans helped to deflect inappropriate development and shape appropriate proposals. However, other planning guidance was equally effective depending on the weight given to it by the local authority and central government.

(ii) The larger sophisticated development interests recognized that plans were only one element in determining a local authority's current policies and development opportunities.

(iii) Development plans were not of direct concern to public sector agencies concerned with the provision of infrastructure.

(iv) Development plan preparation was just one of the ways in which central government sought to influence local planning policy.

(v) Local pressure groups placed considerable weight on development plan policies and the extent to which these were upheld (i.e., assumed that policies in plans could be readily used as a guide to current policy).

The extent to which generalizations about the influence of policies and proposals in development plans on development processes can be made on this basis is limited by the restricted size of the study areas and their common location in the outer South East of England. In addition, this first stage of the research focused on situations characterized by substantial continuity and stability as to policies, powers of implementation, and institutional frameworks. The preparation of structure plans in the 1970s offered an opportunity for the review of established policies rather than policy development; the structure plans thus tended to confirm existing policies for the study areas. As a result of this continuity, there was little discrepancy between the policies that were actually being followed and the policies that were set out in plans. However, further work on the implementation of planning policies in the South East indicates that the policy stability and continuity charted by Healey *et al.* in the period 1974–80 is breaking down as a result of national economic trends and shifts in central government policies and attitudes. This research has investigated the release of land for housing and employment-generating development in the two Hertfordshire districts of Dacorum DC and North Hertfordshire DC, where policies of development restraint apply (see, for instance, McNamara 1982, 1984; Elson, 1983). The effectiveness of restraint policies is seen to be determined by the legal basis of the policy involved, the support received from central government, local public support, clarity of aims, and the strength of the trends the policy seeks to reverse (McNamara, 1984). Also important is the degree of commitment which various implementing bodies have towards the policies in their documents or in those of other (higher) planning authorities. Thus in Hertfordshire the county was able to exert considerable pressure on the districts to accept strategic restraint policies as a consequence of the information/data services which the county could offer to districts anxious (for local political reasons) to prevent potentially vulnerable sites from being released at appeal. In terms of the release of land for *housing* development in a restraint context, a key factor is how central government reacts to the pressures and demands of the housebuilding industry, via representative bodies such as the House Builders' Federation, for a loosening of restraint and greater levels of land release. McNamara (1982) suggests that the latest compromise reaction to these pressures, the five-year land supply agreed in the context of DoE Circular 9/80, is potentially meaningless, particularly in restraint areas, since land outwith the agreed supply is afforded no increased protection. On the contrary, DoE Circulars 22/80 and 15/84 both emphasize that the fact that the housebuilding needs of an area can be met from identified sites is not in itself sufficient reason for refusing planning permission elsewhere (DoE, 1980b, 1984d). Local planning policies seeking to release *only* a five-year supply will be dependent for success upon *explicit* central government support, which following Circular 15/84 is unlikely to be available. For employment-generating development, Elson (1983) documents how the emphasis of restraint policy in Dacorum has shifted in the face of rapid industrial decline in the area since 1980. Structure plan activity-based restraint policies operated reasonably successfully in the period 1974–79, acting as a powerful deterrent to industrial and warehousing development and also effectively restricting development by firms from outside the county. These policies encouraged negotiations over the *principle* of development (as well as over detailed aspects) which started from a set of presumptions against permission; most applicants for new development used local agents to conduct these negotiations. However, during this period the supply of new industrial land in Dacorum was drying up. Approvals of planning applications on non-allocated land demonstrated that (Elson, 1983, p. 139):

statutory planning was lagging behind decisions emerging from the interplay of development interests represented by the Commission for the New Towns, restraint interests represented by the County Council and local pressure groups in Dacorum, and the more pragmatic and facilitatory approach of District Councillors.

From 1980 onwards, macro-economic changes led to shifts in the locally perceived priorities for land use planning and hence to more relaxed criteria for assessing applications and to allocations of new industrial land in the 1982 Dacorum local plan. Elson (1983, p. 148) comments however that formal plans play only a restricted role in identifying new land; rather 'the *process* of plan-making provides a forum where the stronger interests can assert their positions and shift the consensus'. In Dacorum the strength of development interests in the early 1980s partly derived from the concerns of local councillors over rising unemployment and job losses. Changing local attitudes to land release for new employment-generating development and the subsequent adjustments to planning policies thus represented a contingent response to uncertainties over the consequences for the local economy of poor national economic performance.

At the same time as local attitudes towards employment-generating development are changing in traditional areas of restraint, the role of structure and local plans in defining a supply of land for such development is likely to be further weakened by calls from central government for a more liberal attitude towards industrial development (DoE Circular 16/84) and employment generation (DoE Circular 14/85). This advice reflects the view of central government that such developments should be encouraged in order to promote economic recovery. The Circulars emphasize that applications for planning permission for development that is contrary to approved development plans should not be refused permission solely on this basis; that there is always a presumption in favour of allowing applications for development; and that the development plan is only one of the material considerations that must be taken into account in dealing with planning applications. At the same time, it is made clear that central government remains committed to protecting what Circular 14/85 (para. 3) refers to as 'interests of acknowledged importance', i.e. to 'the need to preserve our heritage, to improve the quality of the environment, to protect the Green Belts and conserve good agricultural land'. Protection of Green Belts continues to receive special emphasis (DoE Circular 14/84). Taken together with the views expressed by central government on the availability of land for new housing development in DoE Circulars 22/80 and 15/84, it is clear that central government support for locally-established policies of development restraint is being explicitly reduced unless such policies seek to protect 'interests of acknowledged importance'. This suggests that structure and local plans embodying restrictive policies are likely to become less reliable as predictors of central government attitudes towards appeals against decisions based on these policies; and that the significance of statutory plans will be reduced, since this partly depends on the degree of support that central government is prepared to give them at appeal. The case studies analysed by Healey *et al.* (1982) were characterized by long-standing agreement between levels of government as to the goals of land use planning policy in the South East. In these conditions of certainty over goals, statutory development plans containing policies on the release of land were found to provide an effective means of implementation for such policies. However, these conditions are now changing as central government calls on local authorities to adopt a permissive attitude towards new development proposals, with the traditional policy goals of restraint receiving little attention except in so far as 'interests of acknowledged importance' are involved. In Christensen's terms (1985), a situation of 'known technology, agreed goal' is being eroded as circular advice informs local authorities of present central government policies and attitudes, calls into question the value of development plans (many of which undoubtedly contain restraint policies, reflecting the former consensus among levels of government), and emphasizes the discretion available to local planning authorities in development control.

In contrast to the initial work in the South East, the second stage of the Oxford Polytechnic research on the implementation of development plans deals with much larger case study areas in Greater Manchester and the West Midlands where major policy reorientations (from the regulation of development to development promotion) occurred in the 1970s in the face of changing economic factors and consequent land use and development problems (Healey *et al.*, 1985). The main concern of strategic policy in these areas was urban regeneration through peripheral restraint, with evidence that the latter was being achieved while the former was not. In order to explore how this outcome was being produced, the research focused on a series of case studies of land use change and development within the two conurbations, including city centres, areas on the urban fringe intended for housing and industrial development, open land areas, and major strategic development sites. The role and significance of development plans was found to vary with policy objectives, the local political situation, and the nature of interests in land, leading to the identification of six tasks necessary for the implementation of strategic planning policies:

1 Land allocation.
2 Co-ordination and progression of large development projects.
3 Attraction of resources for investment in environmental change.
4 Organization of investment.
5 Conservation of the environment.
6 Management of small-scale change where strategically important.

To some extent these tasks can be associated with the broad policy goals of urban regeneration (attraction and organization of investment) and peripheral restraint (land allocation, development co-ordination and conservation), with the use of plans varying according to the task being addressed.

1 Land allocation. This task requires policies specifying both *location* and *level* of a given activity or land use, the main difficulty for policy formulation being to achieve and subsequently sustain a balance between control (vital to support agreements reached over land allocations) and flexibility (required by the various activities for which sites are being allocated). Policy formulation involves mediating between and resolving conflicts of interest, and in this respect statutory structure and local plans were found to be useful and powerful tools in urban fringe and open land areas. The various mechanisms for consultation, objection and inquiry enshrined in the formal procedures for the preparation of these plans were judged to be particularly valuable in producing reasonable and justifiable policy frameworks. Such policies were seen by many parties to have considerable strength so long as they continued to receive local and central government support. Plans thus offered an effective and accountable framework for investment, of value to the development industry, public sector infrastructure agencies, and to development control. In particular structure plans were seen as vital to this task and were commonly supplemented by statutory local plans, which themselves could be complemented by statutory local plans defining the Green Belt and by housing land availability studies; these studies did not challenge adopted plans, but in some cases threatened to pre-empt policy choices under discussion as part of local plan preparation.

2 Co-ordination and progression of large development projects. This task requires a policy framework capable of guiding negotiations between the various parties involved and co-ordinating investment by public agencies, landowners and development firms. Statutory plans provided a useful context as a result of their land allocations, while the process of preparing a plan could secure commitments from, for example, public infrastructure agencies. The most important tool for this task however was the development brief, used in urban fringe, inner and city centre areas to guide and record discussion and agreements reached with development interests. In open land areas there were other activity-specific ways of progressing sites, e.g. conditions on mineral sites; management plans and agreements when extraction was underway. Development briefs not only organized agencies involved in complex projects, providing a focus for negotiation and co-operation, but also provided a means of explicitly bringing forward policy objectives and constraints, thereby ensuring that these figured in negotiations over development projects.

3 Attraction of resources for investment. Statutory plans played only a limited role here, since they could not be readily tailored to the investment requirements of individual agencies; however, they were used to some extent to promote sites and opportunities for private development, and in attempts to secure the commitment of public resources to projects or areas.

4 Organization of investment. This task is of particular relevance to the strategic goal of urban regeneration where it relates to the management of change in city centres and inner areas, but it is also relevant to open land areas and the urban fringe. Statutory action area plans were initially intended to provide a policy vehicle for the organization and management of intensive change in urban areas. This approach has been difficult to realise in practice as a consequence of overall limits on public and private resources, and as a result of difficulties in using local plan statements to control or influence the flow of public funds. Spatial co-ordination, which might be achieved after lengthy negotiation, could be recorded in plans or other documents such as local authority budgets, but these rarely subsequently constrained the actions of other municipal departments. Various forms of policy statements were produced, typically linked to spending programmes (e.g. inner city Partnership programmes, corporate plans), but these failed to provide a coherent statement of land use policies and proposals for areas experiencing change. Some districts are now producing inner area statements or studies which attempt to give such a policy overview (e.g. Birmingham City Council). Statutory plans are not incapable of providing frameworks for investment, but both public resource and private investment contexts have typically been too variable and uncertain for local plans to have much of a stabilizing effect. This task has thus been characterized by a degree of agreement over goals (e.g. urban regeneration) even though the means used to achieve these goals have often proved ineffective.

5 *Conservation of the environment.*   This was found to be an important planning concern in all of the case study areas, with statutory plans having an important role in identifying protected sites, areas and buildings and helping to deflect development interest from such localities. Structure plans indicated criteria for conservation of natural and built environments, and most authorities used local plans in the contested and pressured areas of the urban fringe to further specify these policies, perhaps in conjunction with separate Green Belt plans, which provided clear statements of locally important policy concerns. Conservation policies restrict development opportunities and so are likely to be regularly challenged at appeal; hence precise formulation is important.

6 *Management of small-scale change.*   The bulk of planning applications handled by the development control system concern minor development proposals or changes of use which are of little strategic concern. Where development proposals raised the strategic issue of safeguarding the quality of the environment, statutory plans were found to play a crucial role in guiding and supporting development control decisions, particularly where these were subject to challenge. Structure and local plans, however, rarely provided guidance of sufficient detail for development control within urban areas, particularly where intensive change was involved, e.g. city centres or the inner city. Other non-statutory policy documents were used to offer supplementary guidance to development interests, although for the local planning authority this practice raised the problem of ensuring that such policy documents were always up-to-date.

The Oxford research thus suggests that statutory plans offer a useful means of resolving conflicts over land allocations and an effective framework for the implementation of conservation policies. This in turn undoubtedly reflects the weight given by central government to conservation and land allocation as roles for the statutory land use planning system. In other words, development plans have been useful tools for the progression of conservation and land allocation policies not merely as a result of the statutory format *per se*, but also – and perhaps largely – because central government support for these policies has been forthcoming.

Both of these sets of policies or tasks involve the use of negative powers, restricting or preventing development in sensitive locations (conservation) or in areas not identified for development in structure and local plans (land allocation). As regards conservation, the view of the Secretary of State for the Environment in 1981 was that in this area 'the planning system has achieved a very great deal . . . National Parks, the AONBS, the listed building machinery, the SSSIs, the Green Belt . . . here lie the great successes of the planning mechanism . . . [which all] involve the use of negative powers'.[6] Central government continues to emphasize conservation as a land use planning task. As we have seen, DoE Circular 14/85 qualifies the general presumption in favour of allowing applications for development by reference to the need to protect 'interests of acknowledged importance', i.e. the national heritage, Green Belt, environment, agricultural land. Recent circulars also emphasize the facilitation of new housing (Circular 15/84) and industrial development (16/84) through the allocation of sites in plans. Land for housing development may also be identified through the joint studies undertaken between planning authorities and housebuilders. At the same time these circulars make it clear that the allocation or identification of a suitable supply of land is not in itself sufficient reason for refusing planning permission on other sites. This advice therefore seeks to restrict the use of the negative powers of development control in this respect. In short, in recent circulars central government is emphasizing one aspect of the land allocation task for plans (the identification of sites) while explicitly withdrawing support for another aspect (the control of development elsewhere, i.e. on non-allocated sites). There remains however a general presumption against inappropriate development, i.e. defined by Circular 16/84 (para. 9) as that involving losses of countryside, Green Belt and agricultural land.

Central government support has been less forthcoming for the other tasks identified by Healey *et al.* (1985). Where plans attempted to organize investment, co-ordination problems remained, reflecting the complexity of the public sector and the volatile nature of private sector investment in the built environment. Statutory plans as such were found to be neither essential nor appropriate for the co-ordination of large projects, the management of small-scale change, or the attraction of investment. Instead, the requirements of these tasks has typically been met by the use of non-statutory policy guidance such as development briefs (site co-ordination), development control policy notes (management of small-scale change), marketing and promotional documents (investment attraction), informal plans/policy statements or studies (investment organization). These findings echo Farnell's conclusion (1983) that local plans seeking to facilitate development have succeeded, while those whose primary concern was to 'investigate problems and search for solutions' in inner city areas suffering decline, partly through a lack of investment interest, have not. Such plans tend either to be abandoned or reoriented towards development facilitation. The significance of statutory plans for the tasks of land allocation and conservation confirms the earlier conclusion of Healey *et al*. (1982) that development plans are particularly important and significant where the policies concerned relate to the way in which land is to be released and developed. Other types of local planning policy document may be equally or more appropriate in situations where the traditional planning power of refusing applications for permission to develop is of less immediate relevance to the management of environmental change.

8.5.3   Industrial and commercial development

Local planning authorities have typically adopted a favourable and positive stance to industrial development, reflecting a concern with the local economy and unemployment rates at a time of national recession. Central government is also concerned that structure and local plans should act to facilitate appropriate industrial development (DoE, 1984e, 1984g). Local authority policy stances have shifted as industrial decline has continued, even in traditional areas of development restraint such as Dacorum (Hertfordshire). As we have seen (Section 8.4.3), local authority attempts to promote economic development include among other activities the use of local plans to make policy statements establishing a general presumption in favour of industrial development (usually subject to other planning criteria, e.g. access), together with specific allocations of industrial land. Given development pressure and interest, the identification and release of industrial land in plans may be sufficient to secure the implementation of plan proposals, but otherwise more positive activities will have to be undertaken to attract investment and facilitate development. Thus, site allocations can be supplemented by action to ensure that the land involved is truly 'available', e.g. acquisition and marketing, infrastruction provision. These activities and others reflect the recognition that a relatively passive enabling role (as represented by the allocation of sites in a local plan) is often insufficient to ensure implementation by the private sector. A subsidiary policy concern has been to relocate established industrial firms where these presently constitute non-conforming uses.

Recent trends in industrial property development include the interest shown by the financial institutions in industrial property as an investment sector, and the involvement of property companies in the provision of standard estate-type developments. At the same time changing manufacturing practices, most notably a shift towards light manufacturing and assembly, have reduced the need for specialized (owner-occupied) premises. These trends have typically emphasized the production of property catering for light industry, warehousing and distribution, rather than heavy industrial plant, in the form of standardized developments with good motorway access (Boddy and Barrett, 1979; Boddy, 1981). Investment and property development criteria favour developments located in the suburbs or on the urban fringe rather than in the inner areas, for reasons of site size, working and residential environment, strategic access, and investment security. Such trends do not aid the implementation of planning policies designed to promote industrial development in the inner city, although a strong case for institutional investment in the inner areas can be made (Ratcliffe, 1978). Moreover, problems of industrial (manufacturing) land availability are severe in the conurbations and London relative to rural areas and small towns, inner city decline notwithstanding, although it is not clear if this is as a result of constraints on supply or a lack of demand; consistently high prices for industrial land in the conurbations indicates the former (Fothergill, Kitson and Monk, 1983). The emergence of a demand for landscaped 'science parks' catering for high-technology industry and requiring good access to airports and motorways will exacerbate the pressure on the urban fringe and on Green Belt land. It is in these locations where challenges to local planning policies can be expected to be concentrated, with difficult policy conflicts arising as a result, e.g. between industrial development and promotion and the protection of agricultural land or the Green Belt.

Planning policies for retailing established in structure plans and detailed in local plans are characteristically concerned to encourage new shopping development within the existing hierarchy of city, town and district shopping centres. This approach to the locational structure of retail provision reflects an acceptance of central place concepts as well as the operational structure of retailing in the post-war years (Burt, Dawson and Sparks, 1983), but its relevance to modern retailing is less clear (Dawson, 1980). Retailing is an area of rapid change and innovation (Guy, 1982) but the spatial and land use implications of structural changes have seldom been fully addressed in development plans. Policy issues in retail planning have been divided by Burt *et al.* (1983, p. 11) into two groups: those arising from structural change (hypermarket and superstore development, local shop loss, the development of large non-food stores, and the location of retail establishments outside established shopping districts); and those relating to the impact of these changes on existing urban planning policy (the maintenance of a hierarchy of shopping districts, the regeneration of inner city areas, urban conservation, traffic management within established shopping districts, and the maintenance of viable shopping districts). In their survey of retailing policies in structure plans, Burt *et al.* (1983, p. 13) found the following:

1 Few county planning authorities had a comprehensive view of retail planning, most authorities typically responding to two or three of these issues in a reactive manner.
2 Formulated policies in structure plans sought to maintain the *status quo* in retail provision with respect to the existing hierarchy, so that although retail planning has emerged as a coherent element in structure plans 'many of the policy approaches are rather conservative responses to the activities of a very dynamic economic sector'.

Structure plans of course provide the policy context for local plan policies, and there is little indication that local plans are doing anything other than detailing strategic retailing policies at the local level. For example, the draft local plan (1984) for Hereford, a free-standing market town with an extensive agricultural hinterland, contains the following main shopping policies:

- to promote Hereford as a major sub-regional shopping centre;
- to protect the commercial viability of the existing central shopping area (defined on the proposals map) and retain it as the prime shopping area;
- to restrict further major shopping provision within the central area beyond a defined floorspace limit in order to maintain viability; this limit is taken up by commercial redevelopment by the local authority in partnership with the private sector;
- central area shopping developments to have regard to the historic character of the area;
- to promote the location of non-food retail uses such as do-it-yourself stores and bulky goods retailers in central or edge of centre locations, providing the viability of existing shops in the central area is not threatened and subject to other planning considerations;
- to restrict additional shopping floorspace outside the central area to existing local shopping centres or to serve new residential development;
- to restrict further development of new shopping facilities except for small corner shops;
- to restrict the non-retail use of ground-floor premises in defined primary shopping streets with provision for such uses being made in secondary shopping streets.

Given local plan policies along these lines, it is not surprising that as applied to the determination of retail planning applications in development control, their effect has been to facilitate change and growth in town centres while hindering retailing innovation whose space requirements could not be met within the existing hierarchy (Guy, 1980). Planners have taken a sympathetic attitude towards retail interests wishing to expand or develop town-centre sites, a stance balanced by conservation issues in historic centres. Moreover, where local authorities have become involved in town centre redevelopment schemes in conjunction with the private sector, a concern to secure the commercial success of such ventures provides an additional reason for local plan policies supporting the existing hierarchy of shopping provision. At the same time, many innovative retail organizations have locational and land use requirements (for large and cheap sites with good access) which are difficult to meet within existing shopping centres. This has led such operations to seek out-of-centre sites, which has brought them into conflict with local authority planning policies attempting to concentrate new retail investment in existing shopping centres (Jones, 1984). In practice, planning policies prohibiting superstore, hypermarket and retail warehouse developments outside established centres have been supplemented or replaced by the more positive identification of suitable specific sites in local plans or the treatment of individual applications on their merits (Gibbs, 1985). Such compromises reflect a pragmatic realization that large retail developments do have benefits to offer, e.g. job creation, use of vacant land (Jones, 1983). There is also the possibility that such developments will be allowed on appeal where policies are judged overly restrictive, the DoE pointing out that although planning authorities may take account of the likely impact of new development on the viability of established centres, structure and local plan policies and proposals should not attempt to regulate competition between retailers or seek to prevent the evolution of new forms of retail provision (DoE, 1984g, para. 4.22). Nevertheless, Gibbs argues that planners have not compromised on the underlying locational principle of the retail hierarchy. Thus superstores have been directed to established shopping centres or to form the basis of new district centres, perhaps in conjunction with residential development. Here local plans provide a means of establishing planning policies on the location of retail development and of identifying appropriate sites. For instance, the City of Gloucester Local Plan (1983) has core policies indicating that comparison and convenience shopping facilities will continue to be concentrated within the city centre, and allocates separate sites for two DIY stores and a superstore on the periphery of the main shopping area. The overall effect of retail planning policies has been a slowing down of decentralization i.e. of the rate of development of surburban shopping facilities (Guy, 1980), and where planners continue to support the existing shopping hierarchy through policies in structure and local plans this trend is set to continue. Finally, it is worth pointing out that there are many aspects of recent retail change relatively unaffected by planning policies, such as the loss of local shops, even though these are perceived as issues of concern by planners.

The development and provision of offices in Britain is dominated by London's role as a national office centre. Office development in London takes place in a particularly active, volatile and profitable property market, with the growth of user and investment demand for new office space being accommodated in three ways, each of which has resulted in a distinct planning response (Barras, 1984):

1 Intensification of existing central area office stock via redevelopment, reflecting competition to develop the prime sites, where rents and development profitability are highest.
2 Extension of the boundaries of the traditional central office areas by new development on the fringes.
3 Development of new office centres in the suburbs as one manifestation of a wider process of decentralization occurring in response to the costs of locating in the central area.

Intensification of stock is characterized by higher redevelopment densities and the assembly of large individual sites; developments of this scale inevitably put pressure on established physical planning policies seeking to control the scale, design and layout of individual buildings, e.g. policies on building height or plot ratio, as well as on policies seeking to conserve older buildings which may be included as part of a larger site intended for redevelopment (Barras, 1983). Issues of this nature have recently been raised for instance in proposals (turned down at appeal by the Secretary of State) for the redevelopment of London's Mansion House Square.[7] There are also examples of development and landowning interests directly challenging restrictive local planning policies within the traditional central office area, such as the recent successful attempt of Great Portland Estates to quash office policies in the Westminster District Plan (see Chapter 4). The hostile reception given in some quarters to the conservation policies of the draft City of London Local Plan suggests that these pressures on local plan formulation will continue.[8]

The difficulties of finding and assembling sites for large-scale redevelopment within the central area has encouraged office development schemes on the fringes of the City in areas previously in industrial, commercial or residential use. In particular, large development sites are becoming available as a result of the closure of docks and associated warehouses on the South Bank of the Thames (e.g. at Coin Street, Hays Wharf, Surrey Docks) and in London Docklands to the west of the City (e.g. at Canary Wharf, where current development proposals envisage the provision of up to eight million square feet of offices). Development opportunities coming forward in London Docklands also reflect the activities of the Development Corporation established for this area. In some cases, particularly on the South Bank, there is considerable local planning authority opposition to such schemes which tend to conflict with existing local plans proposing a continuation of traditional and small-scale industrial and residential uses appropriate to inner London (Barras, 1984). In contrast, office developments in surburban centres have generated less planning opposition, despite potential conflicts with other land uses, and indeed in some areas have been positively welcomed. The trend to office decentralization which developments in suburban centres manifest is likely to be fostered by the impact of high technology on office automation, encouraging dispersal as automated systems make the traditional central area location unnecessary.

8.5.4 Development plans and land supply for private residential development

Land availability is a crucial aspect of the residential development process. On the one hand, the allocation of sites for residential development in plans is the means by which the planning system (via development control) seeks to influence the scale and location of development. On the other hand, a suitable supply of land is a necessary physical resource for the housebuilder. In addition, stocks of land held by builders in the form of land banks provide them with a means of securing profit from land development as well as from housing production, while land banks also give builders a degree of bargaining power in their dealings with local planning authorities (over planning permission) and with landowners (over future acquisitions). In particular, stocks of land ready for development allow the builder to maintain housing construction even though additional sites or planning permission are not immediately available. This section (1) outlines the objectives and interests of landowners, local planning authorities and housebuilders in residential land availability, and (2) considers how the housebuilding industry gathers the necessary resources to allow private residential development to proceed.

8.5.4.1. *Objectives and interests*

Landowners, local planning authorities and housebuilders participate in the residential development process in order to pursue or fulfil distinctive interests, objectives or policy aims. Landowner, developer and planning policy interests can be expected to vary from locality to locality and as a result of historical circumstance. It is nevertheless possible to characterize in broad terms the perspectives underlying the involvement of these groups in residential land supply and development.

1 Landowners. Relatively little attention has been given to the part played by landowners in the supply of land for residential development, although some information is now becoming available (see, for example, Goodchild and Munton, 1985). The role of the landowner in the development process stems from the fact that development cannot begin until the proprietary rights of landownership are held by the developing agency. For this reason landownership by public or private non-development agencies can act as a supply-side constraint by limiting availability, whatever the supply of land 'available' in planning terms, i.e. with planning permission or allocated in plans. This factor may be significant in the non-implementation of planning permissions for residential development (DoE, 1978b).

The most important action of the landowner in the residential development process is arguably the decision to sell and at what price. Here one of the most important factors is seen to be landowners' reasons for holding land, or their motives of ownership (Clawson, 1962; Kaiser and Weiss, 1970; Goodchild, 1979). At a national level it is clear from the work of Massey and Catalano (1978) that much land is held variously:

1 as the historical basis of a wider role in society (former landed property);
2 as a factor of production, emphasizing use value (industrial land-ownership); and
3 as an alternative investment sector (financial landownership).

None of these roles necessitates a concern with the supply of land for residential development. As Ball (1983, p. 153) comments,

these characteristics of landownership reinforce the economic attractions of holding onto potential development land rather than selling it at the first opportunity . . . [forming] a fundamental base for the power that landowners can exert in the residential land market.

Not only do landowners rarely *have* to sell their land, but there are also often sound social and economic factors encouraging retention. Rydin (1984) points out that for this reason landowners may be supportive of planning restraint policies. Thus, for farmers rural land is an essential factor of production, while rural landowners favour restraint as one means of preserving existing land use and ownership patterns and the social relations which they reflect.

2 *Local planning authorities.*   The basic concern of local planning authorities in the residential development process is to regulate the market in the interests of wider economic, social and environmental needs, and to control individual developments accordingly. Housebuilders often portray land use planning as a supply-side constraint, arguing that the amount of land allocated in development plans is insufficient or that sites are poorly located. Nevertheless land use planning seeks to promote efficient change in the built environment, in which housebuilders have a central interest. There are however inherent contradictions and conflicts in the involvement of local planning authorities in the programming of residential land supply. Planning authorities may allocate sites for housing and control development rights, but other rights of ownership will usually be held privately and the development actually carried out by the private sector. Planning authorities formulate policies and land allocations on the basis of need, while the private sector responds to economic demand. Estimates of housing requirements based on these two approaches may overlap to some extent but are unlikely to be identical. Housebuilders complain for instance that land allocation proposals fail to consider the relationship between site characteristics and market potential (Housing Monitoring Team, 1980). The trend of recent central government advice in circulars is to favour and encourage a greater incorporation of market- and demand-oriented criteria in the land use planning process. Circular 9/80 promoted local land availability studies, undertaken *jointly* between local planning authorities and representatives of the housebuilding industry, so as to 'bring together the housebuilders' assessment of market demand and the development potential of particular sites, with the local planning authority's assessment of planning objectives' (DoE, 1984d, para. 14). The results of such studies are of more than comparative interest since Circular 15/84 promises that they will be treated as a material consideration in determining planning appeals, thereby giving

the housebuilding industry something of a privileged position in the development control process.

3 The housebuilding industry. The structure of the modern speculative housebuilding industry is characterized by great diversity, featuring many small firms together with a limited number of much larger ('volume') house-builders. All speculative housebuilders, however, share a common financial objective, to maximize profits, although in practice this can be achieved or defined in several ways. Both the manner in which firms undertake housing development in pursuit of this objective and the land requirements of firms vary with firm size. Small builders (by output) are reported typically to prefer the upper end of the housing market, a limited number of expensive houses offering better returns than the generally more competitive volume markets. Land requirements for small builders are thus for small and well-serviced sites, usually infilling, in established residential areas. On the other hand, larger companies seeking profit growth have much larger outputs in the main markets over wide geographical areas. For these companies, land requirements comprise future building land, to be held in stock in the form of land banks, as well as land ready for immediate development (Harloe, Issacharoff and Minns, 1974); in general it is the larger housebuilders who carry out land banking (DoE, 1975c).

Land banks are held primarily to ensure a steady flow of suitable and available land for construction to proceed. The holding of sites prior to development thus allows for design, the obtaining of planning permission and the completion of initial site servicing. These technical reasons typically require a land bank cycle (period for which land is held prior to development) of some two to three years. In addition to these factors, however, sites are also held in land banks as a means of acquiring and retaining development gain that would otherwise accrue to the landowner. Development gain can be defined as the difference between house price and construction costs, and is apportioned between the builder as development profit and the landowner as the purchase price of land (capitalized rent). The builder's development profit can be considered as deriving from both land development and housing production. Ball (1983) suggests that it is not practically or even theoretically possible to distinguish between these two components, but it is nevertheless clear that many of the larger housebuilders have successfully integrated profits from landownership into the structure of their productive activities (Massey and Catalano, 1978). It must be emphasized that while speculative housebuilding is a form of land investment and a means of realizing gains from landownership, such realization necessarily requires housing construction. Thus, 'firms in the industry therefore have to be involved in an industrial process (building) rather than being solely holders of an appreciating financial asset (land)' (Ball, 1983, p. 52).

The development gain that accrues to the housebuilder as a result of land banking activities stems from two sources: inflation and planning (Smyth,

1982). Both of these factors imply a relatively long time period for holding land. The recent tendency for land price rises to outstrip the rate of inflation ensures that builders' land costs can be held down relative to house prices if the land is held for several years after purchase but prior to development. However, housing land price rises do not *inevitably* exceed inflation. Land use planning affects land price by the granting of planning permission for development, the value of sites with permission being much higher than the value of sites without, leading to a clear dual market in housing land along these lines (Drewett, 1973b). Hence land without permission held by builders is potentially subject to the most profits deriving from permission being granted. The study of residential land availability in the South East of England by the Economist Intelligence Unit for the DoE (DoE, 1975c, p. 44) in the early 1970s comments that:

where land without outline planning permission is considered for purchase there is a general rule of careful consideration being given to its planning prospects before purchase . . . the larger developers seem fairly confident of their ability to judge correctly the likelihood of obtaining permission on a site and indeed a number of them are willing to put substantial resources behind getting land zoned for development, such as offers of improvement to infrastructure and amenities to the local authority.

Nevertheless, such land constitutes a risk which the builder must assess against the financial advantages involved; prevailing planning policies are an important factor here. In purchasing land without planning permission or land unallocated for residential development in structure and local plans, the builder accepts that the site, subject to holding costs, is likely to be in his land bank, undevelopable, for some time while planning permission is sought. Substantial resources may have to be committed to 'promoting' the site for development, e.g. fighting appeals, resubmission of planning applications; there is always the risk of not obtaining planning permission at all. In an attempt to avoid this eventuality and minimize the risks and costs involved, housebuilders and their representative organizations support an increase in the supply of land allocated for housing over and above that already given in plans, hoping thereby to secure planning permission on holdings of unallocated land. Land releases of this nature also serve to weaken the monopoly position of landowners holding allocated land, while providing builders with a greater choice of sites.

8.5.4.2 *The evolution of policy*
The public statements of housebuilders suggest that as major 'customers' of the planning system they require an adequate supply of land at the right time and place (see, for example, Baron, 1980). This begs questions as to the criteria to be used for example in establishing the location of new residential development, and minimizes the difficulties posed for both housebuilders

and local planning authorities by the private ownership of allocated development land. To say that the housebuilding industry is a 'customer' of the planning system emphasizes the latter's role as a service to developers and underplays the conflicts between need and demand referred to earlier. It is not a concern of local planning authorities to ensure the continuing profitability of speculative housebuilders. Nevertheless, development gain associated with the obtaining of planning permission and captured by housebuilders via land banking contributes to builders development profit. The prescription that housebuilders require suitable land at the 'right' time is perhaps most appropriate for the builder relying on profits from housing construction, whose land needs emphasize turnover, throughput and short land bank cycles to meet technical requirements. However, housing land supply is a more complex matter than merely allocating suitable sites as and when required for construction, since land banking intervenes. Moreover, at least in the past, allocating sites for residential development in plans has not precluded a high proportion of the land subsequently developed for housing from coming forward outwith the development plan system. The DoE/EIU study (DoE, 1975c, p. 6), emphasizing the then tenuous relationship between land available in planning terms and that actually available, found that,

in all four county areas studied less than half the planning permissions granted in the period since October 1972 were on land identified at that date as available in planning terms. This phenomena is the result of the peak demand conditions in the period referred to, and to the inherent difficulty of anticipating which parcels of land are available to come forward for development so that they can be designated in planning terms.

This clearly weakens housebuilders' arguments that land use planning is a significant supply-side constraint on their activities (via a shortfall of allocated housing land), and emphasizes the role of unallocated land held by the housebuilding industry. The *ad hoc* release of land documented by the DoE/EIU study involved infill and redevelopment sites or deliberate departures from policy assumptions (e.g. as a result of appeal decisions on Green Belt sites). *Ad hoc* land release encourages builders to buy unallocated land with low hope value, but in itself is no answer to their land supply requirements since it largely bypasses the role of planning as a facilitating mechanism, e.g. there is no guarantee that such sites will be free from servicing and infrastructure problems. It must be noted that the DoE/EIU study was concerned with a statutory planning context very different from that prevailing today. In 1973, for instance, the DoE (1973b, paras 2 and 4) envisaged that pending structure plan approval and the subsequent adoption of local plans, and in order to meet current housing land requirements,

development will have to be allowed on a good deal of land shown as 'white' on existing Development Plans ... in growth areas planning applications must be considered against the background of a strong general presumption in favour of housing.

Given this advice from central government, the extent of 'white' land receiving residential planning permission in the early 1970s in the South East is not surprising. Structure and local plans today provide a comprehensive framework for assessing needs and demands for housing land (see Home, 1985), and no doubt the likelihood of gaining planning permission for housing on land not allocated for such is much less, Circulars 22/80 and 15/84 notwithstanding. Nevertheless Cuddy and Hollingsworth (1983, p. 175) report that in Wales some 40 per cent (6500 hectares) of the potential housing land held by builders and developers as stock in trade lacked any form of planning approval, and comment that:

The activities of developers in acquiring land unallocated for development is an understandable precaution, and can often produce considerable returns, but the operation of this on a large scale must inevitably be counter-productive. The developer may have a large amount of working capital tied up, the land itself is very often unused or underused and the planning authority have to spend time resisting such development in appeals when they could be proceeding more rapidly with the production of up-to-date plans.

In addition, Ball (1983) estimates that nationally the industry has a land bank equivalent to nine years production at the 1980 starts level; this comprises three-and-a-half years of land with planning permission and five-and-a-half currently without permission. Such holdings of land without planning approval presumably indicate something of builders' perceptions of the likelihood of winning permission at appeal and the status of statutory plans. This in turn raises questions as to how far, and how successfully, the housebuilding industry has pressed its case for land release with central government, and how this national lobbying has worked through at local level.

Since the housebuilding industry comprises a diverse collection of firms, it should not be seen as a single monolithic body acting with a single purpose. The industry features a number of representative organizations,[9] which represent the industry nationally and lobby on its behalf; they are essentially pressure groups promoting the broad interests of their members. Locally however builders are in direct competition with each other, although there are aspects of local co-operation, e.g. joint ventures and consortia. In terms of policy-making at the national level, there appear to be close links between government and the housebuilders' representatives which typically emphasize the interests of the larger firms (Rydin, 1984; Hooper, 1979, 1980; Humber, 1980). The involvement or even incorporation of housebuilders within central government policy-making is fertile ground for conspiracy theorists but is more likely to reflect a common perception of the housing-land supply problem. In any event DoE Circulars 9/80, 22/80 and 15/84 allow the industry to become involved in *local* policy review of available housing land supply and establish a general presumption in favour of development. Thus the identification of a five-year supply of housing-land in accordance with Circular 9/80 does not preclude residential development on sites outwith that supply – each case must still be considered on its merits (Circular 22/80, para. 8, restated in Circular 15/84).

The detailed operation of the local housing land availability studies established by Circular 9/80, involving the joint assessment of specific sites by planners and builders, can be likened to Christensen's (1985) conditions of 'known technology, no agreed goal' and hence seen as a bargaining process. The study methodology is closely prescribed following the initial joint study in Greater Manchester (DoE/House Builders' Federation, 1979), while the two main participants bring very different interests, perceptions, and objectives to bear on the process. As noted the housebuilding industry takes an economic or market orientation to landownership, while the planning system is concerned with public control over land use. These standpoints result in not only different *definitions* but also different *conceptions* of land availability, which currently diverge considerably in locational terms (Hooper, 1980). Land use planning favours peripheral restraint to aid inner area policies, while builders seeking a marketable environment for new houses prefer surburban areas; 'the spatial location of . . . housebuilding operations can only be shifted to places desired by planners if a marketable environment for selling houses can be created at the new locations' (Ball, 1983, p. 265). The housebuilders' profit criteria are incorporated into land availability studies as 'marketability'. Housing-land making up the five years' supply must not only be free from planning and other constraints but must 'also be capable of being economically developed, be in areas where potential house buyers are prepared to live, and be suitable for the wide range of housing types which the housing market now demands' (DoE, 1980a, para. 5). This has allowed housebuilders to challenge planning criteria for site allocation with their own market criteria. In turn, marketing criteria enshrined in the joint study methodology – particularly the market-banding principle – encourage inflationary estimates of land availability and future land needs (Hooper, 1980, 1983), and in this sense the methodology is hardly neutral despite its 'technical' (and therefore implied value-free) appearance.[10]

Each of the study participants is dependent on resources controlled by the other in order to achieve their aims *vis-à-vis* residential development. There is as a result a basic mutual inter-dependence underlying participation in the studies. Local planning authorities control the granting of planning permission, but this is primarily a negative power of veto, despite the commonplace informal negotations between applicant and authority. Housebuilders control initiation of the production process, but the strength of this 'resource' may be reduced by holdings of unallocated land on which they desire to secure planning permission. On the other hand, local planning authorities may be dependent on builders for implementation of planning policies on residential development, a dependency increased by reductions in expenditure on municipal housing construction. The builder obviously depends on the planning authority for the permission to develop; this is significant if the builder already holds the land, but less so if he is not committed to a specific site. In any case, the local authority's 'monopoly' of planning permission is not absolute, since the builder can (at some cost) appeal to central government. The resultant decision might threaten wider planning strategies for the area via precedent, while for the builder, land release in this manner is unpredictable. This situation of mutual inter-dependency over housing-land encourages negotiated settlement and consensus where 'the final agreed assessment of land supply represents the bargain struck between the parties rather than an objective assessment of future land supply' (Cuddy and Hollingsworth, 1983, p. 176). This is only possible where 'availability' comprises a set of negotiated criteria rather than an objective or technical definition. As Christensen points out, the aim of bargaining where goals are conflicting or uncertain is to accommodate multiple preferences. Housing land availability studies seek to do this through the production of a consensus as to land supply which is seen as legitimate by both groups of participants. However, this does not remove underlying differences in orientation towards the residential development process.

One of the options available to housebuilders as an alternative, or in addition, to detailed site-by-site assessments carried out in conjunction with local planning authorities is to mount explicit attacks on local planning policies with the hope of winning permission at appeal. This is in effect the strategy adopted by Consortium Developments Limited (CDL) in their attempt to force the release of housing land in the South East. CDL, composed of ten leading volume housebuilders,[11] and therefore a spectacular example of co-operation in the industry, propose a series of new settlements in the South East outside London. This strategy for major urban expansion within rural areas is contrary to established planning policies of restraint in the region, typified by the Green Belt. It is proposed that each settlement accommodate between 13,000 and 18,000 people in 5000 to 7000 houses, requiring a land take of 850 to 1000 acres, and mainly comprising private housing for sale to be developed by Consortium members and local housebuilders (Bennett, 1984). Settlements could also include some local authority special needs' housing and housing association development for rent, together with shops, industry and offices, schools, recreation and other community facilities. In total, new settlements would be developed as comprehensive schemes. The first settlement to be announced (May 1985) is 'Tillingham Hall', a 'new country town' of 14,000 population to be located on Green Belt land in Essex; local planning authority (Thurrock BC) opposition is expected,[12] with local communities also mobilizing to contest the proposals (Wolley and Rosborough, 1985).

The case for the provision of such new settlements by the speculative housebuilding industry has been put by Shostak and Lock (1984, 1985) in terms of a selection of planning policy objectives and demographic trends. An overriding concern is the gap that is said to exist in the South East between structure plan housing provision and forecast increase in households. The conclusion that 'put brutally, the effect of current structure plans in the South East is to create a housing shortage' (Shostak and Lock, 1985, p. 19) is clearly an effective political point, albeit one which plays down the role of agreed regional and sub-regional strategies in setting structure plan contexts. Also, household increase may well generate additional housing need, but it is surely unlikely that this will be converted in a straightforward fashion into effective housing demand, to which housebuilders respond. The CDL proposals are being promoted as a solution to housing shortage, although there is no reason to assume that new settlements developed by private housebuilders represent the only or indeed the most appropriate response to housing provision in the South East. One alternative for instance would be to develop a 'centripetal' rather than 'centrifugal' urban strategy in order to overcome substantial under-use of existing urban housing stock and land (Merrett, 1984). Wray (1985), giving qualified support to Shostak and Lock, accepts much of their analysis if not their prescription, arguing instead for a planned resurrection of the new town and town-expansion programmes. In terms of regional strategy, the CDL proposals represent an *ad hoc* rather than planned response. However, it is a mistake to assume that CDL is reacting in an altruistic manner to housing shortage; rather, the new settlements represent a response by the volume housebuilders to their own difficulties over land supply in the region. Forecast housing shortage provides the political justification for land release. It is indeed acknowledged that before 1991 'the position will become critical *for the industry*' (Bennett, 1984, p. 4, our emphasis). The response of CDL members to this problem has partly taken the form it has because as Healey suggests volume housebuilders need to sustain volume output in order to maintain profits.[13] This is especially relevant to those builders who emphasize turnover in their operations, and whose land banks, though large, are rapidly depleted. Large sites also offer continuity of land supply and housing production, together with economies of scale. The new settlements reflect the land requirements of Consortium members in seeking to respond to effective regional housing demand rather than flowing from existing public development strategies; hence the proposals are for 'communities of pre-determined size and character looking for somewhere to happen' (Savage, 1984, p. 4). In seeking the release of large sites *within the Green Belt*, CDL is also looking for a high-quality local environment to aid marketing – ironically, an environment maintained by the very planning policies now under attack. Individual builders can be expected to be supportive of planning policies of restraint and environmental protection when these aid marketing of the finished product. Equally, they are unlikely to be supportive when such policies prevent *them* from building in the area. The logic of the CDL strategy requires both *release* of Green Belt land and the continuing *retention* of these policies outwith the new settlements.

In securing the planning release of required sites, the CDL 'carrot' is the promise that the developers will provide necessary infrastructure and community facilities together with superior landscaping and estate layout. It is said that the costs of these elements will be met from the relatively low land costs of the new settlements, although 'ultimately of course, the acquisition costs and costs of basic infrastructure will determine whether any public investment in a new settlement will be required' (Bennett, 1984, p. 7). The success of CDL in acquiring sites at less than full development value is crucial here, but there will still be revenue implications for public authorities (Savage, 1984). The CDL 'stick' is the threat that sites will be released following any subsequent appeal by the Consortium to central government. In fighting such appeals no doubt the quality and low public cost of the proposals will again be emphasized. The promotion of the new settlements as well-defined and comprehensive schemes embodying high-quality design echoes earlier successful challenges to Green Belt policy on the basis of integrated layout plans reported by Gracey (1973). However, the result of appeals by CDL is far from a foregone conclusion; local opposition seems likely and may unite a range of interests to produce a broad coalition between farmers, rural landlords, suburban residents and the environmental movement in the manner envisaged by Rydin (1984); Green Belt policy has strong and widespread (if uncritical) public support; and the government, while confirming its commitment to a continued growth in owner occupation, has also recently reaffirmed the objectives of Green Belt policy.

8.5.4.3 *Local planning, local plans and residential land availability*
The land use planning system and the housebuilding industry interact first over the supply of land for residential development, which may be set out in plans or established by the granting of outline planning permission, and second over detailed features of proposed developments, such as layout, design or materials. Planning policies on such latter aspects usually constitute supplementary planning guidance. The concern of the housebuilding industry is not so much with policies as expressed in plans or other statements *per se*, but with how the local planning authority exercises its central power of granting or witholding planning permission. In seeking to secure a supply of 'suitable' land (the definition of which varies from firm to firm) the builder must assemble ownership rights from the landowner and secure planning permission from the local planning authority. The response of the larger builders in particular to this dependency has been to extend their control over the necessary resource of land by landbanking. In the process of bringing land forward for development, plans and policy statements are of

interest as *guides* to future development control decisions, and as such may affect the timing of an application or the purchase decision. How individual housebuilders view plan allocations depends on their existing land holdings in the area and company strategy. Builders concentrating on housing construction rather than land development can be expected to place more emphasis on the usefulness of plans in identifying readily available sites in planning, ownership and infrastructure terms, and which are located in suitable (i.e. marketable) areas. These are factors addressed by joint housing-land availability studies. On the other hand, builders undertaking land development activities will be seeking the allocation of their sites for residential development, and the subsequent granting of planning permission, in order to realize an element of development gain. To the extent that local plans exclude such sites, calls and pressures for more land release can be expected to continue, a five-year supply of housing-land agreed locally notwithstanding. Builders are unlikely to see policies and proposals as contained in emerging or operative statutory local plans as definitive or immutable. The principal options open to housebuilders seeking to challenge such policy statements, apart from negotiations with the local planning authority, are to object during plan preparation, e.g. at the PLI, to apply (following adoption) to the High Court under s.244 of the 1971 Act for the plan to be quashed in whole or in part,[14] or to appeal to central government following the subsequent refusal of a planning application by the planning authority. Local planning policies expressed in a statutory local plan are thus open to formal challenge during plan preparation *and* implementation. These ways of challenging policy locally are in addition to national representations made by the industry to central government, which may if heeded influence the policy context in which local planning policies are formulated and operated.

8.5.5 Local planning, local plans and the private sector

This review of the relationships between private development interests and local planning authorities suggests that local planning policies are primarily of relevance to such interests in so far as they are applied in development control by local planning authorities and upheld at appeal by central government. In particular, private sector interests will be concerned with local plan proposals for land release for residential, industrial and commercial development (Capner, 1982). Plans (or rather policies in plans) represent a partial guide to the outcome of development control, and development interests are dependent upon the granting of planning permission. The primary concern of development interests is with the current policies of local authorities, evident from decisions on planning applications and from policy statements (plans) in so far as these are up-to-date. Development plans inform development control and so can be used by private developers as a source of information on likely future policy. Central government's disapproval of non-statutory material seeks to establish statutory local plans as the main local planning instrument in this respect. Housebuilders may thus consult structure and local plans in order to identify growth areas, and to assess the likelihood of gaining residential planning permission on particular sites, allocated or not. How they use the information thus gleaned depends on company strategy and the nature of their land requirements. Many development interests, aware of the potential discrepancy between policies in plans and current policies, use published planning documents as an initial guide to planning policy, which is then developed through consultations with the planning authority (to determine the present status of policy statements) and subsequently through negotiations over particular proposals. Local plans thus constitute an important element in the development game played between developers and planners, and to this extent patterns of development on the ground can be attributed at least in part to the expression of policies in plans (e.g. see Blyth's 1984 study of the implementation of the Northwich District Plan). Plans represent provisional statements of intention as to how the local planning authority will exercise its powers of development control. They are not (and cannot be) a comprehensive guide, since development control must also have regard to other material considerations.

Plans may also be used selectively by private development interests in fighting appeals; the policy statements of other tiers of government are also important here if they can be used to support a developer's case, e.g. circulars, structure plans. In general, the strength of development plans in development control depends on explicit central government support for their provisions. In situations where policies have been applied consistently by different levels of government and there has been stability over time, as in the South East during the 1970s, statutory plans are likely to provide an important tool for carrying through the policies within them, particularly where these policies relate to the central land use planning power of land allocation and release. However, recent central government advice calls for a more liberal approach to planning applications for housing and industrial developments and emphasizes the significance of other material considerations rather than development plan provisions (DoE Circulars 15/84, 16/84, and 14/85). Although the facilitation and co-ordination of private sector development continues to be regarded as an important task for structure and local plans, particularly as regards land supply, it is made clear that the identification of sites in plans does not preclude other sites from being brought forward. At the same time, central government remains committed to such objectives as conservation and the protection of the environment, objectives which will often be detailed at the local level in development plans, e.g. Green Belt. In short, it is clear that central government support for development plan policies is being explicitly reduced, with the exception of policies seeking to protect what the government regards as 'interests of acknowledged importance', i.e. the national heritage, Green Belt, agricultural land, the environment (Circular 14/85, para. 3). Development interests seeking the release of sites which have not been allocated for development in plans will benefit from Circular advice by using the formal processes of development control to appeal stage (where central government attitudes supposedly take effect) rather than being satisfied with negotiations with the local planning authority. Alternatively, if local planning authorities accede to the exhortations of circulars, this will initially encourage a drift of policies away from those enshrined in plans, and emphasize the role of consultation, negotiation and discussion between the authority and development interests as the latter seek to determine current local authority policy and its implications for their own development proposals and requirements.

8.6 Conclusions

There are few indications that statutory local plans or informal local planning documents *per se* are of central concern to 'implementing' agencies in the public or private sectors, whatever the hierarchical model of policy formulation and implementation underpinning the British land use planning system suggests. Many plans contain proposals and site allocations which are intended to be carried out or taken up by private sector development interests. However, beyond issuing development briefs or in a limited number of cases entering into partnership with private developers, there is little the local planning authority can do to bring about development in line with plans. Site allocations in local plans for example do not automatically resolve landownership or development finance issues. The planning authority can of course prevent development occurring that is contrary to plan policies – as long as these are supported at appeal – and statutory local plans are likely to be especially useful here given present central government attitudes to informal plans and policies. The ability of the local planning authority to refuse planning permission is a key factor in the development process to which all private development interests must have regard; this dependency is the foundation of the bargaining strength of the planning authority. Statutory local plans and other statements of local planning policy give an indication of the likely outcome of development control decisions by formulating and expressing policy and providing a means of carrying such policies through. Hence, policies of development restraint may be usefully incorporated into statutory plans where these provide a mechanism for carrying this strategy through into decisions on planning applications. On the other hand, statutory plans are likely to be less appropriate for establishing policies of development promotion, especially where development pressure is slack. The bargaining power of local planning authorities is relatively limited where local plan policies and proposals are dependent upon the private sector for implementation. More active forms of involvement in the development process may be appropriate here, e.g. site assembly, infrastructure provision, but in general resources for such intervention are limited. In the public sector, co-operation seems to be most successful between local planning authorities and infrastructure and development agencies, but is not necessarily based on plans as such. Instead, it is more likely to relate to informal channels of communication, perhaps established through a concern on the part of the planning authorities with implementation (as in Berkshire) or as a result of the collaborative and consultative efforts required of public agencies during local plan preparation. Infrastructure agencies have an important role in ensuring allocated land is genuinely available, at least in terms of services, while the various development agencies and the resources they command are also important for plan implementation where policies relate to their area of operations. An explicit concern to carry out positive planning, aimed at implementing local planning policies (whether in plans or not), is not a central part of the remits of these agencies, but they can be expected to operate within the framework of adopted planning policy in plans. Where the nationalized industries are involved in land and property development, technical or commercial criteria outweigh the small area concerns of local plans. Sectoral social and community service provision and funding has the effect of fragmenting policy areas, with planning departments having little effective influence over the timing of associated developments. However, the land use implications of *service* development plans can where available be included in local plans in an attempt to ensure co-ordination.

The essential point here is that few public agencies or private sector interests organize their land holding and development activities primarily with reference to local planning policies. In other words, development decisions are seldom plan-based in that policies and proposals are followed because they have been expressed in local plans, so that these cannot be said to guide development (Healey, 1983). Reflecting this, few statutory local plans contain details of how their policies and proposals are to be implemented, beyond an indication of whether the private or public sector is responsible and a review of committed expenditure as regards the latter (see Chapter 4). In short, it seems that the legislative framework within which statutory local plans are prepared and used provides an insufficient basis for the effective management of change. Instead of development decisions being typically plan-based, where local plans are taken account of by private sector development interests and public-sector agencies engaged in infrastructure provision and development, the relationship is characterized by negotiation and bargaining, whereby 'development interests and planning authorities are appraised of each other's interests and intentions in various ways through an interactive process' (Healey, 1983, p. 234). Local plans, statutory and informal, are one of the available channels for informing interested parties of local planning authority policies. Policies and proposals in plans can be regarded as bargaining counters to be deployed as required in a bid to control or at least to influence the outcome of negotiations. Here, the discretionary nature of development control ensures room for the flexible interpretation of local plan statements, opening up scope for a negotiative relationship between applicant and authority. Public development agencies are likely to take up projects as they are put forward by planning authorities through processes of consultations and programming, while co-ordination with infrastructure agencies can relate to planning policies but be expressed and developed through informal or semi-formal channels of consultation and communication, ie. be derived from continuous interaction rather than being plan-based. The importance of these negotiative relationships between local planning authorities and development interests suggests that an important planning task is to attempt to oversee the activities of public agencies and private interests so as to manage rather than control change in the built environment in the pursuit of policy aims. This requires planners to act as facilitators; to mediate between interests, to promote development and bargain with actors involved in the development process in addition to producing plans, which will still be prepared as required, e.g. to support restraint policy. However, the effectiveness with which statutory local plans carry out their formal roles within the land use planning hierarchy (i.e. to bring local planning issues before the public, to detail structure plan policies and proposals, and to provide a detailed basis for development control) is threatened where the relationships between local planning authorities and development interests are based on continuous interaction rather than plans. In turn, this raises issue of local public accountability and of the conformity of local planning policies as being pursued to structure plan strategy.

Finally, this review of the ways in which local planning policies influence or seek to influence development activity emphasizes the complexity and variety of development interests and processes. Much development activity, particularly in the private sector, is also highly volatile and unpredictable. Development processes are both dynamic and uncertain, reflecting the operation of wider economic forces which give rise to demands for changes in the built environment. Planners have not always successfully anticipated such demands, or readily responded to them (for instance, large-scale retail developments). Development processes and the land and development issues that arise from these processes have many of the characteristics of complex and 'wicked' problems as identified by Mason and Mitroff (1981), i.e. interconnectivity, uncertainty, ambiguity and conflict. In the process of finding acceptable solutions to development pressures, local planners have developed negotiative relationships with both public and private sector development interests. Bargaining is a key activity here, with the main resource available to the planning authority being the ability to grant or refuse planning permission. Negotiation and bargaining carried out at the local level can be seen as a pragmatic and contingent response to the complexity of development processes, allowing and facilitating the management of environmental change. Nevertheless, there is still a need for such interaction between interests and agencies to be guided by policy aims. The British land use planning system does have a loose hierarchical structure based on plans as statements of policy, but this framework for defining and progressing policy is potentially weakened where the local planner's role in reality emphasizes bargaining and management rather than the plan-oriented control of change. Local negotiative relationships deciding land and development issues, however useful and expedient in practice, must be explicitly located within the wider policy context if higher-level strategies for socio-economic change are to be satisfactorily translated into development on the ground.

Notes

1 Particularly as regards the extensive holdings of the Forestry Commission. See *Chartered Survey* or (1980) Selling surplus public property, supplement to Vol. 112, No. 9, April. There are also provisions for the registration of surplus public sector land under Part X of the *Local Government, Planning and Land Act 1980*, and the Secretary of State can direct (under ss.98 and 99 of that Act) that registered sites be sold off to the private sector. See 'Register sites in disposal direction', *Planning*, No. 596, 23 November 1984, p. 1; 'More councils hassled to release register land', *Planning*, No. 625, 5 July 1985, p. 16. There are no equivalent provisions for the identification of unused or underused private sector land, e.g. land held in the land banks of housebuilders.

2 See 'Surplus NHS land to be sold', *Guardian*, 13 February 1982; 'TUC turns back to ACAS in health dispute', *Guardian*, 29 June 1982; and 'Green belt health drive', *Planning*, No. 504, 4 February 1983, p. 1.

3 For instance, the regularly updated Department of Health and Social Security/Welsh Office statement of policy on land matters, 'Handbook on Land Transactions'.

4 See 'Public Inquiry ordered by the Ministry of Agriculture into the disposal of land at Crichel Down', Cmnd. 9176, Ministry of Agriculture and Fisheries, June 1954, and 'Disposal of Government land to former owners', *Journal of Planning and Environment Law*, February 1982, p. 67.

5 See 'Secret coal-powered gas works', *Planning Bulletin*, 45/80, 28 November 1980.

6 Secretary of State's Address to the Town and Country Planning Summer School, 1981.

7 See 'Impact on area demolishes case for glass stump', *Planning*, No. 620, 31 May 1985, p. 1.

8 See 'Commerce blows a gale on draft plan', *Planning*, No. 604, p. 9; 'City draft local plan crushed in tank attack', *Planning*, No. 611, 29 March 1985, p. 16.

9 The Federation of Master Builders, the main interest group representing smaller firms; the House Builders' Federation; the National Federation of Building

Trades Employers; and the Volume Builders' Study Group. The latter comprises nine of the largest housebuilders by output.

10　See the recent debate on the methodology prescribed by Circular 15/84 (Annex A) in *Planning*: 'Strange side effects from central prescription', *Planning*, No. 610, 22 March 1985, p. 8; Letters to Editor, *Planning*, No. 612, 5 April 1985 to No. 615, 26 April 1985; No. 617, 10 May 1985, and No. 618, 17 May 1985.

11　Barratt Developments plc, Bovis Homes Ltd., Broseley Estates Ltd., Christian Salvesen (Properties) Ltd., Comben Group plc, Tarmac plc, New Ideal Holdings plc, Wilcon Homes Ltd., Wimpey Homes Holdings plc, and Leech Homes Ltd.

12　See 'Consortium village troubles council', *Planning*, No. 598, 7 December 1984, p. 16, and 'Thurrock green belt earmarked for new village', *Planning*, No. 615, 26 April 1985, p. 1.

13　See Healey, P., 'New villages aimed at wrong target', Letter to the Editor, *Planning*, No. 621, 7 June 1985, p. 2.

14　For example, see Fourth Investments Ltd. *v.* Bury Metropolitan Borough Council and the Secretary of State for the Environment (Queen's Bench Division, McCullough J., 23 July 1984), reported in *Journal of Planning and Environment Law*, March 1985, pp. 185–188.

9 Local planning in practice: conclusions and implications

9.1 Introduction

The theoretical perspectives on local planning reviewed in Chapter 2 are set out as if they are free-standing, largely for clarity of presentation. In reality they are highly inter-related and all point in slightly different ways to the complex inter-related nature of planning problems and the need for a contingency approach to be adopted in dealing with 'wicked' public policy issues, e.g. the characteristics of wicked public policy problems identified by Mason and Mitroff (1981) give emphasis to uncertainty and the impact of societal constraints (or contingent factors) in managing such problems; Christensen (1985, p.69) in her treatment of uncertainty shows that '. . . planning processes can be understood as addressing different conditions of uncertainty . . . . By tailoring planning to real world conditions, the planner is acting contingently.' The *Law of Requisite Variety* by implication points to the need to adopt a contingent approach, whilst the view of planning as the management of change in the environment similarly reflects this view.

Against this background, this final chapter:

1 considers the relevance of the theoretical perspectives reviewed in Chapter 2 for local planning in practice;
2 evaluates the achievements of statutory local plans against their established roles, i.e. to develop and detail structure plan policies and proposals; to provide a detailed basis for development control; to provide a detailed basis for the co-ordination of the development and other use of land, and to bring local planning issues before the public (DoE, 1984g, para. 3.1);
3 reviews the resultant implications for the practice of local planning.

9.2 Theory and local planning in practice

9.2.1 Local planning, complex 'wicked' problems, and variety

In the words of Mason and Mitroff (1981, p. 4) '. . . every real world policy problem is related to every other real world problem', and local planning problems are no exception. The Planning Advisory Group in reviewing the future of development plans in 1965 recognized this when it stated 'the problems of physical development and redevelopment . . . are often highly complex, involve investment decisions of great magnitude and extend across many fields of policy. Many different agencies, public and private, are involved and many different interests are affected' (PAG, 1965, p. 3). Goldsmith with the advantage of hindsight explicitly spells out the complex inter-relationships between housing policies intended to provide security of tenure in the private-rented sector, and the physical deterioration of much private-rented housing (Goldsmith, 1980, p. 23). Similarly the inter-relationships between farm price support programmes and the inner city problem in the USA, as reported by Mason and Mitroff (1981, pp. 8–9), reinforce this point.

At a more local level the Southtown local plan (see Chapter 4, Section 4.7.2) and the proposed industrial and commercial developments at Miskin (see Chapter 7, Section 7.5) emphasize the inter-related nature of local planning problems. Thus in Southtown the proposed economic regeneration of the area was inextricably linked with issues of housing policy, traffic, conservation, the redistribution of retail turnover and the relative power of the District and County Councils. In the case of Miskin, the proposed industrial and warehouse development showed the inter-relationship between the wider issue of employment generation, and conservation, sewage disposal, the loss of agricultural land, and the protection of industrial sites already agreed in the relevant structure plans in other parts of South Wales.

Despite these recognized complex inter-relationships, the statutory land use planning system is restricted to a concern with the development and other use of land in a way which makes it difficult to take full account of these inter-relationships. Whilst the development plan established by statute may take account of wider social and economic issues in so far as they affect the development and other use of land, the form, content and procedures associated with the production of the development plan give almost exclusive emphasis to the land use aspects of these inter-relationships. At the same time, the limited range of statutory land use plans available to practice (the structure plan and three types of local plan) ensure that only a limited range of public policy areas can be handled adequately by the statutory system. Whilst the Hertfordshire local planning authorities find the statutory planning system an effective means of controlling growth and new development, the authorities responsible for the planning and development of inner city areas find the statutory planning system largely irrelevant to their needs. In short, the variety of problem situations encountered in the practice of land use planning is not matched by variety in the system established to deal with these problems.

Compounding this problem is the fact that British public policy is established and implemented by specialist and segmented sets of interlocking institutions, each consisting of politicians, professional experts and interest groups. Inevitably these institutions constrain what gets debated and how policy goals and objectives (ends) and the means of achieving those objectives are agreed. The specialist institutions in, for example, housing, transport, agriculture, and social services fix their agreed goals largely independently and then proceed to consider a narrow range of alternative means which must be acceptable to the politicians, the professional experts and the vested interests in the speciality. The inter-related and complex nature of planning problems is ignored in a concern to agree compartmentalized ends and means. Rarely, for example, is housing expenditure increased at the expense of transport, or vice versa. As a result, in addition to the statutory structure and local land use plans, a range of compartmentalized but statutory local planning instruments are also operated in the areas of:

- *housing*, including GIAs, HAAs, and HIPs;
- *environment and conservation*, including National Park Plans, AONBs, Heritage Coasts, Nature Reserves, SSSIs, conservation areas, Operation Groundwork and waste disposal plans;
- *urban development and regeneration*, including IIAs, Partnership arrangements, the Urban Programme, Enterprise Zones, Urban Development Areas and Corporations, and Task Forces;
- *transport*, including TPPs and PTPs (Chapter 5).

It can be argued that this range of local planning instruments reflects the variety of problem situations encountered in practice and thus accords with the concepts associated with the *Law of Requisite Variety*. However the problems of successfully using these instruments arise out of the fact that they are largely compartmentalized and are operated without an over-riding strategic framework. In the absence of such a guiding framework the system of planning and implementation which emerges in practice is inevitably disjointed and incremental. As Solesbury (1981) has illustrated in the fields of regional development, housing and health, development on the ground resulting from day-to-day decisions often bears little relation to any overall strategy.

9.2.2 Strategic planning, contingency and the management of change

Management theory suggests that one way of handling problems of complexity and inter-connectivity is to structure policy and decision-making in a hierarchical fashion, which reflects a military distinction between strategy and tactics, i.e. a 'top-down' approach to the policy–action relationship. This classical organizational principle has been supplemented in recent years by a recognition within the field of management of the important role of contextual or contingent factors which define the characteristics of particular problem situations. This contingency approach indicates that managers should be able to select the approach and course of action which is judged to be most appropriate at the time for dealing with specific issues within their particular contexts. There is no single approach applicable at all times to all problems in all situations. Applying these ideas to strategic planning, it can be argued that any framework for planning and managing change in the environment should be capable of the following:

1 Ensuring that higher order policies or objectives for social and economic change are transmitted vertically through the framework in a way which leads to a co-ordination of plans and their implementation at different levels.

2 Allowing actors at different levels in the framework to develop approaches and solutions to particular problems in the light of prevailing contingent factors – which will include higher level policies and decisions.

These two 'prescriptions' can be used to develop an idealized framework for the definition and progression of policies for environmental change, against which the British planning system can be evaluated. This exercise also points up deficiencies and omissions in the theoretical approaches themselves.

A hierarchical model of policy control and formulation is relevant since the planning system has a recognizable hierarchical structure, even though certain elements are currently given little weight (e.g. regional planning) or are subject to change (e.g. the abolition of the metropolitan counties and the introduction of unitary development plans). Certainly this model underlies the development plan legislation and informs the DoE's view of the system in operation, e.g. development plan schemes and certification can be seen to embody the strategic planning principle that policies set out in the level above should influence those developed in the level below (Chapter 3). A review of the statutory local plan against an idealized hierarchical framework points to various difficulties with such documents as presently constituted, e.g. there are problems in expressing local socio-economic policies as a result of the area of concern of statutory local plans being restricted to matters of land use; there are difficulties in establishing a meaningful basis for the co-ordination of development given the range of autonomous vertically organized public agencies; there are the problems of implementing plan policies through development control, and limitations on the extent to which the necessary resources can be controlled or even influenced by the plan itself. Local plans can readily be used to define land use policies. However the problems outlined here cast doubts on their value for policy implementation, i.e. on their effectiveness as instruments for the control and management of change, in such a way that change takes place in accordance with planning policy. In turn, these shortcomings highlight the discrepancies between the British planning system and that suggested by strategic planning principles (Chapter 4). At the same time, these difficulties can be related to the shortcomings of the hierarchical approach itself, which takes an explicitly 'top-down' view of the process of putting policy into effect. Key assumptions are that lower-level implementers are constrained by higher-level policy-makers, and that all policy is formulated at the top and is an essential pre-condition for action. However, action can precede policy, where policy emerges as a response to problems and solutions devised on the ground. Moreover, many agencies whose actions are relevant to the implementation of planning policies operate in a relatively autonomous fashion, at least as far as the local planning authority is concerned. For instance, the local activities of a nationalized industry can be crucial to attaining local plan objectives (e.g. land release, or plant closure may wreck a local plan strategy) but are unlikely to be amenable to local planning authority influence. These criticisms of the hierarchical model and the assumptions it embodies lead to a different view of the policy–action relationship, as an essentially political process characterized by negotiation, bargaining and compromise between the groups involved (Bruton, 1980; Barrett and Hill, 1984). From this perspective, according to Bruton (1983b, p. 42):

the development plan is an essential and integral element in the bargaining process which is development control – it is the element in the process which establishes the boundaries of authority of planning policy and legitimizes the role of the local planning authorities. It is the element of the process against which the applicant (and appellant) use their counter arguments, power, influence, and resources in an attempt to implement development which would otherwise not be permitted.

Nevertheless, the problems for local plans and local planning identified above (e.g. the difficulties of securing effective inter-agency co-ordination, or of implementing plan policies through development control) remain whether we view these as shortcomings of the statutory plan and organizational issues within the context of the hierarchical approach, or as central features of public policy only reconcileable through processes of negotiation and bargaining. More generally, a comprehensive view of local planning is more likely to emerge from a combination of approaches rather than a single perspective applied in isolation.

The contingency approach emphasizes the significance of the policy and operational contexts within which local planning takes place and as a consequence (a) anticipates the wide range of variety in local planning strategies and (b) suggests that detailed prescriptive advice issued by central government will in general be of limited relevance. Whilst the formal contexts for local planning emerge from strategic or structure plan policy, the development plan legislation and subsequent interpretation by the DoE, and the statutory procedural requirements for local plan preparation and adoption, these form only part of the range of contingent factors influencing the selection or evolution of local planning strategies by local planning authorities. Other factors impinging on the local planning process include the nature of inherited policies, required strength at appeal, the perceived need for policy frameworks in terms of timescale (urgency), topic and area coverage, the nature of prevailing county–district relations, the local political situation, and the planning issues involved. For instance, factors acting to dictate the need for a statutory plan, e.g. as a firm basis for restraint policy, may outweigh associated drawbacks such as the length of the preparation process. The contingency approach suggests that the use by planning practice of a wide range of informal as well as statutory means of progressing policies is a reflection of the range and local variety of local planning contexts, which recognizes that the prime determinant of any planning strategy is likely to be the local context or environment which contains the very issues the strategy must deal with. The range of planning strategies adopted in practice reflects the principles of the *Law of Requisite Variety* (Ashby, 1964), whereby complex systems can only be guided by equally complex systems of control. Thus the complexity of real-world planning issues has in part been matched by the evolution of mechanisms intended to manage and control these issues which display an equivalent variety. The fact that practice uses both statutory and non-statutory policy documents implies that the formal local plan is too restricted an instrument, in terms of form, content and procedures, to match the variety of local planning issues faced by local planning authorities throughout the country. Authorities have taken up whichever form of planning instrument, statutory or informal, appears to them to be the most appropriate means of coping with a particular problem or problems in particular circumstances. Hence the variety of statutory plans and informal documents in use in practice, whatever DoE advice has had to say, and hence the variety which characterizes statutory local plan content within the prescribed limits.

This inevitably leads to the recognition that various aspects of planning strategies such as the choice of statutory or informal documents are largely influenced by or contingent upon local contextual circumstances. In turn, this has the effect of putting into perspective those contingent factors associated with the formal hierarchy of policy control, and of emphasizing the importance of local authority discretion as to choice of strategy. In particular, guidance and advice contained in circulars from the DoE on local planning matters is only one of a number of contingent factors to which local planning authorities have regard in choosing local planning strategies. In this respect, contingent factors produce a 'unique solution' for individual authorities, hence, 'it is not appropriate to offer universal guidance on the sort of approach that should be taken in different circumstances' (Fudge *et al.*, 1983, p. 128). However, herein lies the main drawback of the contingency approach. It is difficult to proceed much beyond a recognition of the importance of local operational circumstances in order to generalize about how any given set of contingency factors identified as relevant will affect the local planning strategy taken up. Indeed, until such generalizations are formulated, prescriptive statements as to, for instance, the appropriate strategy to adopt in any given circumstances, will lack precision and detail. For example, central government advises local authorities to prepare statutory local plans where needed, but cannot define such situations other than in the most general terms (DoE, 1984g, paras 3.10–3.11).

The hierarchical model of policy control implies that the process of planning and managing complex and 'wicked' issues is logical and straightforward, even if present administrative arrangements are highly complex. However, as we have seen, this approach gives little attention to power relations and conflicts of interest between policy-makers and those 'responsible' in some sense for implementation. Addressing these issues leads to the recognition that decision-making takes place in a social and political context, where issues of power and influence and the ability to bargain can have a significant effect on the outcome. Fixed ideas of 'policy' and 'implementation' lose their meaning when the policy–action relationship is seen as an interactive negotiative process rather than a logical step-by-step sequence. On the basis of the work of such writers as Schelling (1960) and Bacharach and Lawler (1981), generalizations can be made about the nature of the processes of bargaining which arise from the role of land use planning as an inherently distributional activity taking place in a socio-political context. The contexts for negotiations, which provide the 'rules of the game' and make bargaining parties unequal or limit the scope for negotiations, are less amenable to analysis but no less important (Strauss, 1978; Barrett and Hill, 1984). Nonetheless, frameworks specifying the nature and characteristics of bargaining processes have been found to be useful and highly relevant in analysing negotiations in development control and local plan preparation. For example, such policy devices as development plan schemes and certification, which in one sense reflect strategic planning principles, can also be seen as 'bargaining counters' used in a process of distributional bargaining to establish relative levels of responsibility for plan-making between tiers of local government. These ideas also provide a perspective on the development process which leads to a very different view of the role of local plans and local planning policies in this process than that suggested by the hierarchical model (see Chapter 8 and Bruton, 1980, 1983b). For example, policies in plans may be used as tactical statements of position, which are subsequently open to amendment as negotiations between the interests involved proceed. In this way the land use planning tasks of plan preparation and implementation form an essential part of the development and maintenance of interactive relationships between interests and agencies, relationships characterized by bargaining and negotiation over access to the various resources involved. It is these relationships which in practice require the local planning authority to proceed beyond the plan-based control of change and to attempt the more active management, organization and motivation of public agencies and private interests in pursuit of policy objectives.

In essence, given the complex and inter-related nature of public policy problems; given the uncertainty which is integral to any planning system; given the significance of contextual or contingent factors in dealing with complex public policy issues; given the inherent bargaining and conflicts of interest in any system which aims to distribute or redistribute scarce resources; and given the need for an over-riding guiding framework to ensure that policy and implementation are vertically related; then land use planning must be concerned explicitly to manage change in the environment through the application of the following four activities:

- *Planning*, which involves the establishment of goals and objectives for the system, i.e. the establishment of a strategic framework;
- *Organizing*, which includes the provision of resources needed to achieve the strategic objectives, and the definition of (*a*) the tasks of bodies and individuals in the system and (*b*) the relationships between them;
- *Controlling*, which is concerned to ensure that implementation conforms to the strategic objectives and that the relationships between bodies and individuals are established and working satisfactorily; and
- *Leading*, which is the process of motivating and influencing others so that they contribute to the achievement of the strategic goals and objectives set by the system.

Currently, the legislation relating to the development plan quite unrealistically prescribes a role for the land use planner which is restricted to the production of land use plans and the negative control of development. Explicit in this legislation is a hierarchical relationship between elements of the development plan – the structure plan and local plans. Less explicit is the hierarchical relationship between the development plan and (*a*) higher-order policies for social and economic change, and (*b*) the means of implementing policies set out in the development plan. At the same time, land use planning in practice, both at local authority and central government level, is gradually coming to accept that the resolution of complex planning problems involves bargaining and conflicts of interest; it is attempting to adjust to the realities of the situation by adopting a contingency approach without a clearly established over-riding strategic framework, e.g. local planning authorities are producing a wide range of non-statutory local planning instruments which are responsive to their own local contingent factors; central government has introduced a range of free-standing planning initiatives, such as HAAs and UDCs, which are intended to cope with what are seen as local issues influenced by local contingent factors.

This short summary of the strengths and weaknesses of the various theoretical perspectives used in the foregoing chapters to analyse local planning practice indicates that all of these approaches offer by themselves a partial view of the activity of local planning and of the context in which it takes place. Different approaches it seems are suitable for specific concerns and issues:

- *The hierarchical model of policy control* emphasizes the statutory and organizational setting and the system of powers and resources within which local land use planning takes place. It is most useful as a framework for identifying the formal complexity of the organization of public-sector policies with a bearing on land use and development;
- *Contingency* approaches draw attention to the possibly diverse determinants of individual local planning strategies, thus relating these to the complexity and variety of planning issues, problems and operating contexts;
- *Bargaining and negotiation* characterize specific aspects of such strategies and the way they are formulated and implemented; an analysis of the role of local plans in the development process emphasizes the importance of negotiative relationships and continuous interaction rather than the value of a fixed and static policy framework which is subsequently 'implemented'. Plans however do provide a resource to be deployed in the management of such relationships, either as bargaining counters or in recording agreements which have been reached. As such they provide a useful means of managing change in the local environment.

An eclectic rather than unitary approach is thus suggested. Christensen (1985) provides an example here in suggesting that in dealing with the uncertainty which characterizes all planning problems, planners should act contingently by selecting appropriate planning styles for prevailing situations or 'problem conditions'; suggested styles include bargaining in order to accommodate multiple preferences.

9.3 The roles of statutory local plans in practice

The theoretical perspectives summarized above provide a framework against which the use of statutory local plans and their contribution in practice to the achievement of local planning policies can be assessed. However, it must be noted that plan production, whatever the emphasis of legislation, is essentially only one means of securing the co-ordination and management of environmental change. Put simply, statutory local plans serve as a mechanism for establishing land use policies and proposals and provide a means of policy implementation via the operation of development control. In assessing these broad functions, a useful starting point is the four statutory local plan roles recognized by the DoE, i.e.,

1 to develop and detail structure plan policies and proposals;
2 to provide a detailed basis for development control;
3 to provide a detailed basis for the co-ordination and direction of the development and other use of land; and
4 to bring local planning issues before the public.

A review of plan content and procedures on the basis of these four roles shows how statutory local plans have actually sought to carry out these functions, and indicates how effective plans have been (or can be expected to be) in these respects.

1 The development and detailing of structure plan policies and proposals

This first role explicitly defines the relationship of the local plan to the structure plan, and so emphasizes the position of the local plan as a link in the hierarchy of policy control. The structure-local plan link is reinforced by such integrating and co-ordinating devices as development plan schemes (which have been useful in resolving questions of the relative responsibilities and planning roles of county and district authorities), certification and local plan briefs. There are instances of conflicts between structure plan strategy and local plan proposals, usually focused around these devices, but these instances are relatively rare and in the main it seems that the working relationship between these two planning tiers is now characterized by consultation and co-operation. Conformity is thus generally ensured through the normal processes of consultation during plan preparation, with local plans reflecting the strategic context in a variety of ways. Difficulties have arisen in using local plans to establish social and economic as well as land use and development policy; this limitation also applies to structure plans and is aggravated by the absence or poor expression of national and regional socio-economic policy in the British planning system.

2 The provision of a detailed basis for development control

In developing the policies and proposals of the structure plan, local plans, as part of the development plan for an area, are concerned to provide a suitably detailed policy context for the operation of development control. Again this role emphasizes the statutory local plan as an important link in the policy hierarchy. The relationships between local plans and development control on the one hand and between local plans and structure plans on the other together embody the strategic planning principle that the level above should constrain the planning of the level below, while itself being constrained by the level above. The link with development control is likely to be of particular importance for local plans, especially as regards the implementation of proposals involving the private sector. However, development control is a highly discretionary and complex decision-making process in terms of establishing exactly what constitutes relevant 'other material' considerations for individual applications and as a result of the implicit or explicit processes of bargaining involved. Statutory local plans, as part of the development plan, represent only one 'input' or factor in development control, and their significance varies in relation to others such as central government policy guidance, other local authority policies, officer–councillor relationships, 'facts' related to a particular development, and bargaining. It is clear that the importance of statutory plans in this process is at times only marginal. Nonetheless, Healey *et al.* (1982) show that given certain strategic conditions such as high demand for land and policy stability, the policies in statutory plans may be effectively implemented where these require the controlled release of land, with the preparation of plans giving legitimacy to policies.

3 The provision of a detailed basis for the development and other use of land

The local plan proposals map provides a basic indication of how local plan policies and proposals are related spatially. In addition, several other methods and techniques, such as summary chapters and area statements, have been used in a bid to emphasize the relationships between topic-based plan policies. Less attention has been given to the linked questions of resource availability and the attitudes of public- and private-sector imple- menting agencies, reflecting the local plan's lack of control or influence over these issues. Local plans give the planning authority little effective control over the local activities of public agencies, some of which can be expected to work within or have regard to planning policies. However, their activities are as likely to reflect processes of consultation and communication as being plan-based. The implementation of proposals involving the private sector depends on the discretionary process of development control in the first instance and subsequently on other development process factors such as the resolution of landownership difficulties. Local plans may succeed in provid- ing a detailed policy basis for future land use and development within their areas, but have typically been less successful as mechanisms for managing change in the environment. This generalization applies more to areas of low or no growth, rather than to situations where pressure for land release is high and policy implementation can be secured through the traditional land use planning powers of land allocation and the granting of planning permission on identified sites. The role of the local planning authority in the develop- ment process is based primarily on its ability to prevent development occur- ring, and local plans provide a relatively precise indication of how this power is to be used. Policy statements in plans are nonetheless subsequently open to interpretation in the light of particular development proposals. As a result, it is probably more appropriate to regard the relationship between planning policy in local plans and development control and development processes as characterized by negotiative interaction rather than being a technical process of policy definition and subsequent implemen- tation. Hence this role of providing a detailed set of policies and proposals leading to the co-ordination of development activity can be understood as supplying an initial indication of policy around which public and private development interests can negotiate with the local planning authority.

4 The bringing of local planning issues before the public

Much effort has been expended on public participation exercises in local plan preparation, but local plans are not ideal vehicles for such exercises e.g.

the difficulty of securing a representative response to participation exercises, and the constraints imposed by the need to maintain a general conformity to the structure plan (Bruton and Lightbody, 1980b). In any case it is likely that in carrying out such exercises authorities will be concerned not so much to identify some notional public interest as to legitimate the policies involved as a formal basis for local land use planning. Public participation is likely to be regarded as less of an end in itself as a means to plan and policy legitimation. The production of a local plan, by virtue of its site-specific nature, is likely to lead to conflicts of interest which can only be resolved by complex and necessarily lengthy processes of negotiation and bargaining. Hence it is not surprising that the timescale of local plan production is usually numbered in years. In many cases this investment of time and resources is necessary in order to establish an effective base for land use planning, and in the face of complex and conflicting interests is likely to be judged worthwhile. The public local inquiry offers a means of arbitration to resolve conflicts of interest generated by local plan preparation, and as a result is a key stage in the process of conflict resolution. This inquiry also offers a further channel for policy legitimation i.e. the review and testing of policies in an open public forum by an independent third party in order to establish whether the plan is reasonable, fair and competent. In these respects the credibility of the inquiry is threatened by the ability of the local authority to reject Inspector's recommendations, although the way in which the local authority accommodates any suggested modifications to the plan can offset this danger. The successful resolution of conflicts of interest goes hand in hand with the legitimacy of local plan policies and proposals; both of these aspects could be weakened if the public participation and inquiry stages of local plan production were significantly reduced (Bruton and Lightbody, 1980b).

These points suggest that while statutory local plans have sought to carry out their formal roles within the hierarchy, problems remain in respect of each. For example, it is difficult to express the local implications of strategic socio-economic policies (where such policies are available) since local plans cannot define policies on non land use matters (though such matters may be referred to in the reasoned justification, and local plans can deal with the land use implications of socio-economic policies). In formulating their policies and proposals on the development and other use of land, local plans have attempted to inform development control but with varying success. Local plans may undoubtedly attempt to provide a meaningful basis for development co-ordination, and can potentially provide an integrated spatial base for development activity, but are hindered by autonomous public agencies and the difficulties of implementing plan policies through development control. A further linked problem is the limited extent to which resources can be controlled or influenced by plan allocations and proposals. Public funds for investment in the built environment for instance are typically under the control of other local authority spending departments or of other public agencies. Such resources are often linked to other area-based initiatives, the local plan not having a monopoly within the public sector as to the formulation of policy affecting the built environment, or are deployed through expenditure programmes organized by policy topic rather than local plan areas. A wide range of formal initiatives, plans and prog-rammes dealing with a variety of matters has been introduced into the planning system on an *ad hoc* basis, often as a political response to particular issues and problems. These initiatives and the resources they command have significant land use and development implications, but nonetheless they are commonly established and progressed largely independently of the statutory local plan. For example, policies on future change in areas facing acute social, economic and physical problems are increasingly being taken in advance of or apart from the process of local plan production–Partnership, enterprise zones, urban development areas and corporations, task forces, urban development grants. The danger here is that 'local plans will . . . come to seem peripheral to other means of managing continuous change in those localities which are suffering greatest stress or most rapid physical change' (Wannop, 1981, p. 18). In particular, the ability of local plans to secure the effective co-ordination of these measures at the local level is threatened by their free-standing nature, by the tendency to focus on specific 'problem' issues, by the practice of limiting proposed solutions to specific areas or zones, and by the short-term and action-orientated timescale usually involved. Such difficulties however are wider problems of integration which cannot be wholly accounted for in terms of the shortcomings of one particu-lar policy vehicle. There are difficulties in using statutory local plans to establish and progress local land use policies, but these must be related to the wider context of the organization and operation of both public-sector agen-cies and policies and private sector development interests.

Many authorities, partly in response to the perceived problems and short-comings of the statutory framework, have taken to non-statutory means of defining their local planning policies. None of the roles prescribed for local plans by the DoE actually *requires* a statutory format in any case. Informal documents offer ease of preparation and amendment, although it is clear that many are subjected to preparation procedures analogous to those laid down in the statutory regulations. Indeed, they are used to carry out a wide range of local planning roles and appear to be taken into account at appeal (Bruton and Nicholson, 1984a, c). In some cases their use can be directly linked to the difficulties associated with the statutory format, i.e. as an alternative and less restricted means of stating social and economic policies and their land use implications; as a mechanism for linking resources to local planning policies. In choosing or evolving local planning strategies authorities appear to take a pragmatic approach where the main considera-tion is to formulate policies which reflect contingent factors and deal effectively with local situations, problems and issues. This local operating context varies from authority to authority with the relative importance of a wide variety of contingent factors. Authorities can be expected to use whatever form of local planning instrument appears to them to be most suitable. It seems that statutory local plans are most appropriate where complex and conflicting interests are involved; where public support and strength at appeal are important; and where the policies concerned relate to the central planning responsibility of the allocation of land and the control of its release. The timescale of statutory plan preparation may only be a relative disincentive, judged unimportant when compared to the need for a formal basis for land use planning. Elsewhere, for instance in rural areas where the main planning issues relate to the management of land use rather than development, structure plan policies may be seen as adequate for development control, perhaps in conjunction with supplementary planning guidance. In other situations such as inner cities, where the public sector is the main agent for change and private development pressure is limited, informal plans may be established to direct or review spending programmes. The contingency approach suggests that the range of situations faced by local authorities has produced an equivalent range of local planning responses, and this in turn implies that the attempts of the DoE to prevent planning authorities expressing land use policies through vehicles other than the statutory local plan are unlikely to succeed. Since individual authorities are seeking to deal with very different planning issues and problems, *detailed* central advice must inevitably be limited in its relevance. Such central advice can be seen as one of the factors that planning authorities will have regard to in devising local planning strategies relevant to their needs. Certainly, practice has not reacted favourably to restrictions on local plan content. Indeed this appears to be one of the main reasons for the continued use of non-statutory documents, particularly where land use issues concern decline or management rather than growth and development.

At the same time, it should not be assumed that by taking an informal approach local planning authorities have been able to avoid the various difficulties posed for the progression of policies by the complexity of the organizational context. At the local authority level, the management of change in the environment in accordance with local planning policies is hindered by the complexity of the relationships between central and local government, between the nationalized industries and other public agencies, and between the public and private sectors. This context of fragmentation has made policy integration at the local level extremely difficult, and these problems face both statutory and informal local planning strategies.

9.4 The implications for practice

A useful way of summarizing these arguments and their implications for practice is to relate them to the approach to uncertainty and contingency in planning outlined by Christensen (1985) and reviewed in Chapter 2. She suggests a four-fold classification of the conditions typifying planning problems which allows a preliminary identification of the different situations in which different planning roles will be appropriate. Particular planning processes should be understood not as predetermined mechanisms but as tools whose selection and use must vary depending on circumstances; planners should act contingently by tailoring their styles and actions to the particular problems, conditions and situations prevailing at the time. By considering to what extent different planning problems are characterized by uncertainty as to *means* ('technology') and/or *ends* ('goal'), Christensen is able to categorize different planning roles according to the problem conditions they address (see Figure 2.4 p. 65). This suggests that planners must first assess the actual conditions of uncertainty characterizing the particular problem at hand, and then select a suitable planning style or role. Christensen's framework provides an initial means of generalizing about how specific local planning strategies are influenced by local contextual or contingent circumstances, which may be understood in terms of the degree of certainty available in defining planning goals and the subsequent methods of implementation. There may still be a 'unique solution' for particular authorities, but Christensen's approach offers the hope that the suitability of particular planning roles (including plan production and development control) can be related – at least broadly – to the nature of the planning context in which they are both formulated and pursued.

The division of planning problems into four 'prototype' conditions is avowedly simplistic; 'the line that divides means and ends often blurs . . . technologies are rarely completely known or completely unknown; . . . a goal may elicit various degrees of disagreement' (Christensen, 1985, pp. 63–4). Moreover, the nature of planning problems is such that it can be difficult to define the degree of certainty and agreement attached to either the goals of planning intervention or the means of achieving those goals. These points reflect the 'wicked' nature of public policy issues which planning addresses, issues which are characterized by complexity, ambiguity and conflict. Nonetheless, the schema usefully summarizes local planning contexts and the nature of the possible planning responses. In British local planning practice, the 'goal' dimension can be roughly understood as referring to local planning policy and proposals (as sometimes established in statutory and non-statutory plans), while 'technology' can be seen as including the development control process as well as the use of development briefs, development partnerships, and other forms of intervention such as the assembly of sites or the provision of infrastructure. This distinction between goal and technology is not necessarily mutually exclusive, e.g. statutory local plans provide a means of establishing land use policy goals and proposals, and a mechanism for their progression through development control.

Problem conditions of *known technology and agreed goal* are typified by relative certainty as to both ends and means, although periodic uncertainty is unavoidable should circumstances change. Statutory local plans appear to be most useful for these conditions, where the planning goals involved can be agreed and supported by different levels of government (district, county and central), and where the policies concerned are achievable through the control of land release and site allocation. Sufficient development pressure to ensure the take-up of allocated sites is also a pre-condition for the implementation of plan policies if the essentially responsive development control process is to be the main means of policy implementation. Statutory plans are thus likely to be most useful where there is a level of agreement over goals *and* where the relevant 'technology' is likely to be effective. In these conditions, planning policies can be established in plans, applied legitimately in development control (perhaps in conjunction with supplementary planning guidance such as policy notes or development briefs), and subsequently supported by central government at appeal. For instance, the study by Healey *et al.* (1982) of the implementation of development plans in the South East of England during the 1970s traces how policies clearly relating to town and country planning powers were successfully progressed in conditions of policy continuity and stability. In conditions where goals are agreed and technology known, the planner acts in Christensen's terms as a *programmer*, defining the scale, location and timing of development and then using proven means to bring this about. These conditions call for traditional rationalist planning skills, i.e. the planner as regulator, analyst and administrator. Examples include new town development or phased land releases for development by the private sector, particularly where large sites are involved. In this situation planning documents may resemble master plans, which in effect record agreements reached over goals as well as the particular commitments and contributions required of individual agencies, formulate a rational and technical land use planning solution, and provide a clear tool or guide for subsequent implementation. The typical local planning task involved here is one of the management of pressures for growth, where the planner acts to facilitate development, providing a service to development interests by reducing uncertainties over land supply through the allocation of sites in plans. The statutory format is not essential for development facilitation, but may be associated with greater certainty by development interests as a consequence of the formal preparation procedures. Statutory plans appear to have been successful mechanisms for the facilitation of development, i.e. local planners have matched development needs with the environmental and amenity context within which such needs were to be met (Farnell, 1983, p. 103), although informal plans can also be significant tools. Indeed in the past these have been sanctioned by central government (as in the land release circulars in the early 1970s), while DoE disapproval of non-statutory plans since 1976 has not prevented the Inspectorate accepting such instruments at appeal as valid statements of local planning authority policy (Bruton and Nicholson, 1984c). Currently, central government is emphasizing the role of facilitation of development for statutory local plans, while attempting to establish the statutory local plan (one of a range of vehicles that can be used to establish local planning policy) as the principal means of policy expression legiti-mately available to local planning authorities. The objective of development facilitation is likely to be acceptable to, and supported by, development interests, so that in this respect central government and the private development industry share a common view of the future role of statutory local plans and implicitly of local planning. Also, such objectives may well be supported by district (if not county) authorities concerned with their local economic situation, although facilitation will have to shift to promotion in the absence of development pressure. Ironically, the current (1985) government's view of the role of local plans echoes that of the Planning Advisory Group, for whom local plans were not primarily a control mechanism but rather devices for the guidance and co-ordination of development. However, PAG had assumed an economic context of growth, with consequent urban expansion and renewal; in today's economic climate planning strategies for facilitation look rather less than appropriate or adequate in all areas of the country outside the South East.

Many local planning contexts are characterized by problem conditions of *known technology and no agreed goal* in that while the principal means of policy implementation in the form of development control is prescribed by legislation, the goals of planning intervention often engender conflict rather than consensus. For instance, development interests such as housebuilders can be expected to take a hostile or ambiguous attitude towards policies concerned to restrain development. Such policies include for example structure plan targets and local plan allocations of land for housing, based on forecast needs and requirements rather than likely effective demand; policies intended to conserve and protect the national heritage. In the case of the housebuilding industry, opposition to such policies reflects the distinctive role of land supply and landholdings in the overall structure of housebuilders' productive activities (Chapter 8). However, as long as these restraint and conservation policies are agreed and supported by central government the local planner can still act successfully to control development, notwithstanding the opposition of development interests. Statutory local plans are thus useful as part of the hierarchy of land use plans, and can be regarded as bringing certainty to a small part of the development process. The statutory local plan preparation procedures, allowing for consultations, objections and the resolution of conflicts of interest through public local inquiries have an important role here in producing what are seen as legitimate policy frameworks. However, the policy consensus between levels of government in traditional areas of restraint such as the South East is currently under threat. In particular, more liberal attitudes to development proposals are evident in recent central government circulars dealing with housing (Circular 15/84), industrial development (16/84), and development and employment (14/85), although support for conservation policies such as Green Belt remains strong (14/84). The move to a situation where central government support for local policies of restraint is being redefined suggests a necessary shift in local planning style away from the plan-based 'programming' of change to one of *bargaining*. Here plans are important in so far as they can be used to predict local (if not central) responses to planning applications, i.e. they can indicate how the resource of development control might be deployed, and provide an *initial* statement of policy as a basis for bargaining. The room for these negotiations is provided by the normal flexibility and discretion which characterizes development control, and this has been emphasized by circulars stressing the importance of 'other material considerations' in the development control decision-making process at the expense of the development plan (see for example Circular 14/85). Much day-to-day local planning explicitly employs political bargaining processes in order to accommodate divergent or conflicting interests, the role of the planner in this respect principally deriving from the power to refuse planning permission. The planner can also bargain on the basis of control of other powers (such as those of compulsory purchase) or resources (such as land-ownership. The significance of the ability to decide planning applications is contingent upon the likely extent of central government support for development control decisions at appeal, and central government policy statements in circulars form an important part of the context for local bargaining situations. Greater certainty over goals could be introduced into the planning process by national and regional level policy statements, perhaps along the lines of Scotland's National Planning Guidelines (Diamond, 1979) and Regional Reports (McDonald, 1977), but such policy guidance for the development plan system in England and Wales is at present only poorly expressed (Bruton and Nicholson, 1985a).

Conditions of *unknown technology and agreed goal* are also common in implementing local planning policies given the essentially negative and responsive nature of development control and the difficulties of co-ordinating public sector land and development activities and resources. There may be a broad measure of policy agreement around general goals, such as development promotion or urban regeneration, but much less consensus about either the nature of the problem or the most appropriate response. Statutory local plans can be used to formulate policies and proposals on the development and other use of land, though in the absence of development pressure or in conditions of decline the usual means of policy implementation, the attempts of the development control process to guide development proposals to allocated sites, is not usually suitable or appropriate. Where goals are sufficiently specific and small scale and are agreed, planning styles of *experimentation* are more suitable, perhaps involving the production of promotional development briefs to publicise or market sites, or other forms of positive planning, e.g. development Partnerships. These initiatives are generally linked to or based on policies in statutory or informal plans. Local planners it seems are also prepared to respond to and utilize any opportunities for policy implementation which may arise. Central government policy initiatives, such as Enterprise Zones or urban development grants for example, may be significant and valuable at the local level, given that such measures often have additional resources and powers associated with them. Finance for local projects may also be available from sources such as the EEC (e.g. the European Regional Development Fund). Successfully to tap such initiatives and sources may entail an entreprenurial or opportunist style of local planning, which in an ideal situation should be informed by a framework of local planning policies and objectives.

In situations of *unknown technology and no agreed goal* the nature of the problem is itself hard to define, leading to prevailing confusion as a result of uncertainty over both means and ends. Inner city decline has been a particularly intractable problem in this regard, although political initiatives have periodically attempted to define goals *and* solutions, e.g. the 'total approach' of Partnership. The linked social, economic and environmental issues affecting the inner city form a classic example of the 'wicked problem' (Rittel and Webber, 1973). Statutory local plans have been initiated in a bid to carry out what Farnell (1983, p. 11) terms 'problem investigation and solution search' in response to the complex problems of the inner city, but these have proved difficult to prepare and (once adopted) difficult to implement, partly because of the lack of resources under the direct control of local planners. An initial planning task is to generate order, but this has often meant no more than the production of summaries of the policy intentions and committed investments to be made by other public agencies. Some authorities such as Birmingham City are beginning to produce various policy-linked devices in a bid to overcome this problem. Birmingham's inner area studies, designed to co-ordinate major change led by the public sector, have thus been characterized as a local planning response to uncertainty where 'the aim is to shift the emphasis of local planning away from development control and longer-term land use towards annual rolling investment programmes' (Wenban-Smith, 1983, p. 205). Authorities faced with these complex planning issues have found both the form and land use remit of the statutory local plan too restrictive, and have developed more flexible policy vehicles capable of addressing socio-economic issues directly, as well as their land use implications. In addition, informal documents in some areas show a concern with resource and investment programming matters missing from many statutory local plans.

To summarize, these points suggest that statutory local plans as statements of policy are likely to be relatively successful, where policy goals are well established and supported by other levels of the policy hierarchy if not by development interests; where the means of implementation relate to the central statutory planning power of development control, i.e. to the control of land release, and where there is sufficient development pressure to ensure the take-up of allocated sites. This is a situation of relative certainty and agreement over both means and ends. The support of development interests can be expected to the extent that policies are concerned with development facilitation and management rather than restraint or conservation. Statutory plans thus have a clear role where change is to be achieved through the guiding of private sector development pressures by means of the development control system. This role is typically seen in urban fringe locations where the key land use planning issues are the allocation of land for development in line with structure plan targets, coupled with restraint elsewhere, e.g. protection of the Green Belt. Statutory plans are less appropriate in urban and metropolitan areas where, given an absence of development pressure, initiatives are required to offset decline or promote development, or where the stability of existing land use patterns means that little change is anticipated, e.g. surburban areas. In situations where major change is proposed but where 'traditional' means of policy implementation through development control are not successful or are inappropriate other means will be sought, such as development briefs, promotional and publicity documents. Finally, where goals are not agreed the task is one of problem identification and definition as a preliminary to the setting of policy goals.

From this analysis it can be seen that local planning styles or roles are likely to vary according to the task in hand. No single methodology or prescribed approach will be appropriate for all situations or 'problem conditions' in the management of environmental change. Moreover, these varying roles for planners require different sets of planning instruments or tools; this is evidenced by practice in its commitment to both statutory plans and informal documents. The statutory local plan, however flexibly used within the confines of regulations, is only one of the available tools for policy definition and implementation. In broad terms this implies that DoE advice, focusing on the legislative basis of the town and country planning system, is over-restrictive, both as to the *type* of vehicle that can be used to establish local planning policy, and as to the *role* of that vehicle in practice. Following DoE exhortations, and no doubt mindful of the likely future status of informal local plans and policies at appeal if the threats of circulars that such documents can have little weight for development control purposes are put into effect, there is some evidence to suggest that local planning authorities are becoming more enthusiastic about adopting statutory local plans. If such a choice of strategy reflects central advice rather than conditions deriving from the local planning context, then the logic of this decision must be questioned. Most local planning contexts are characterized by uncertainty as to goals and/or methods of implementation, while the statutory plan is arguably best suited to conditions of growth and relative certainty as to both ends and means. The danger is that under the impact of central government advice, statutory local plans will be produced with insufficient attention being given to the extent to which the context for such plans is typified by uncertainty as to goals and/or methods of implementation. Local plan written statements in any case imply agreement rather than conflict over the policies and proposals they establish, while means of implementation are seldom discussed in any detail, even though these issues are highly problematic. In conditions of uncertainty, assumptions made during the preparation of plans about the degree of consensus and agreement surrounding policies and proposals and the ease and means of carrying policies through are likely sooner or later to break down, leaving adopted policies in disarray. In short, the risk is that statutory local plans will be prepared on the basis of assumptions of premature consensus over both goals and 'technology' (Christensen, 1985). To avoid this risk, local planning must begin by assessing local contexts, the sort of role most appropriate, and hence the suitability of alternative ways of establishing and progressing local planning policies. In other words, strategy or approach 'must be developed from a thorough *understanding of the planning department's operational context* – i.e. the circumstances and pressures within its area and the powers, responsibilities and resources (broadly defined) available to it' (Fudge *et al.*, 1983, p. 129). Planners must face rather than ignore conditions of uncertainty and react accordingly by adopting roles appropriate to actual planning problems and contexts. This requires the planner to recognize his or her broad role as a manager of environmental change before assessing the contribution that statutory powers and duties – the production of plans and the control of development – can make to the planning styles required by problem conditions. In terms of the planning documents judged necessary to support such roles, the variety of policy vehicles actually in use by practice suggests that many authorities have indeed weighed up the pros and cons of different strategies, giving due weight to the legislative provisions for local plans but also regarding informal approaches as viable and justifiable in the light of local conditions. What has happened is that local authorities are exercising local discretion in the manner envisaged in 1965 by the Planning Advisory Group, but the resultant strategies and the documents involved are being drawn from both statutory provision and informal experimentation and innovation, as the statutory local plan has become circumscribed by regulations and circular advice. This situation seems set to continue given the emphasis being placed by central government on statutory plans as facilitators of development and the range of complex problem conditions, characterized by uncertainty and stemming from economic recession, being faced by local planning authorities throughout the country.

References

Ackoff, R. L. (1970) *A Concept of Corporate Planning*, John Wiley, New York.

Adcock, B. (1979) 'The effectiveness of local plans in the inner urban areas', *Local Planning Practice*, PTRC, London, pp. 37–50.

Adcock, B. (1984) 'Regenerating Merseyside Docklands', *Town Planning Review*, Vol. **55**, No. 3, pp. 265–289.

Alder, J. (1979) *Development Control*, Sweet and Maxwell, London.

Alderton, R. (1984) 'Urban Developments Grants – Lessons from America', *The Planner*, Vol. **70**, No. 12, pp. 19–21.

Allan, C. F. (1974) 'The Inspector's criteria', *Planning Inquiry Practice*, Sweet and Maxwell, London, pp. 3–9.

Allan, M. S. (1978) 'The inner city partnership arrangements – the Lambeth approach', *Structure and Regional Planning Practice and Local Planning Practice*, PTRC, London, pp. 101–117.

Allen, M., Shipley, H., Thirlaway, J. and Worters, J. H. (1983) 'Derwenthaugh Industrial Improvement Area', *The Planner*, Vol. **69**, No. 3, pp. 98–99.

Alonso, W. (1964) *Location and Land Use*, Harvard University Press, Cambridge, Massachusetts.

Ambrose, P. and Colenutt, B. (1975) *The Property Machine*, Penguin, Harmondsworth.

Amos, F. J. C. (1982) *Manpower Requirements for Physical Planning*, Institute of Local Government Studies, University of Birmingham.

Anderson, M. A. (1981) 'Planning policies and development control in the Sussex Downs AONB', *Town Planning Review*, Vol. **52**, No. 1, pp. 5–25.

Angell, R. and Taylor, N. (1985) 'Unitary development plans – an initial appraisal', *The Planner*, Vol. **71**, No. 12, pp. 21–23.

Angus, I. (1985) *Outside the Law? Non-statutory Local Planning in Scotland*, Strathclyde Papers in Planning, No. 5, Department of Urban and Regional Planning, University of Strathclyde, Glasgow.

Ansoff, H. I. (1969) 'Towards a strategic theory of the firm', in H. I. Ansoff (ed.), *Business Strategy*, Penguin, London, pp. 11–40.

Arnstein, S. (1971) 'A ladder of citizen participation in the USA', *Journal of the Town Planning Institute*, Vol. **57**, No. 4, pp. 176–182.

Ashby, W. R. (1964) *An Introduction to Cybernetics*, Methuen, London.

Ashworth, W. (1954) *The Genesis of Modern British Town Planning*, Routledge and Kegan Paul, London.

Bacharach, S. B. and Lawler, E. J. (1981) *Bargaining*, Jossey-Bass, San Francisco.

Bajaria, N. (1982) 'A budget for local plans to lever actions by others', *The Planner*, Vol. **68**, No. 5, pp. 152–153.

Ball, M. (1983) *Housing Policy and Economic Power*, Methuen, London.

Barnard, T. (1980) 'A review of local plans', in C. Fudge (ed.), *Approaches to Local Planning (2)*, Working Paper 17, School for Advanced Urban Studies, University of Bristol, pp. 42–76.

Baron, T. (1980) 'Planning's biggest and least satisfied customer', *Report of Proceedings of the 1980 Town and Country Planning Summer School*, Royal Town Planning Institute, London, pp. 34–40.

Barras, R. (1983) 'Development of profit and development control: the case of office development in London', in S. Barrett and P. Healey (eds., 1985), *Land Policy: Problems and Alternatives*, Gower, Aldershot, pp. 93–105.

Barras, R. (1984) 'The office development cycle in London', *Land Development Studies*, Vol. **1**, No. 1, pp. 35–50.

Barrett, S. (1980) 'CLA: did it fail?', *Planning*, No. 373, 20 June, pp. 6–7.

Barrett, S. and Fudge, C. (1981) 'Examining the policy–action relationship', in S. Barrett and C. Fudge (eds.), *Policy and Action*, Methuen, London, pp. 3–38.

Barrett, S. and Hill, M. (1984) 'Policy, bargaining and structure in implementation theory: Towards an integrated perspective', *Policy and Politics*, Vol. **12**, No. 3, pp. 219–240.

Barrett, S., Stewart, M. and Underwood, J. (1978) *The Land Market and Development Process: A Review of Research and Policy*, Occasional Paper No. 2, School for Advanced Urban Studies, University of Bristol.

Barrett, S. and Whitting, G. (1980) *Local Authorities and the Supply of Development Land to the Private Sector*, Working Paper 19, School for Advanced Urban Studies, University of Bristol.

Barrett, S. and Whitting, G. (1983) *Local Authorities and Land Supply*, Occasional Paper No. 10, School for Advanced Urban Studies, University of Bristol.

Bayliss, D. (1975) 'TTPs and structure plans', *The Planner*, Vol. **61**, No. 9, pp. 334–335.

Bell, D. and Held, V. (1969) 'The community revolution', *Public Interest*, No. 16, pp. 142–177.

Bennett, A. (1984) 'New settlements: the role of Consortium Developments', *Housing and Planning Review*, Vol. **39**, No. 5, pp. 4–7.

Benyon, J. (1982) 'Dual loyalty of planners and councillors – the case of Beverley Hills', *Local Government Studies*, Vol. **8**, No. 4, pp. 53–63.

Berkovitch, I. (1977) *Coal on the Switchback*, George Allen and Unwin, London.

Berlo, D. K. (1960) *The Process of Communication: An Introduction to Theory and Practice*, Rinehart Press, San Francisco.

Blacksell, M. and Gilg, A. (1977) 'Planning control in an Area of Outstanding Natural Beauty', *Social and Economic Administration*, Vol. **11**, No. 3, pp. 206–215.

Blacksell, M. and Gilg, A. (1981) *The Countryside: Planning and Change*, George Allen and Unwin, London.

Blowers, A. (1980) *Limits of Power*, Pergamon, Oxford.

Blyth, R. J. (1984) *The Implementation of Local Plans*, unpublished dissertation, Department of Civic Design, University of Liverpool.

Boaden, N. (1982) 'Urban development corporations – threat or challenge?', *Local Government Studies*, Vol. **8**, No. 4, pp. 8–13.

Boddy, M. (1981) 'The property sector in late capitalism: the case of Britain', in M. Dear and A. J. Scott (eds.), *Urbanisation and Urban Planning in Capitalist Society*, Methuen, London, pp. 267–286.

Boddy, M. and Barrett, S. (1979) *Local Government and the Industrial Development Process*, Working Paper 6, School for Advanced Urban Studies, University of Bristol.

Botham, R. and Lloyd, G. (1983) 'The political economy of enterprise zones', *National Westminster Bank Quarterly Review*, May, pp. 24–32.

Bowen, E. and Yates, S. (1974) 'Comments', *The Planner*, Vol. **60**, No. 1, p. 502.

Bowie, D. (1979) *Housing policy and capital expenditure in Oxford: A case-study of the operation of the Housing Investment Programme System*, Oxford Shelter Group, Oxford.

Bramley, G., Leather, P. and Murie, A. (1980) *Housing Strategies and Investment Programmes*, Working Paper 7, School for Advanced Urban Studies, University of Bristol.

Briscoe, B. (1985) 'Certification of local plans', *The Planner*, Vol. **71**, No. 3, p. 17.

Bristow, R. (1985) 'Some questions on unitary development plans — a plain man's guide?', *Regional Studies*, Vol. **19**, No. 3, pp. 263–268.

British Steel Corporation (1981) *Annual Report and Accounts 1980–81*, the British Steel Corporation, London.

Bruton, M. J. (1980) 'Public participation, local planning and conflicts of interests', *Policy and Politics*, Vol. **8**, No. 4, pp. 423–442.

Bruton, M. J. (1982) 'The Malaysian planning system: A review', *Third World Planning Review*, Vol. **4**, No. 4, pp. 315–334.

Bruton, M. J. (1983a) 'Local plans, local planning and development plan schemes', *Town Planning Review*, Vol. **54**, No. 1, pp. 4–23.

Bruton, M. J. (1983b) *Bargaining and the Development Control Process*, Papers in Planning Research No. 60, Department of Town Planning, UWIST, Cardiff.

Bruton, M. J. (1983c) *Legislation and the Role of the Town Planner in Society*, Papers in Planning Research No. 61, Department of Town Planning, UWIST, Cardiff.

Bruton, M. J. (1984) 'Strategic planning and inter-organizational relationships' in M. J. Bruton (ed.), *The Spirit and Purpose of Planning*, Hutchinson, London, pp. 78–94.

Bruton, M. J., Crispin, G. and Fidler, P. M. (1980) 'Local plans: Public local inquiries' *Journal of Planning and Environment Law*, pp. 374–385.

Bruton, M. J., Crispin, G. and Fidler, P. M. (1982) 'Local plans: The role and status of the public local inquiry', *Journal of Planning and Environment Law*, pp. 276–286.

Bruton, M. J., Crispin, G. and Fidler, P. M. (1983) 'The conduct and content of local plan inquiries' *Journal of Planning and Environment Law*, May, pp. 279–287.

Bruton, M. J., Crispin, G., Fidler, P. M. and Hill E. A. (1982) 'Local Plans PLIs in Practice, 2', *The Planner*, Vol. **68**, No. 2, pp. 50–51.

Bruton, M. J., Crispin, G., Fidler, P. M. and Perry, M. (1984) *Local Planning Authorities' Reaction to Inspectors' Recommendations on Local Plans*, Stage One Interim Report, April.

Bruton, M. J. and Gore, A. (1980) *Vacant Urban Land in South Wales*, Department of Town Planning, UWIST, Cardiff.

Bruton, M. J. and Lightbody, A. J. (1979) *Southtown Leamington Spa: A Case Study of Public Participation in Local Planning*, Working Paper No. 8, School of Planning and Landscape, City of Birmingham Polytechnic.

Bruton, M. J. and Lightbody, A. J. (1980a) *County–District Relations in Local Planning: A Case Study of Warwick District Council 1975–79*, Working Paper No. 11, School of Planning and Landscape, City of Birmingham Polytechnic.

Bruton, M. J. and Lightbody A. J. (1980b) *Final Report to the DoE on Public Participation in Local Planning: A Case Study of Warwick District Council, December 1975–December 1979*. City of Birmingham Polytechnic.

Bruton, M. J. and Nicholson, D. J. (1983a) 'Non-statutory local plans and supplementary planning guidance', *Journal of Planning and Environment Law*, July, pp. 432–443.

Bruton, M. J. and Nicholson, D. J. (1983b) *Local Planning in South Wales*, Papers in Planning Research 69, Department of Town Planning, UWIST, Cardiff.

Bruton, M. J. and Nicholson, D. J. (1984a) *Local Planning in England*, Papers in Planning Research 74, Department of Town Planning, UWIST, Cardiff.

Bruton, M. J. and Nicholson, D. J. (1984b) *Local Planning and Development Briefs: A Case Study of Warwick District Council*, Papers in Planning Research 79, Department of Town Planning, UWIST, Cardiff.

Bruton, M. J. and Nicholson, D. J. (1984c) 'The use of non-statutory local planning instruments in development control and Section 36 Appeals: I', *Journal of Planning and Environment Law*, August, pp. 552–565.

Bruton, M. J. and Nicholson, D. J. (1984d) 'The use of non-statutory local planning instruments in development control and Section 36 Appeals: II. The future of supplementary planning guidance', *Journal of Planning and Environment Law*, September, pp. 633–638.

Bruton, M. J. and Nicholson, D. J. (1985a) 'Strategic land use planning and the British development plan system', *Town Planning Review*, Vol. **56**, No. 1, pp. 21–41.

Bruton, M. J. and Nicholson, D. J. (1985b) 'Supplementary planning guidance and local plans' *Journal of Planning and Environment Law*, December, pp. 837–844.

Bryant, C. R., Russwurm, L. H. and McLellan, A. G. (1982) *The City's Countryside: Land and its Management in the Rural–Urban Fringe*, Longman, London.

Buchanan, C. D. (1968) 'Presentation of the gold medal of the TPI', *Journal of the Royal Town Planning Institute*, Vol. **54**, No. 2, pp. 49–55.

Buchanan, C. (1972) *The State of Britain*, Faber and Faber, London.

Burgess, E. W. (1925) 'The growth of the city' in R. E. Park, E. W. Burgess and R. D. McKenzie (eds.), *The City*, Chicago.

Burt, S., Dawson, J. A. and Sparks, L. (1983) 'Structure plans and retailing policies', *The Planner*, Vol. **69**, No. 1, pp. 11–13.

Byrne, D. F. (1978) 'The need for a comprehensive approach to local planning', *Structure and Regional Planning Practice and Local Planning Practice*, PTRC, London, pp. 181–191.

Cadman, D. and Austin-Crowe, L. (1978) *Property Development*, E. and F. Spon Ltd., London.

Capner, G. (1982) 'A private sector view', *The Planner*, Vol. **68**, No. 5, p. 149.

Cardiff City Council (1976) *Thornhill District Plan, Written Statement*, City Planning Officer, Cardiff.

Cardiff City Council and South Glamorgan County Council (1975) *District Plan Written Statement: St. Mellons and Trowbridge Mawr Development*, Cardiff City and South Glamorgan County Councils, Cardiff.

Carnworth, R. (1982) *Blundell and Dobry's Planning Appeals and Inquiries* (3rd ed), Sweet and Maxwell, London.

Carter, H. (1981) *The Study of Urban Geography* (3rd ed), Edward Arnold, London.

Chadwick, G. (1971) *A Systems View of Planning*, Pergamon, Oxford.

Chandler, A. D. (1962) *Strategy and Structure: Chapters in the History of the Industrial Enterprise*, MIT Press, Cambridge, Massachusetts.

Chape, A. T. (1978) 'The inner city Partnership: Its background and approach in Liverpool', *Structure and Regional Planning Practice and Local Planning Practice*, PTRC, London, pp. 118–133.

Chapin, F. S. and Kaiser, E. J. (1979) *Urban Land Use Planning* (3rd ed), University of Illinois Press, Urbana.

Chartered Surveyor (1980) *Selling Surplus Public Property*, supplement to Vol. **112**, No. 9, April.

Cherry, G. E. (1974) *The Evolution of British Town Planning*, Leonard Hill, Leighton Buzzard.

Child, J. (1977) *Organization, A Guide to Problems and Practice*, Harper and Row, London.

Christensen, K. S. (1985) 'Coping with uncertainty in planning', *Journal of the American Planning Association*, Vol. **51**, No. 1, pp. 63–73.

Clawson, M. (1962) 'Urban sprawl and speculation in surburban land', *Land Economics*, Vol. **38**, No. 2, pp. 99–111.

Clawson, M. (1971) *Surburban Land Conversion in the United States: An Economic and Governmental Process*, Johns Hopkins University Press, Baltimore.

Clawson, M. and Hall, P. (1973) *Planning and Urban Growth: An Anglo-American Comparison*, Resources for the Future, Johns Hopkins University Press, Baltimore.

Cloke, P. J. (1983) *An Introduction to Rural Settlement Planning*, Methuen, London.

Collins, C. A., Hinings, C. R. and Walsh K. (1978) 'The officer and the councillor in local government', *Public Administration Bulletin*, No. 28, pp. 38–41.

Cook, O. (1967) *The Stansted Affair: A Case for the People*, Pan Books, London.

Cooke, P. N. and Rees, G. (1981) *The Industrial Restructuring of South Wales*, Papers in Planning Research No. 25, Department of Town Planning, UWIST, Cardiff.

Cornwall County Council (1983) *Countryside Local Plan*, Cornwall County Council, Truro.

Couch, C. (1978) 'Local Planning Practice', *Planning*, No. 289, 13 October 1978, pp. 8–10.

Countryside Commission (1978) *Areas of Outstanding Natural Beauty: A Discussion Paper*, CCP 116, Countryside Commission, Cheltenham.

Countryside Commission (1983) *Areas of Outstanding Natural Beauty: A Policy Statement*, CCP 157, Countryside Commission, Cheltenham.

Cox, A. (1980) 'Continuity and discontinuity in Conservative urban policy', *Urban Law and Policy*, Vol. **3**, pp. 269–292.

Craven, E. (1969) 'Private residential expansion in Kent 1956–64: A study of pattern and process in urban growth', *Urban Studies*, Vol. 6, No. 1, pp. 1–16.

Cuddy, M. and Hollingsworth, M. (1983) 'The review process in land availability studies: bargaining positions for builders and planners' in S. Barrett and P. Healey (eds., 1985), *Land Policy: Problems and Alternatives*, Gower, Aldershot, pp. 160–178.

Cullen, P. (1981) *An Evaluation of the Heritage Coast Programme in England and Wales*, CCP 155, Countryside Commission, Cheltenham.

Damer, S. and Hague, C. (1971) 'Public participation in planning: Evolution and problems', *Town Planning Review*, Vol. 42, pp. 217–224.

Dashwood, R. (1980) 'Selling 79,000 acres of railway land for £226m', *Chartered Surveyor*, Vol. 112, No. 9, pp. 19–21.

Davidoff, P. (1965) 'Advocacy and pluralism in planning', *Journal of the American Institute of Planners*, Vol. 31, No. 4, pp. 331–338.

Davidson, J. (1983) 'Breaking new ground', *Town and Country Planning*, Vol. 52, No. 9, pp. 232–233.

Davies, H. W. E. (1980) 'The relevance of development control', *Town Planning Review*, Vol. 51, No. 1, pp. 7–17.

Davies, H. W. E., Edwards, D. and Rowley, A. R. (1986) *The Relationship between Development Plans, Development Control and Appeal*. Working Papers in Land Management and Development, Environmental Policy, No. 10, Department of Land Management and Development, University of Reading.

Davies, J. G. (1972) *The Evangelistic Bureaucrat*, Tavistock, London.

Dawson, J. (1980) *Retail Geography*, Croom Helm, London.

Dearlove, J. (1973) *The Politics of Policy in Local Government*, Cambridge University Press, London.

Denman, D. R. and Prodano, S. (1972) *Land Use: An Introduction to Proprietary Land Use Analysis*, George Allen and Unwin, London.

Dennier, D. A. (1978) 'National Park Plans', *Town Planning Review*, Vol. 49, No. 2, pp. 175–183.

Dennier, D. A. (1980) 'National Park Plans' in A. Gilg (ed.), *Countryside Planning Yearbook 1*, Geo Books, Norwich, pp. 49–66.

Dennis, N. (1970) *People and Planning*, Faber and Faber, London.

Dennis, N. (1972) *Public Participation and Planner's Blight*, Faber & Faber, London.

Department of Economic Affairs (1964) *The National Plan 1964*, HMSO, London.

Department of the Environment (1971) *Management Networks: A Study for Structure Plans*, HMSO, London.

Department of the Environment (1972) *Land Availability for Housing*, Circular 102/72, HMSO, London.

Department of the Environment (1973a) *Local Government Act 1972: Town and Country Planning: Co-operation between Authorities*, Circular 74/73, HMSO, London.

Department of the Environment (1973b) *Land Availability for Housing*, Circular 122/73, HMSO, London.

Department of the Environment (1974a) *Local Government Act 1972: Town and Country Planning: Development Plan Provisions*, Circular 58/74, HMSO, London.

Department of the Environment (1974b) *Local Government Act 1972: National Parks*, Circular 65/74, HMSO, London.

Department of the Environment (1974c) *Structure Plans*, Circular 98/74, HMSO, London.

Department of the Environment (1975a) *Housing Act 1974: Renewal Strategies*, Circular 13/75, HMSO, London.

Department of the Environment (1975b) *Housing Act 1974: Parts IV, V and VI: Housing Action Areas, Priority Neighbourhoods and General Improvement Areas*, Circular 14/75, HMSO, London.

Department of the Environment (1975c) *Housing Land Availability in the South East*, consultants' study by Economist Intelligence Unit Ltd., HMSO, London.

Department of the Environment (1975d) *Vacant Land*, Liverpool Inner Area Study, report of consultants to the Steering Committee, DoE, London.

Department of the Environment (1976) *Development Plan Schemes and Local Plans*, Local Plans Note 1/76, DoE, London.

Department of the Environment (1977a) *Inner Area Studies: Liverpool, Birmingham and Lambeth: Summaries of Consultants' Final Reports*, HMSO, London.

Department of the Environment (1977b) *Historic Buildings and Conservation Areas – Policy and Procedure*, Circular 23/77, HMSO, London.

Department of the Environment (1977c) *Housing Strategies and Investment Programmes: Arrangements for 1978–79*, Circular 63/77, HMSO, London.

Department of the Environment (1977d) *Policy for the Inner Cities*, Cmnd. 6845, HMSO, London.

Department of the Environment (1977e) *The Town and Country Planning Act 1971 (Part II as amended by the Town and Country Planning (Amendment) Act 1972 and the Local Government Act 1972): Memorandum on Structure and Local Plans*, Circular 55/77, HMSO, London.

Department of the Environment (1977f) *Local Government and the Industrial Strategy*, Circular 71/77, HMSO, London.

Department of the Environment (1977g) *Nature Conservation and Planning*, Circular 108/77, HMSO, London.

Department of the Environment (1978a) *Form and Content of Local Plans*, Local Plans Note 1/78, DoE, London.

Department of the Environment (1978b) *Land Availability: A Study of Land with Residential Planning Permission*, consultants' study by Economist Intelligence Unit Ltd., HMSO, London.

Department of the Environment (1979a) *The Town and Country Planning Act 1971 (Part II as amended by the Town and Country Planning (Amendment) Act 1972, the Local Government Act 1972 and the Inner Urban Areas Act 1978): Memorandum on Structure and Local Plans*, Circular 4/79, HMSO, London.

Department of the Environment (1979b) *Central Government Controls over Local Authorities*, Cmnd. 7634, HMSO, London.

Department of the Environment (1980a) *Land for Private Housebuilding*, Circular 9/80, HMSO, London.

Department of the Environment (1980b) *Development Control – Policy and Practice*, Circular 22/80, HMSO, London.

Department of the Environment (1981a) *Historic Buildings and Conservation Areas*, Circular 12/81, HMSO, London.

Department of the Environment (1981b) *Enterprise Zones*, DoE, London.

Department of the Environment (1981c) *Local Government, Planning and Land Act 1980, Town and Country Planning: Development Plans*, Circular 23/81, HMSO, London.

Department of the Environment (1982) *Draft Memorandum on Structure and Local Plans*, DoE, London.

Department of the Environment (1983) *Planning Appeals: A Guide*, HMSO, London.

Department of the Environment (1984a) *Simplified Planning Zones: A Consultation Paper*, DoE, London.

Department of the Environment (1984b) *The Reallocation of Planning Functions in the Greater London (GLC) and Metropolitan County Council (MCC) Areas, Revised Proposals Paper*, DoE, London.

Department of the Environment (1984c) *Green Belts*, Circular 14/84, HMSO, London.

Department of the Environment (1984d) *Land for Housing*, Circular 15/84, HMSO, London.

Department of the Environment (1984e) *Industrial Development*, Circular 16/84, HMSO, London.

Department of the Environment (1984f) *Crown Land and Crown Development*, Circular 18/84, HMSO, London.

Department of the Environment (1984g) *Memorandum on Structure and Local Plans: The Town and Country Planning Act 1971: Part II (as amended by the Town and Country Planning (Amendment) Act 1972, the Local Government Act 1972, and the Local Government, Planning and Land Act 1980)*, Circular 22/84, HMSO, London.

Department of the Environment (1984h) *Local Plans: Public Local Inquiries. A Guide to Procedure*, HMSO, London.

Department of the Environment (1985a) *Development and Employment*, Circular 14/85, HMSO, London.

Department of the Environment (1985b) *Chief Planning Inspector's Report for 1984*, DoE, London.

Department of the Environment and Department of Transport (1977) *Residential Roads and Footpaths*, Design Bulletin No. 32, HMSO, London.

Department of the Environment and House Builders Federation (1979) *Study of the Availability of Private House-Building Land in Greater Manchester 1979–1981*, 2 vols., HMSO, London.

Department of Health and Social Security (1976) *The NHS Planning System*, HMSO, London.

Department of Transport (1981a) *Road Layout and Geometry: Highway Link Design*, Departmental Standard TD 9/81, DoT, London.

Department of Transport (1981b) *Junctions and Accesses: The Layout of Major/Minor Junctions*, Departmental Advice Note TA 20/81, DoT, London.

Diamond, D. (1979) 'The uses of strategic planning: the example of the National Planning Guidelines in Scotland', *Town Planning Review*, Vol. **50**, No. 1, pp. 18–25.

District Planning Officers' Society (1978) *Local Plans*, DPOS.

District Planning Officers' Society (1982) *The Local Plan System: The Need for Radical Change*, discussion paper, DPOS.

Donnison, D. (1974) 'Policies for Priority Areas', *Journal of Social Policy*, Vol. **3**, No. 2, pp. 127–135.

Dowrick, F. E. (1974) 'Public ownership of land-taking stock 1972–1973', *Public Law*, Spring, pp. 10–24.

Drake, M., McLoughlin, B., Thompson, R., and Thornley, J. (1975) *Aspects of Structure Planning in Britain*, Research Paper 20, Centre for Environmental Studies, London.

Drewett, R. (1973a) 'The developers: Decision processes' in P. Hall, H. Gracey, R. Drewett and R. Thomas (1973), *The Containment of Urban England, Volume Two: The Planning System*, George Allen and Unwin, London, pp. 169–193.

Drewett, R. (1973b) 'Land values and the surburban land market' in P. Hall, H. Gracey, R. Drewett and R. Thomas (1973), *The Containment of Urban England, Volume Two: The Planning System*, George Allen and Unwin, London, pp. 197–245.

Duc, T. (1979) *Local Plans: The Role of East Sussex County Council*, unpublished project report, City of Birmingham Polytechnic.

Duerden, B. (1978) 'Local Planning in Liverpool' in C. Fudge (ed.), *Approaches to Local Planning*, Working Paper No. 3, School for Advanced Urban Studies, University of Bristol.

Dunleavy, P. (1977) 'Protest and quiescence in urban politics: a critique of some pluralist and structuralist myths', *International Journal of Urban and Regional Research*, Vol. **1**, No. 2, pp. 193–218.

Edwards, M. (1980) 'Land value theories and policies: the need for reappraisal', *Built Environment*, Vol. **6**, No. 4, pp. 272–276.

Elkins, S. E. (1974) *Politics and Land Use Planning: The London Experience*, Cambridge University Press, Cambridge.

Elson, M. (1983) 'Containment in Hertfordshire: Changing attitudes to land release for new employment-generating development' in S. Barrett and P. Healey (eds., 1985), *Land Policy: Problems and Alternatives*, Gower, Aldershot, pp. 127–150.

Elson, M. J. (1986) *Green Belts: Conflict Mediation on the Urban Fringe*, Heinemann, London.

Emery, F. W. and Trist, E. L. (1965) 'The causal texture of organisational environments', *Human Relations*, Vol. **18**, pp. 21–32.

Farnell, R. (1983) *Local Planning in Four English Cities*, Gower, Aldershot.

Field, B. G. (1983) 'Local plans and local planning in Greater London', *Town Planning Review*, Vol. **54**, No. 1, pp. 24–40.

Field, B. G. (1984) 'Theory in practice: The anatomy of a borough plan', *Planning Outlook*, Vol. **27**, No. 2, pp. 68–78.

Finney, J. and Kenyon, J. S. (1976) 'Relationship between structure planning and local plans in a metropolitan area', *Local Planning, Structure Planning, Social Policy and Planning Games*, PTRC, London, pp. 57–64.

Foley, D. L. (1972) *Governing the London Region: Reorganization and Planning in the 1960s*, University of California Press, Berkeley.

Fortune, P. H. (1981) 'Economic regeneration in Cardiff, the role of local planning', *Local Planning Practice*, PTRC, London, pp. 67–74.

Fothergill, S., Kitson, M. and Monk, S. (1983) 'The supply of land for industrial development' in S. Barrett and P. Healey (eds., 1985), *Land Policy: Problems and Alternatives*, Gower, Aldershot, pp. 27–40.

Friend, J. K. and Jessop, W. M. (1969) *Local Government and Strategic Choice*, Tavistock, London.

Friend, J. K., Power, J. M. and Yewlett, C. J. L. (1974) *Public Planning: The Intercorporate Dimension*, Tavistock, London.

Frith, D. W. (1976) 'District planning', *Local Planning, Structure Planning, Social Policy and Planning Games*, PTRC, London, pp. 42–46.

Fudge, C. (1976) 'Local plans, structure plans and policy planning', *The Planner*, Vol. **62**, No. 6, pp. 174–176.

Fudge, C. (1984a) 'London Borough of Camden' in C. Fudge and P. Healey, *Local Planning in Practice (3): Camden and Manchester*, Working Paper 32, School for Advanced Urban Studies, University of Bristol, pp. 1–67.

Fudge, C. (1984b) 'Sheffield City Council' in C. Fudge and P. Healey, *Local Planning in Practice (4): Hampshire and Sheffield*, Working Paper 33, School for Advanced Urban Studies, University of Bristol, pp. 92–143.

Fudge, C., Lambert, C., Underwood, J. and Healey, P. (1983), *Speed, Economy and Effectiveness in Local Plan Preparation and Adoption*, Occasional Paper No. 11, School for Advanced Urban Studies, University of Bristol.

Galloway, T. D. and Mahayni, R. G. (1977) 'Planning theory in retrospect: The process of paradigm change', *Journal of the American Institute of Planners*, Vol. **43**, No. 1, pp. 62–71.

Gault, I. (1981) *Green Belt Policies in Development Plans*, Working Paper No. 41, Department of Town Planning, Oxford Polytechnic.

Gibbs, A. (1985) 'Planners and retail innovation', *The Planner*, Vol. **71**, No. 5, pp. 9–10.

Gilbert, T., MacIntyre, M., Manning, K. and Summall, I. (1982) 'Consulting with statutory bodies', *The Planner*, Vol. **68**, No. 5, pp. 143–144.

Glasson, J. (1978) *An Introduction to Regional Planning*, (2nd ed), Hutchinson, London.

Gloucester City Council (1983) *City of Gloucester Local Plan*, Gloucester City Council, Gloucester.

Goldsmith, M. (1980) *Politics, Planning and the City*, Hutchinson, London.

Goodchild, R. (1979) *The Supply of Development Land: The Role of the Landowner*, unpublished PhD thesis, University of Cambridge.

Goodchild, R. and Munton, R. (1985) *Development and the Landowner*, George Allen and Unwin, London.

Goodhall, P. (1985) 'Urban Development Grant – An early assessment of regional implications', *Planning Outlook*, Vol. **28**, No. 1, pp. 40–42.

Gracey, H. (1973) 'The Planners: control of new residential development' in P. Hall, H. Gracey, R. Drewett and R. Thomas, *The Containment of Urban England, Volume Two: The Planning system*, George Allen and Unwin, London, pp. 126–145.

Grant, M. (1978) 'Developers' contributions and planning gain: Ethics and legalities', *Journal of Planning and Environment Law*, pp. 8–15.

Grant, M. (1982) *Urban Planning Law*, Sweet and Maxwell, London.

Greater London Council (1985) *Erosion of the Planning System: A GLC Analysis*, GLC London.

Green, G. (1978) 'Birmingham's partnership: participation excluded?', *The Planner*, Vol. **64**, No. 3, p. 76.

Green, P. (1982) 'Central Leicester District Plan', *The Planner*, Vol. **68**, No. 5, p. 149.

Gregory, D. (1977) 'Green Belt Policy and the conurbation' in F. Joyce (ed.), *Metropolitan Development and Change*, University of Aston in Birmingham.

Griffiths, J. (1978) 'Planning for Hertfordshire' in C. Fudge (ed.), *Approaches to Local Planning*, School for Advanced Urban Studies, University of Bristol, pp. 18–36.

Groundwork Trust (1983) *Operation Groundwork, Making Good between Town and Country*, Groundwork Trust, St. Helens.

Gunn, L. A. (1978) 'Why is implementation so difficult?', *Management Services in Government*, November, pp. 169–176.

Guy, C. M. (1980) *Retail Location and Retail Planning in Britain*, Gower, Farnborough.

Guy, C. M. (1982) 'Push-button shopping and retail development', Papers in Planning Research 49, Department of Town Planning, UWIST, Cardiff.

Gwent County Council (1978) *Gwent Structure Plan, Written Statement of Policies and Proposals*, Gwent County Council, Cwmbran.

Gyford, J. (1976) *Local Policies in Britain*, Croom Helm, London.

Haig, R. M. (1926) 'Toward an understanding of the metropolis', *Quarterly Journal of Economics*, Vol. **40**, pp. 179–208.

Hall, P. (1984) 'Enterprises of great pith and moment?', *Town and Country Planning*, Vol. **53**, No. 11, pp. 296–297.

Hall, P., Gracey, H., Drewett, R. and Thomas, R. (1973a) *The Containment of Urban England, Volume One: Urban and Metropolitan Growth Processes*, George Allen and Unwin, London.

Hall, P., Gracey, H., Drewett, R. and Thomas, R. (1973b) *The Containment of Urban England, Volume Two: The Planning System*, George Allen and Unwin, London.

Hambleton, R. (1976) 'Local planning and area management', *The Planner*, Vol. **62**, No. 6, pp. 176–179.

Hambleton, R. (1977) 'Policies for areas', *Local Government Studies*, Vol. **3**, No. 2, pp. 13–29.

Hambleton, R. (1983) 'Symposium on expenditure-based planning systems: Developments in planning systems', *Policy and Politics*, Vol. **11**, No. 2, pp. 161–165.

Hamilton, R. N. D. (1977) 'Legal issues on the preparation of structure and local plans . . . I', *Local Government Chronicle*, 20 May, No. 5747, pp. 417–419.

Hammersley, R. and Williamson, J. (1985) 'Industrial and commercial improvement areas: An assessment', *The Planner*, Vol. **71**, No. 11, pp. 8–10.

Hampton, W. (1977) 'Research into public participation in structure planning' in W. R. D. Sewell and J. T. Coppock (eds.), *Public Participation in Planning*, John Wiley and Sons, London, pp. 27–42.

Hargreave, R. B. and Kirkpatrick, D. (1982) 'Stockport's alternative to the Enterprise Zone', *Report of Proceedings of the 1982 Town and Country Planning Summer School*, Royal Town Planning Institute, London, pp. 34–40.

Harloe, M., Issacharoff, R. and Minns, R. (1974) *The Organization of Housing*, Heinemann, London.

Hatch, S. and Sherrott, R. (1973) 'Positive discrimination and the distribution of deprivation', *Policy and Politics*, Vol. **1**, No. 3, pp. 223–240.

Hawke, J. M. (1981a) 'Planning agreements in practice: I', *Journal of Planning and Environment Law*, January, pp. 8–14.

Hawke, J. M. (1981b) 'Planning agreements in practice: II', *Journal of Planning and Environment Law*, February, pp. 86–97.

Hayes, M. G. (1981) 'Local planning and the inner city: The Liverpool Case', *Local Planning Practice*, PTRC, London, pp. 75–84.

Healey, P. (1979a) 'Statutory local plans: their evolution in legislation and administrative interpretation', Working Paper No. 36, Department of Town Planning, Oxford Polytechnic.

Healey, P. (1979b) 'Central–local relations in Green Belt local plans', in M. Elson (ed.), *Perspectives on Green Belt Local Plans*, Working Paper 38, Department of Town Planning, Oxford Polytechnic, pp. 3–33.

Healey, P. (1983) *Local Plans in British Land Use Planning*, Pergamon, Oxford.

Healey, P. (1984) 'Manchester City Council', in C. Fudge and P. Healey, *Local Planning in Practice (3): Camden and Manchester*, Working Paper No. 32, School for Advanced Urban Studies, University of Bristol.

Healey, P. and Elson, M. (1981) 'Development plans and development investment', *Structure and Regional Planning Practice*, PTRC, London, pp. 45–54.

Healey, P., Davis, J., Wood, M. and Elson, M. (1982) *The Implementation of Development Plans*, Report of an exploratory study for the DoE, Department of Town Planning, Oxford Polytechnic.

Healey, P., Doak, A. J., McNamara, P. F. and Elson, M. (1985) *The Implementation of Planning Policies and the Role of Development Plans*, Report to the DoE, Department of Town Planning, Oxford Polytechnic.

Heap, D. (1982) *An Outline of Planning Law* (8th ed), Sweet and Maxwell, London.

Heap, D. and Ward, A. (1980) 'Planning bargaining – the pros and cons', *Journal of Planning and Environment Law*, pp. 631–637.

Hebbert, M. and Gault, I. (eds., 1978) *Green Belt Issues in Local Plan Preparation*, Working Paper No. 34, Department of Town Planning, Oxford Polytechnic.

Hereford City Council (1984) *Hereford Local Plan, Draft Written Statement*, Hereford City Council, Hereford.

Hewitt, E. (1985) 'Local planning in East Lindsey – 'patchwork quilt on a bed of nails'', *The Planner*, Vol. 71, No. 12, pp. 18–20.

Hickling, A. (1974) *Managing Decisions: the Strategic Choice Approach*, Mantec, Rugby.

Hickling, A. (1978) 'AIDA and the levels of choice in structure plans', *Town Planning Review*, Vol. 49, No. 4, pp. 459–475.

Hickling, A., Friend, J. and Luckman, J. (1979) *The Development Plan System and Investment Programmes*, Centre for Organisational and Operational Research, DoE, London.

Higgins, J. C. (1980) *Strategic and Operational Planning Systems: Principles and Practice*, Prentice-Hall, London.

Himsworth, K. H. (1980) *A Review of Areas of Outstanding Natural Beauty*, Countryside Commission, CCP 140, Cheltenham.

Historic Buildings Council for England (1981) *Annual Report 1979–80*, HMSO, London.

Hollingsworth, M. J. and Cuddy, M. (1979) 'The Land Authority for Wales: Its mediating role in planning and implementation', *Local Planning Practice*, PTRC, London, pp. 1–13.

Home, R. K. (1985) 'Forecasting housing land requirements', *Land Development Studies*, Vol. 2, pp. 19–34.

Hookway, R. (1978) 'National Park Plans, A milestone in the development of planning', *The Planner*, Vol. 64, No. 1, pp. 20–22.

Hooper, A. (1979) 'Land availability', *Journal of Planning and Environment Law*, November, pp. 752–756.

Hooper, A. (1980) 'Land for private house-building', *Journal of Planning and Environment Law*, December, pp. 795–806.

Hooper, A. (1983) 'Land availability studies and private housebuilding', in S. Barrett and P. Healey (eds., 1985), *Land Policy: Problems and Alternatives*, Gower, Aldershot, pp. 106–126.

Hough, H. T. (1950) 'The Liverpool Corporate Estate', *Town Planning Review*, Vol. 21, No. 3, pp. 236–252.

Housing Monitoring Team (1980) *The Role of the Local Authority in Land Programming and the Process of Private Residential Development*, Research Memorandum 80, Centre for Urban and Regional Studies, University of Birmingham.

Humber, J. R. (1980) 'Land availability – another view', *Journal of Planning and Environment Law*, January, pp. 19–23.

Humphreys, H. C. (1980) *Vacant Urban Land and the Development Process*, unpublished MSc. thesis, Department of Town Planning, UWIST, Cardiff.

Hunter, A. and Trinnaman, J. (1983) 'Policy making and implementation', *Local Government Policy Making*, November, pp. 78–84.

Hurd, R. M. (1924) *Principles of City Land Values*, New York.

Jacobs, J. (1985) 'UDG: The Urban Development Grant', *Policy and Politics*, Vol. 13, No. 2, pp. 191–199.

James, J. R. (1965) 'The future of development plans' *Report of Proceedings of the Town and Country Planning Summer School 1965*, Royal Town Planning Institute, London, pp. 16–30.

Joint Unit for Research on the Urban Environment (1977) *Planning and Land Availability*, University of Aston in Birmingham.

Jones, P. (1983) 'DIY and home improvement centres: the planning issues', *The Planner*, Vol. 69, No. 1, pp. 13–15.

Jones, P. (1984) 'Retail warehouse developments in Britain', *Area*, Vol. 16, No. 1, pp. 41–47.

Jowell, J. (1977) 'Bargaining in development control', *Journal of Planning and Environment Law*, July, pp. 414–433.

Jowell, J. and Noble, D. (1981) 'Structure plans as instruments of social and economic policy', *Journal of Planning and Environment Law*, pp. 466–480.

Kaiser, E. J. and Weiss, S. F. (1970) 'Public policy and the residential development process', *Journal of the American Institute of Planners*, Vol. 36, No. 1, pp. 30–37.

Keeble, L. (1964) *Principles and Practice of Town and Country Planning* (3rd ed), Estates Gazette, London.

Lamarche, F. (1976) 'Property development and the economic foundations of the urban question', in C. G. Pickvance (ed.), *Urban Sociology: Critical Essays*, Methuen, London, pp. 85–118.

Lambert, C. and Underwood, J. (1983) *Local Planning in Practice (2): Kingswood, Leicester, Newcastle-under-Lyme*, Working Paper 31, School for Advanced Urban Studies, University of Bristol.

Land Authority for Wales (1981) *Strategy and Programme 1981–86*, LAW, Cardiff.

Leach, S. and Moore, N. (1979) 'County–district relations in shire and metropolitan counties in the field of town and country planning: A comparison', *Policy and Politics*, Vol. 7, No. 2, pp. 165–179.

Leather, P. (1983) 'Housing (Dis?)Investment Programmes', *Policy and Politics*, Vol. 11, No. 2, pp. 215–229.

Lebas, E. (1977) 'Movement of capital and locality: Issues raised by the study of local power structures', in M. Harloe (ed.), *Urban Change and Conflict*, Conference Paper 19, Centre for Environmental Studies, London, pp. 251–279.

Leclerc, R. and Draffan, D. (1984) 'The Glasgow Eastern Area Renewal Project', *Town Planning Review*, Vol. 55, No. 3, pp. 335–351.

Lee Valley Regional Park Authority (1985) *Lee Valley Park Plan, Consultative Draft*, LVRPA, Waltham Abbey.

Lloyd, M. G. (1984a) 'Enterprise Zones: the evaluation of an experiment', *The Planner*, Vol. 70, No. 6, pp. 23–25.

Lloyd, M. G. (1984b) 'Policies in search of an opportunity', *Town and Country Planning*, Vol. 53, No. 11, pp. 299–301.

Lloyd, M. G. (1985) 'Privatisation, liberalisation and simplification of statutory land use planning in Britain', *Planning Outlook*, Vol. 28, No. 1, pp. 46–49.

Loughlin, M. (1984) *Local Needs Policies and Development Control Strategies*, Working Paper 42, School for Advanced Urban Studies, University of Bristol.

Mabey, R. and Craig, L. (1976) 'Development plan schemes', *The Planner*, Vol. 62, No. 3, pp. 70–72.

McAuslan, J. P. W. B. and Bevan, R. G. (1977) 'The influence of officers and councillors on procedures in planning – A case study', *Local Government Studies*, Vol. 3, No. 3, pp. 7–21.

McAuslan, P. (1975) *Land, Law and Planning*, Weidenfeld and Nicolson, London.

McAuslan, P. (1981) 'Local government and resource allocation in England: Changing ideology, unchanging law', *Urban Law and Policy*, Vol. 4, No. 3, pp. 215–268.

McDonald, I. and Howick, C. (1981) 'Monitoring the Enterprise Zones', *Built Environment*, Vol. 7, No. 1, pp. 31–37.

McDonald, S. T. (1977) 'The Regional Report in Scotland: A Study of change in the planning process', *Town Planning Review*, Vol. 48, No. 3, pp. 215–232.

McGilp, N. (1981) 'Success for single sheet plan format?', *Planning*, No. 427, 17 July, p. 8.

McLoughlin, J. B. (1973) *Control and Urban Planning*, Faber, London.

MacMurray, T. (1974) 'Local planning, corporate management and the public: Strengthening our approach', *The Planner*, Vol. 60, No. 1, pp. 493–495.

McNamara, P. (1982) *Land Release and Development in Areas of Restraint: Housing in Dacorum and North Hertfordshire: Restraint Policy and Development Interests*, Working Paper No. 76, Department of Town Planning, Oxford Polytechnic.

McNamara, P. (1983) 'Towards a classification of land developers', *Urban Law and Policy*, Vol. 6, pp. 87–94.

McNamara, P. (1984) *Land Release and Development in Areas of Restraint: Restraint*

Policy in Action: Housing in Dacorum and North Hertfordshire, Working Paper No. 77, Department of Town Planning, Oxford Polytechnic.

McNamara, P. (1986) 'Statutory Plans for City Centres', *The Planner*, Vol. 72, No. 6, pp. 43–45.

Manchester and Salford Inner Area Partnership Research Group (1978) *Land*, Subject Paper 5, MSIAP Research Group, DoE North West Regional Office, Manchester.

Mandelker, D. R. (1962) *Green Belts and Urban Growth*, University of Wisconsin Press, Madison.

Marsh, G. (1983) 'The local plan inquiry: Its role in local plan preparation', *Progress in Planning*, Vol. 19, Part 2, pp. 91–167.

Mason, M. C. and Whitney, D. K. (1985a) *Urban Development Grants: Evolution and Performance*, Brunswick Environmental Papers No. 53, School of Environmental Studies, Leeds Polytechnic.

Mason, M. C. and Whitney, D. K. (1985b) *Urban Development Grants: A Case Study of Practice in Yorkshire and Humberside*, Brunswick Environmental Papers No. 54, School of Environmental Studies, Leeds Polytechnic.

Mason, R. U. and Mitroff, I. I. (1981) *Challenging Strategic Planning Assumptions: Theory, Cases and Techniques*, John Wiley and Sons, New York.

Massey, D. and Catalano, A. (1978) *Capital and Land*, Edward Arnold, London.

Merrett, S. (1984) 'Villages which have an appetite for land', *Town and Country Planning*, Vol. 53, No. 5, pp. 140–142.

Merrett, S. with Gray, F. (1982) *Owner Occupation in Britain*, Routledge and Kegan Paul, London.

Middleton, M. R. (1982) 'Looking after the land', *The Planner*, Vol. 68, No. 5, pp. 150–151.

Milligan, J. (1972) 'Something the matter with Glasgow', *Official Architecture and Planning*, Vol. 35, No. 1, pp. 18–24.

Ministry of Housing and Local Government (1955) *Green Belts*, Circular 42/55, HMSO, London.

Ministry of Housing and Local Government (1957) *Green Belts*, Circular 50/57, HMSO, London.

Ministry of Housing and Local Government (1962a) *Town Centres, Approach to Renewal*, Planning Bulletin No. 1, HMSO, London.

Ministry of Housing and Local Government (1962b) *Residential Areas: Higher Densities*, Planning Bulletin No. 2, HMSO, London.

Ministry of Housing and Local Government (1963) *Town Centres: Cost and Control of Development*, Planning Bulletin No. 3, HMSO, London.

Ministry of Housing and Local Government (1964) *Planning for Daylight and Sunlight*, Planning Bulletin No. 5, HMSO, London.

Ministry of Housing and Local Government (1967) *Town and Country Planning*, Cmnd. 3333, HMSO, London.

Ministry of Housing and Local Government (1969) *People and Planning*, Report of the Committee on Public Participation in Planning, the Skeffington Report, HMSO, London.

Ministry of Housing and Local Government (1970) *Development Plans: A Manual on Form and Content*, HMSO, London.

Ministry of Housing and Local Government and Ministry of Transport (1965) *Parking in Town Centres*, Planning Bulletin No. 7, HMSO, London.

Ministry of Transport (1966) *Roads in Urban Areas*, HMSO, London.

Muchnick, D. M. (1970) *Urban Renewal in Liverpool*, Occasional Papers in Social Administration No. 33, Bell, London.

Munton, R. (1983) *London's Green Belt: Containment in Practice*, George Allen and Unwin, London.

Murie, A. and Leather, P, (1977) 'Developments in housing strategies', *The Planner*, Vol. 63, No. 6, pp. 167–169.

Nabarro, R. (1980) 'Inner city Partnerships. An assessment of the first programmes', *Town Planning Review*, Vol. 51, No. 1, pp. 25–38.

Nabarro, R. (1981) 'The Urban Development Corporations: Partnership or Autarky?', in S. Markowski (ed., 1982), *Land Policies for the 1980s*, Occasional Paper 1/82, Department of Estate Management, Polytechnic of the South Bank, London.

Nature Conservancy Council (1983) *Eighth Report*, Nature Conservancy Council, London.

Nature Conservancy Council, City of Swansea (1982) *Crymlyn Action Plan*, NCC and Swansea City Council, Swansea.

Newby, J. (1985) 'Local area plans in Camden', *The Planner*, Vol. 71, No. 12, pp. 27–28.

Newham Docklands Forum (1983) *The People's Plan for the Royal Docks*, Newham Docklands Forum, London.

Newstrom, J. W., Reif, W. E. and Monczka, R. M. (1975) *A Contingency Approach to Management Readings*, McGraw-Hill, New York.

Nicholson, D. J. (1979) *Vacant and Cleared Land*, unpublished Diploma dissertation, Department of Town Planning, UWIST, Cardiff.

Norman, P. (1971) 'Corporation town', *Official Architecture and Planning*, Vol. 34, No. 5, pp. 360–362.

Norman, P. (1972) 'A derelict policy', *Official Architecture and Planning*, Vol. 35, No. 1, pp. 29–32.

Nott, S. M. and Morgan, P. H. (1984) 'The significance of Department of the Environment Circulars in the planning process', *Journal of Planning and Environment Law*, September, pp. 623–632.

O'Connor, J. (1973) *The Fiscal Crisis of the State*, St. Martin's Press, New York.

Orchard-Lisle, P. (1980) 'Identifying £100m of new town assets', *Chartered Surveyor*, Vol. 112, No. 9, pp. 14–15.

Page, S. (1985) 'From dingy docks to desirable docklands', *Town and Country Planning*, Vol. 54, No. 11, pp. 327–328.

Pahl, R. (1970) *Whose City?*, Longman, London.

Paris, C. (1974) 'Urban renewal in Birmingham, England – An institutional approach', *Antipode*, Vol. 6, pp. 7–15.

Paris, C. (1979) 'HIPs and underspending: The case of Oxford', *CES Review*, No. 6, pp. 43–49.

Parker, D. J. and Penning-Rowsell, E. C. (1980) *Water Planning in Britain*, George Allen and Unwin, London.

Parkinson, M. H. and Wilks, S. R. M. (1983) 'Managing urban decline – The case of the inner city Partnerships', *Local Government Studies*, Vol. 9, No. 5, pp. 23–39.

Parsons, T. and Schills, E. A. (1951) *Towards a General Theory of Action*, Harvard University Press, Cambridge, Massachusetts.

Patterson, D. T. (1978) 'The Berkshire Development Programme', *Structure and Regional Planning Practice/Local Planning Practice*, PTRC, London, pp. 58–69.

Payne, B. J. (1978) *Water Authorities and Planning Authorities: A Study of Developing Relationships*, Occasional Paper No. 1, Department of Town and Country Planning, University of Manchester.

Pearce, G. R. and Tricker, M. J. (1977) 'Land availability for residential development' in F. Joyce (ed.), *Metropolitan Development and Change*, University of Aston, Birmingham, pp. 253–270.

Penning-Rowsell, E. C. (1982) 'Planning and water services: keeping in step', *Town and Country Planning*, Vol. **51**, No. 6, pp. 150–152.

Perry, J. (1974) 'Introduction: Approaches to local planning', *The Planner*, Vol. **60**, No. 1, p. 492.

Perry, M., Bruton, M. J., Crispin, G. and Fidler, P. (1985) 'Local plans: The process of adoption after an inquiry', *Journal of Planning and Environment Law*, August, pp. 521–529.

Pickvance, C. G. (1977) 'Physical planning and market forces in urban development', *National Westminster Bank Quarterly Review*, August, pp. 41–50.

Planning Advisory Group (1965) *The Future of Development Plans*, HMSO, London.

Pountney, M. T. and Kingsbury, P. W. (1983a) 'Aspects of development control. Part I: The relationships with local plans', *Town Planning Review*, Vol. **54**, No. 2, pp. 138–154.

Pountney, M. T. and Kingsbury, P. W. (1983b) 'Aspects of development control. Part II: The applicant's view', *Town Planning Review*, Vol. **54**, No. 3, pp. 285–303.

Property Advisory Group (1981) *Planning Gain*, HMSO, London.

Purton, P. J. and Douglas, C. (1982) 'Enterprise Zones in the United Kingdom: A successful experiment', *Journal of Planning and Environment Law*, July, pp. 412–422.

Ratcliffe, J. (1978) *An Introduction to Urban Land Administration*, Estates Gazette, London.

Ratcliffe, J. (1982) *Land Acquisition and Disposal*, SSRC Planning Review No. 3, Capital Planning Information, Edinburgh.

Ratcliff, R. V. (1949) *Urban Land Economics*, New York.

Regan, D. E. (1978) 'The pathology of British land use planning', *Local Government Studies*, Vol. **4**, No. 2, pp. 3–23.

Rhymney Valley District Council (1984) *Waste Disposal Plan, Draft Consultative Document*, RVDC, Hengoed.

Rittel, H. W. J. and Webber, M. M. (1973) 'Dilemmas in a general theory of planning', *Policy Sciences*, Vol. **4**, pp. 155–169.

Rodwin, L. (1981) 'On the illusions of planners', in M. Honjo (ed.), *Urbanization and Regional Development*, Maruzen, Asia, Hong Kong, pp. 363–387.

Roger Tym and Partners (1982) *Monitoring Enterprise Zones: Year One Report, Main Report*, Department of the Environment, London.

Roger Tym and Partners (1984) *Monitoring Enterprise Zones: Year Three Report*, Department of the Environment, London.

Roger Tym and Partners, Franklin Stafford Partnership, Richard Barrett Traffic and

Transport Associates (1979) *Time for Industry, Evaluation of the Rochdale Industrial Improvement Area for the Department of the Environment*, HMSO, London.

Rogers, A. (1967) 'Theories of intraurban spatial structure: A dissenting view', *Land Economics*, Vol. **43**, pp. 108–112.

Rose, R. (1974) *Politics in England Today*, Faber & Faber, London.

Roweis, S. T. and Scott, A. J. (1978) 'The urban land question', in K. Cox (ed.), *Urbanization and Conflict in Market Societies*, Methuen, London, pp. 38–75.

Royal Town Planning Institute (1971) *Town Planners and their Future*, RTPI, London.

Royal Town Planning Institute (1972) 'Memorandum on the Local Government Bill', *Journal of the Royal Town Planning Institute*, Vol. **58**, No. 4, pp. 151–152.

Royal Town Planning Institute (1973) *Town Planners and their Future: A Further Discussion Paper*, RTPI, London.

Royal Town Planning Institute (1979) *Land Values and Planning in the Inner Areas*, RTPI, London.

Royal Town Planning Institute (1980) *Planning Officers as Witnesses at Inquiries*, Planning Advice Note No. 1, RTPI, London.

Rydin, Y. (1984) 'The struggle for housing land: A case of confused interests', *Policy and Politics*, Vol. **12**, No. 4, pp. 431–446.

Sant, M. (1985) 'Joint enterprise for Welsh wasteland', *Town and Country Planning*, Vol. **54**, No. 11, pp. 329–331.

Saunders, P. (1981) *Social Theory and the Urban Question*, Hutchinson, London.

Saunders, P. and Couch, C. (1979) 'Delay and uncertainty in local planning', *Local Planning Practice*, PTRC, London, pp. 51–68.

Savage, R. (1984) 'New settlements, questions from a local government planner', *Housing and Planning Review*, Vol. **39**, No. 6, pp. 4–7.

Schaffer, F. (1970) *The New Town Story*, MacGibbon and Kee, London.

Schelling, T. C. (1960) *The Strategy of Conflict*, Harvard University Press, Cambridge, Massachusetts.

Scott, A. J. (1980) *The Urban Land Nexus and the State*, Pion, London.

Scottish Development Department (1984) *Local Planning*, Planning Advice Note 30, SDD, Edinburgh.

Shaw, J. M. (1982) 'Influencing rural change – the planning role', *Report of Proceedings of the 1982 Town and Country Planning Summer School*, Royal Town Planning Institute, London, pp. 56–57.

Shields, R. M. C. (1983) 'Effective local plan making', *Report of Proceedings of the 1983 Town and Country Planning Summer School*, Royal Town Planning Institute, London, pp. 28–29.

Shostak, L. and Lock, D. (1984) 'The need for new settlements in the South East', *The Planner*, Vol. **70**, No. 11, pp. 9–13.

Shostak, L. and Lock, D. (1985) 'New country towns in the South East: a planned response to a regional crisis', *The Planner*, Vol. **71**, No. 5, pp. 19–22.

Shucksmith, M. (1981) *No Homes for Locals*, Gower, Aldershot.

Simmie, J. S. (1974) *Citizens in Conflict*, Hutchinson, London.

Simon, H. A. (1971) 'Decision making and organizational design', in D. S. Pugh (ed.), *Organisation Theory*, Penguin Education, London, pp. 189–212.

Slack, A. (1983) 'Residential and local employment issues in villages – a local plan approach', *Making Development Plans Work*, PTRC, London, pp. 5–15.

Smyth, H. (1982) *Land Banking, Land Availability and Planning for Private House Building*, Working Paper 23, School for Advanced Urban Studies, University of Bristol.

Social Science Research Council (1982) *Review of Research and Education in Planning: Report to SSRC*, SSRC, London.

Solesbury, W. (1981) 'Strategic planning: metaphor or method?', *Policy and Politics*, Vol. **9**, No. 4, pp. 419–437.

Southerton, L. and Noble, T. (1982) 'The corporate approach to local planning', *The Planner*, Vol. **68**, No. 5, pp. 151–152.

Stewart, M. (1983) 'The inner area planning system', *Policy and Politics*, Vol. **11**, No. 2, pp. 203–214.

Stoddart, R. S. (1983) 'Structure plans in Berkshire – theory and practice', in D. T. Cross and M. R. Bristow (eds.), *English Structure Planning*, Pion, London, pp. 150–176.

Stones, A. (1972) 'Stop slum clearance – now', *Official Architecture and Planning*, Vol. **35**, No. 2, pp. 107–110.

Stones, A. (1977) 'Liverpool now: inner city wasteland', *Built Environment Quarterly*, Vol. **3**, No. 1, pp. 47–50.

Strauss, A. (1978) *Negotiations*, Jossey-Bass, New York.

Stuart, R. I. and Beaumont, K. (1984) 'Local plans: the better for change', *Report of Proceedings of the 1984 Town and Country Planning Summer School*, Royal Town Planning Institute, London, p. 41.

Susskind, L. (1981) 'The importance of citizen participation and consensus in the land use planning process', in J. de Neufville (ed.), *The Land Use Policy Debate*, Plenum Press, New York.

Swansea City Council (1981) *Swansea Enterprise Park*, SCC, Swansea.

Tayler, M. (1982) 'Industrial improvement areas, problems and progress', *The Planner*, Vol. **68**, No. 3, pp. 80–82.

Taylor, S. (1981) 'The politics of Enterprise Zones', *Public Administration*, Vol. **59**, Winter, pp. 421–439.

Thomas, D. (1970) *London's Green Belt*, Faber & Faber, London.

Thompson, R. (1977a) 'Making local plans effective: the Camden experience', *Structure Planning Practice, Local Planning Practice*, PTRC, London, pp. 145–155.

Thompson, R. (1977b) 'Camden's local plan: a district-wide approach', *The Planner*, Vol. **63**, No. 5, pp. 145–147.

Thornley, A. (1977) 'Theoretical perspectives on planning participation', *Progress in Planning*, Vol. **7**, No. 1.

Town Planning Institute (undated, pre-1964) *Town Planning as a Career*, TPI, London.

Town Planning Institute (1967) *Progress Report on Membership Policy – Revised Scheme for the Final Examination*, TPI, London.

Townsend, P. (1976) 'Area deprivation policies', *New Statesman*, Vol. **92**, No. 2368, pp. 168–171.

Underwood, J. (1981) 'Development control: A case study of discretion in action', in S. Barrett and C. Fudge (eds.), *Policy and Action*, Methuen, London, pp. 143–161.

University of Aston Management Centre (1985) *Five Year Review of the Birmingham Inner City Partnership*, Public Sector Management Research Unit, University of Aston Management Centre, University of Aston in Birmingham.

Urwin, J. C. and Wenban-Smith, A. (1983) 'Local planning in inner Birmingham', *Making Development Plans Work*, PTRC, London, pp. 43–53.

Wannop, U. (1981) 'The future for development planning', *The Planner*, Vol. **67**, No. 1, pp. 14–18.

Ward, A. (1982) 'Planning bargaining: Where do we stand?', *Journal of Planning and Environment Law*, pp. 74–84.

Warren Evans, R. (1981) 'The Lower Swansea Valley Enterprise Zone', *Built Environment*, Vol. **7**, No. 1, pp. 20–30.

Warwick District Council (1983) *Leamington Town Centre Local Plan*, WDC, Leamington Spa.

Warwickshire County Council (1975) *Development Plan Scheme*, WCC, Warwick.

Warwickshire County Council and West Midlands County Council (1975) Warwickshire Structure Plan 1975, WCC/WMCC, Warwick.

Warwickshire County Council (1979) *Warwick, Leamington and Kenilworth Urban Structure Plan, Written Statement*, WCC, Warwick.

Warwickshire County Council (1983) *Warwickshire Green Belt Subject Plan*, WCC, Warwick.

Weaver, W. (1948) 'Science and complexity', *American Scientist*, Vol. **36**, p. 538.

Welsh Development Agency (1977) *A Statement of Policies and Programmes*, WDA, Treforest.

Wenban-Smith, A. (1983) 'Local plans and the inner cities', *Town and Country Planning*, Vol. **52**, No. 7/8, pp. 203–205.

Williams, R. (1978) 'Statutory local plans: Progress and problems', *Planning Outlook*, Vol. **21**, No. 2, pp. 22–27.

Willis, K. (1982) 'Planning agreements and planning gain', *Planning Outlook*, Vol. **24**, No. 2, pp. 55–56.

Wilson, G. H. (1977a) 'Inflexibility delays progress', *Municipal and Public Services Journal*, 15 July, p. 687.

Wilson, G. H. (1977b) 'Not much scope for reducing delays', *Municipal and Public Services Journal*, 12 August, pp. 779–780.

Wisdom, A. S. (1975) *Local Authorities' Powers of Purchase – A Summary*, Barry Rose, Chichester.

Wolley, R. and Rosborough, L. (1985) 'Tillingham people find their voices', *Town and Country Planning*, Vol. **54**, No. 7, pp. 237–238.

Woodruffe, B. J. (1976) *Rural Settlement Policies and Plans*, Oxford University Press, London.

Wray, I. (1985) 'The false trail to private new towns', *The Planner*, Vol. **71**, No. 2, pp. xiii–xiv.

Young, K. and Kramer, J. (1978) *Strategy and Conflict in Metropolitan Housing*, Heinemann, London.

Index

441